Lecture Notes in Computer Science 12619

More information about this subseries at http://www.springer.com/series/7409

Jan Mazal · Adriano Fagiolini ·
Petr Vasik · Michele Turi (Eds.)

Modelling and Simulation for Autonomous Systems

7th International Conference, MESAS 2020
Prague, Czech Republic, October 21, 2020
Revised Selected Papers

Editors
Jan Mazal
NATO M&S COE
Rome, Italy

Petr Vasik
Brno University of Technology
Brno, Czech Republic

Adriano Fagiolini
University of Palermo
Palermo, Italy

Michele Turi
NATO M&S COE
Rome, Italy

ISSN 0302-9743 ISSN 1611-3349 (electronic)
Lecture Notes in Computer Science
ISBN 978-3-030-70739-2 ISBN 978-3-030-70740-8 (eBook)
https://doi.org/10.1007/978-3-030-70740-8

LNCS Sublibrary: SL3 – Information Systems and Applications, incl. Internet/Web, and HCI

This Springer imprint is published by the registered company Springer Nature Switzerland AG
The registered company address is: Gewerbestrasse 11, 6330 Cham, Switzerland

Preface

This volume contains selected papers presented at the MESAS 2020 Conference: Modelling and Simulation for Autonomous Systems, held on October 21, 2020, in Prague. The initial idea to launch the MESAS project was introduced by the NATO Modelling and Simulation Centre of Excellence in 2013, with the intent to bring together the Modelling and Simulation and the Autonomous Systems/Robotics communities and to collect new ideas for concept development and experimentation in this domain. From that time, the event has gathered together (in keynote, regular, poster, and way ahead sessions) fully recognized experts from different technical communities in the military, academia, and industry. The main topical parts of the 2020 edition of MESAS were "Future Challenges of Advanced M&S Technology," "M&S of Intelligent Systems," and "AxS in Context of Future Warfare and Security Environment." The community of interest submitted 38 papers for consideration. Each submission was reviewed by three Technical Committee members or selected independent reviewers. In this review process, the committee decided to accept 20 papers to be presented (in four sessions), and 19 of these papers were accepted for inclusion in the conference proceedings.

December 2020 Jan Mazal

MESAS 2020 Logo

MESAS 2020 Organizer

NATO Modelling and Simulation Centre of Excellence
(NATO M&S COE)

The NATO M&S COE is a recognized international military organization activated by the North Atlantic Council in 2012, and does not fall under the NATO Command Structure. Partnering nations provide funding and personnel for the centre through a memorandum of understanding. The Czech Republic, Italy, the United States, and Germany are the contributing nations, as of this publication. The NATO M&S COE supports NATO transformation by improving the networking of NATO and nationally owned M&S systems, promoting cooperation between nations and organizations through the sharing of M&S information, and serving as an international source of expertise.

The NATO M&S COE seeks to be a leading world-class organization, providing the best military expertise in modelling and simulation technology, methodologies, and the development of M&S professionals. Its state-of-the-art facilities can support a wide range of M&S activities including but not limited to: education and training of NATO M&S professionals on M&S concepts and technology with hands-on courses that expose students to the latest simulation software currently used across the alliance; concept development and experimentation using a wide array of software capability and network connections to test and evaluate military doctrinal concepts as well as new simulation interoperability verification; and the same network connectivity that enables the COE to become the focal point for NATO's future distributed simulation environment and services.

https://www.mscoe.org/

Organization

General Chairs

Michele Turi	NATO M&S COE, Italy
Adriano Fagiolini	University of Palermo, Italy
Stefan Pickl	Universität der Bundeswehr München, Germany
Agostino Bruzzone	University of Genoa, Italy
Alexandr Štefek	University of Defence, Czech Republic

Technical Committee Chair

Petr Stodola	University of Defence, Czech Republic

Technical Committee

Ronald C. Arkin	Georgia Institute of Technology, USA
Özkan Atan	Van Yüzüncü Yil University, Turkey
Richard Balogh	Slovak University of Technology in Bratislava, Slovakia
Yves Bergeon	CREC St-Cyr, France
Marco Biagini	NATO M&S COE, Rome, Italy
Antonio Bicchi	University of Pisa, Italy
Dalibor Biolek	University of Defence, Czech Republic
Arnau Carrera Viñas	NATO STO CMRE, Italy
Josef Casar	University of Defence, Czech Republic
Erdal Çayirci	University of Stavanger, Norway
Massimo Cossentino	National Research Council of Italy, Italy
Andrea D'Ambrogio	University of Rome Tor Vergata, Italy
Frédéric Dalorso	NATO ACT Autonomy Program, USA
Radek Doskocil	University of Defence, Czech Republic
Jan Faigl	Czech Technical University in Prague, Czech Republic
Luca Faramondi	Università Campus Bio-Medico di Roma, Italy
Jan Farlik	University of Defence, Czech Republic
Pavel Foltin	University of Defence, Czech Republic
Petr Frantis	University of Defence, Czech Republic
Jakub Fučík	University of Defence, Czech Republic
Corrado Guarino Lo Bianco	University of Parma, Italy
Karel Hajek	University of Defence, Czech Republic
Kamila Hasilová	University of Defence, Czech Republic
Vaclav Hlavac	Czech Technical University in Prague, Czech Republic
Jan Hodicky	NATO HQ SACT, USA
Jan Holub	Czech Technical University in Prague, Czech Republic

Jaroslav Hrdina	Brno University of Technology, Czech Republic
Thomas C. Irwin	Joint Force Development, DoD, USA
Shafagh Jafer	Embry-Riddle Aeronautical University, USA
Sebastian Jahnen	Universität der Bundeswehr München, Germany
Jason Jones	NATO M&S COE, Italy
Piotr Kosiuczenko	Military University of Technology, Poland
Tomáš Krajník	Czech Technical University in Prague, Czech Republic
Miroslav Kulich	Czech Technical University in Prague, Czech Republic
Václav Křivánek	University of Defence, Czech Republic
Jan Leuchter	University of Defence, Czech Republic
Pavel Manas	University of Defence, Czech Republic
Michele Martelli	University of Genoa, Italy
Jan Mazal	NATO M&S COE, Rome, Italy
Vladimír Mostýn	Technical University of Ostrava, Czech Republic
Pierpaolo Murrieri	Leonardo S.p.A., Italy
Andrzej Najgebauer	Military University of Technology, Poland
Jan Nohel	University of Defense, Czech Republic
Petr Novák	Technical University of Ostrava, Czech Republic
Gabriele Oliva	Università Campus Bio-Medico di Roma, Italy
Lucia Pallottino	University of Pisa, Italy
Luca Palombi	NATO MS COE, Italy
Václav Přenosil	Masaryk University, Czech Republic
Libor Přeučil	Czech Technical University in Prague, Czech Republic
Dalibor Procházka	University of Defence, Czech Republic
Josef Procházka	University of Defence, Czech Republic
Paolo Proietti	MIMOS, Italy
Jan Rohac	Czech Technical University in Prague, Czech Republic
Milan Rollo	Czech Technical University in Prague, Czech Republic
Marian Rybansky	University of Defence, Czech Republic
Martin Saska	Czech Technical University in Prague, Czech Republic
Vaclav Skala	University of West Bohemia, Czech Republic
Marcin Sosnowski	Jan Dlugosz University in Czestochowa, Poland
Julie M. Stark	Office of Naval Research Global, USA
Vadim Starý	University of Defence, Czech Republic
Jiří Štoller	University of Defence, Czech Republic
Peter Stütz	Universität der Bundeswehr München, Germany
Andreas Tolk	The MITRE Corporation, USA
Petr Vašík	Brno University of Technology, Czech Republic
Jiří Vokřínek	Czech Technical University in Prague, Czech Republic
Přemysl Volf	Czech Technical University in Prague, Czech Republic
Luděk Žalud	Brno University of Technology, Czech Republic
Jan Zezula	University of Defence, Czech Republic
Fumin Zhang	Georgia Institute of Technology, USA
Radomír Ščurek	Technical University of Ostrava, Czech Republic

Contents

Future Challenges of Advanced M&S Technology

Quantum Computing Based on Quantum Bit Algebra QBA

Jaroslav Hrdina and Radek Tichý(✉)

Institute of Mathematics, Faculty of Mechanical Engineering, Brno University of Technology, Technická 2896/2, 616 69 Brno, Czech Republic
hrdina@fme.vutbr.cz, Radek.Tichy@vutbr.cz

Abstract. We describe quantum bit algebra (QBA) as an algebra for quantum formalism. We represent the qubits as a vectors in QBA and the gates as a conjugations. We describe the algebra of their infinitesimal isomorphisms and discuss their relations to orthogobnal Lie algebra $\mathfrak{so}(n)$. We show that QBA can be seen as a model of a hyperbolic quantum computing instead of the classical one.

Keywords: Geometric algebra · Geometric algebra computing · GAALOP · Quantum computing · Hyperbolic quantum mechanics · Quantum bit algebra

1 Introduction

The geometric algebras have played significant role in the engineering applications. The main tool in the scientific engineering computations lies in conformal geometric algebra, for more informations see the classical books [3,6,16,18] or engineering papers [9,10,12,19]. In the context of specific applications, the panorama of geometric algebras becomes more and more popular, for example we can find some applications of the compass ruler algebra in the book [7] or the geometric algebra of conics in the paper [11] or the others algebras in the paper [13].

In the quantum computing based on classical quantum mechanics [2] a qubit can be seen as a superposition of the bits $|0\rangle$ and $|1\rangle$ in the vector in two dimensional vector space over complex numbers $\mathbb{C}^2$:

$$|\psi\rangle = a_0|0\rangle + a_1|1\rangle, \tag{1}$$

where $a_0, a_1 \in \mathbb{C}$ together with the normalization condition $|a_0|^2 + |a_1|^2 = a_0 a_0^* + a_1 a_1^* = 1$, where z^* is a complex conjugation of the complex number z. This is the usual representation of a superposition of the two states $|0\rangle$ and $|1\rangle$ in the classical notation (see, e.g., [2,8,17]). We can see $|a_0|^2$ as the probability that the quantum bit indicates the state $|0\rangle$, while $|a_1|^2$ is the probability that the quantum bit will be found in the state $|1\rangle$.

The first author was supported by the grant no. FSI-S-20-6187. The second author was supported by solution grand FV20-31 science Fund of the FME 2020 at Brno University of Technology.

J. Mazal et al. (Eds.): MESAS 2020, LNCS 12619, pp. 3–14, 2021.
https://doi.org/10.1007/978-3-030-70740-8_1

2 Geometric Algebra

Geometric algebra is a distributive associative algebra over quadratic space $\mathbb{R}^{p,q}$ with the quadratic form of signature (p, q) and with the neutral element $1 \in \mathbb{R}$. The quadratic form defines relations for a scalar product on basis vectors, i.e. basis elements of the quadratic space denoted by $\{\sigma_1, \sigma_2, \ldots, \sigma_{p+q}\} \in \mathbb{R}^{p,q} \subset \mathbb{G}_{p,q}$, in such a way that

$$\sigma_i \cdot \sigma_j = \begin{cases} +1, & 1 \leq (i = j) \leq p, \\ -1, & p < (i = j) \leq p + q, \\ 0, & i \neq j. \end{cases}$$

The product of the algebra called geometric product satisfies the condition

$$a^2 = a \cdot a = Q(a), \quad \forall a \in \mathbb{R}^{p,q}.$$

Thus the geometric product can be defined by relations:

$$\sigma_i \sigma_j = \begin{cases} +1, & 1 \leq (i = j) \leq p, \\ -1, & p < (i = j) \leq p + q, \\ -\sigma_j \sigma_i, & i \neq j. \end{cases}$$

The conclusion of the last case of the geometric product relations is that the geometric algebra is a graded algebra, i.e. the product of a various number of distinct basis vectors represents an element of the algebra. For instance $\sigma_i \sigma_j$, $1 \leq i < j \leq p + q$ is a basis element of the algebra called basis bivector, $\sigma_i \sigma_j \sigma_k$, $1 \leq i < j < k \leq p + q$ is basis 3-vector etc. The element of the highest grade, i.e. the geometric product of $p+q$ different vectors is called pseudoscalar. This element is denoted by I.

3 Geometric Algebra $\mathbb{G}_3$

In the geometric algebra settings the spin states (states of quantum system) can be written in $\mathbb{G}_3$ framework. Let us recall that, the algebra $\mathbb{G}_3$ is an algebra over real numbers $\mathbb{R}$ based on three dimensional real vector space $\mathbb{R}^3$ equipped with the Euclidean scalar product of signature $(+, +, +)$. The vectors correspond to elements σ_1, σ_2 and σ_3 together with identities $\sigma_1^2 = \sigma_2^2 = \sigma_3^2 = 1$. The identification between $\mathbb{C}^2$ and the algebra of even grade elements $\mathbb{G}_3^+$ is based on isomorphism

$$|\psi\rangle = \begin{pmatrix} a_0 + ia_3 \\ -a_2 + ia_1 \end{pmatrix} \leftrightarrow \psi = a_0 + \sum_{k=1}^{3} a_k I \sigma_k, \tag{2}$$

where $I = \sigma_1\sigma_2\sigma_3$. In particular, the identification $|0\rangle \leftrightarrow 1$, $|1\rangle \leftrightarrow -I\sigma_2$ implies a complex multiplication $i_{\mathbb{C}}|\psi\rangle \leftrightarrow \psi I\sigma_3$, which one can see by straightforward computation

$$\begin{aligned}(a_0 + a_3 i_{\mathbb{C}})|0\rangle + (-a_2 + a_1 i_{\mathbb{C}})|1\rangle &\leftrightarrow 1(a_0 + a_3 I\sigma_3) - I\sigma_2(-a_2 + a_1 I\sigma_3)\\ &= a_0 + a_3 I\sigma_3 + I\sigma_2 a_2 - I\sigma_2 a_1 I\sigma_3\\ &= a_0 + a_3 I\sigma_3 + a_2 I\sigma_2 + a_1\sigma_2\sigma_3\\ &= a_0 + a_3 I\sigma_3 + a_2 I\sigma_2 + a_1 I\sigma_1\end{aligned}$$

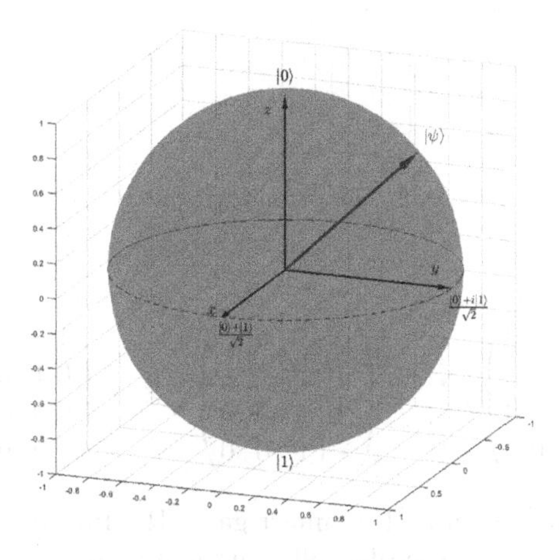

Fig. 1. Bloch's sphere generated by Matlab. Blue circle denotes all qubits with the same probability of states $|0\rangle$ and $|1\rangle$ (Color figure online).

The total probability of the system is $\langle\psi|\psi\rangle = 1$, it follows the rule $|a_0|^2 + |a_1|^2 = 1$ and we can write $|\psi\rangle$ in spherical coordinates $(\gamma, \theta, \varphi)$ as an element of the three dimensional Bloch sphere (see Fig. 1)

$$|\psi\rangle = e^{i\gamma}\left(\cos\left(\frac{\theta}{2}\right)|0\rangle + e^{i\varphi}\sin\left(\frac{\theta}{2}\right)|1\rangle\right),$$

which is equivalent to a group of unitary quaternions and because the probability of the state $|\psi\rangle$ is $\langle\psi|\psi\rangle$ and $\langle e^{-i\gamma}\psi|e^{-i\gamma}\psi\rangle = e^{-i\gamma}e^{i\gamma}\langle\psi|\psi\rangle = \langle\psi|\psi\rangle$ we can use three dimensional sphere parametrisation

$$|\psi\rangle = \cos\left(\frac{\theta}{2}\right)|0\rangle + e^{i\varphi}\sin\left(\frac{\theta}{2}\right)|1\rangle. \tag{3}$$

4 Quantum Gates

Let us introduce the list of basis quantum gates, see Fig. 2. We can introduce Pauli X-gate, Pauli Y-gate and Pauli Z-gate. We can see two dimensional complex vector space $\mathbb{C}^2$ as a four dimensional real vector space $\mathbb{R}^4$. In this context, one can see Pauli matrices as following 4×4 matrices

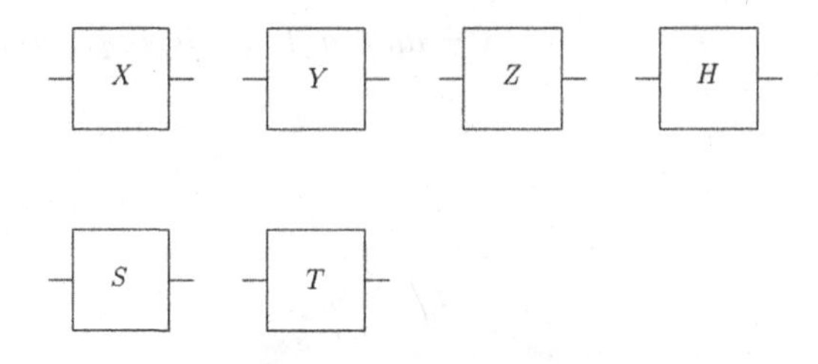

Fig. 2. Quantum logic gates.

$$X = \begin{pmatrix} 0 & 0 & 1 & 0 \\ 0 & 0 & 0 & 1 \\ 1 & 0 & 0 & 0 \\ 0 & 1 & 0 & 0 \end{pmatrix}, \; Y = \begin{pmatrix} 0 & 0 & 0 & 1 \\ 0 & 0 & -1 & 0 \\ 0 & 1 & 0 & 0 \\ -1 & 0 & 0 & 0 \end{pmatrix}, \; Z = \begin{pmatrix} 1 & 0 & 0 & 0 \\ 0 & 1 & 0 & 0 \\ 0 & 0 & -1 & 0 \\ 0 & 0 & 0 & -1 \end{pmatrix}.$$

The next gates are for example Hadamart gate (H), Phase gate (S) and $\frac{\pi}{8}$ gate (T). These gates are represented by following matrices:

$$H = \frac{1}{\sqrt{2}} \begin{pmatrix} 1 & 0 & 1 & 0 \\ 0 & 1 & 0 & 1 \\ 1 & 0 & -1 & 0 \\ 0 & 1 & 0 & -1 \end{pmatrix}, \; S = \begin{pmatrix} 1 & 0 & 0 & 0 \\ 0 & 1 & 0 & 0 \\ 0 & 0 & 0 & 1 \\ 0 & 0 & -1 & 0 \end{pmatrix}, \; T = \begin{pmatrix} 1 & 0 & 0 & 0 \\ 0 & 1 & 0 & 0 \\ 0 & 0 & \cos\frac{\pi}{4} & -\sin\frac{\pi}{4} \\ 0 & 0 & \sin\frac{\pi}{4} & \cos\frac{\pi}{4} \end{pmatrix}.$$

We can define the quantum equivalents to classical computing protocol, for example one can represent the not gate by transformation

$$|0\rangle \to |1\rangle$$
$$|1\rangle \to |0\rangle$$

Fig. 3. X-gate (NOT)

which corresponds to X-Pauli matrix. In the geometric algebra $\mathbb{G}_3$ gates are represented by the geometric product of algebra elements with a qubit. For example Pauli gates X, Y, Z are represented as

$$\psi \xrightarrow{X} \sigma_1 \psi \sigma_3,$$
$$\psi \xrightarrow{Y} \sigma_2 \psi \sigma_3,$$
$$\psi \xrightarrow{Z} \sigma_3 \psi \sigma_3,$$

where ψ is a qubit, see equation (2). Finally, one can formulate the quantum protocol as a circuit over quantum gates. For example in Fig. 4 one can see linear transformation based on matrix

$$\begin{pmatrix} 0&0&1&0\\ 0&0&0&1\\ 1&0&0&0\\ 0&1&0&0 \end{pmatrix} \begin{pmatrix} 1&0&0&0\\ 0&1&0&0\\ 0&0&0&1\\ 0&0&-1&0 \end{pmatrix} \begin{pmatrix} 0&0&1&0\\ 0&0&0&1\\ 1&0&0&0\\ 0&1&0&0 \end{pmatrix} = \begin{pmatrix} 0&0&0&1\\ 0&0&-1&0\\ 1&0&0&0\\ 0&1&0&0 \end{pmatrix} \begin{pmatrix} 0&0&1&0\\ 0&0&0&1\\ 1&0&0&0\\ 0&1&0&0 \end{pmatrix} = \begin{pmatrix} 0&1&0&0\\ -1&0&0&0\\ 0&0&1&0\\ 0&0&0&1 \end{pmatrix}.$$

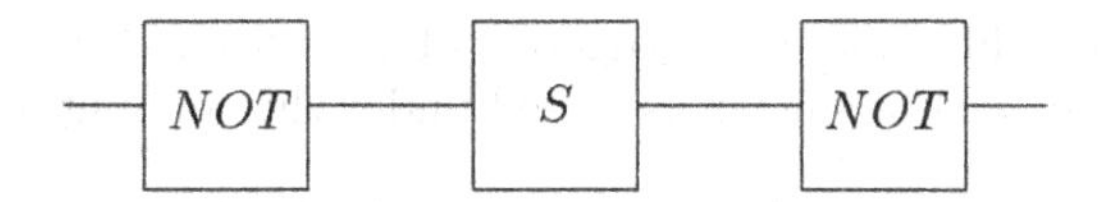

Fig. 4. Example of quantum protocol.

5 QBA

A different possibility how to implement qubit in the geometric algebra framework has been introduced in paper [4]. The authors proposed to model a qubit by unit elements of Clifford algebra $Cl(2, 2)$ with an indefinite bilinear form of the signature $(+, -, +, -)$. We denote the basis of the algebra by vectors

$$(\gamma_0^x, \gamma_0^t, \gamma_1^x, \gamma_1^t),$$

where $(\gamma_0^x)^2 = (\gamma_1^x)^2 = 1$ and $(\gamma_0^t)^2 = (\gamma_1^t)^2 = -1$. A qubit can be defined as

$$\psi = c_0^x \gamma_0^x + c_0^t \gamma_0^t + c_1^x \gamma_1^x + c_1^t \gamma_1^t, \tag{4}$$

where $c_0^x, c_0^t, c_1^x, c_1^t \in \mathbb{R}$. The elements $c_0^x \gamma_0^x + c_0^t \gamma_0^t$, such that $(c_0^x)^2 - (c_0^t)^2 = 1$, define the state $|0\rangle$ and the elements $c_1^x \gamma_1^x + c_1^t \gamma_1^t$, such that $(c_1^x)^2 - (c_1^t)^2 = 1$, define the state $|1\rangle$. It implies that qubits belong to a 3-dimensional surface defined by the identity $\langle \psi | \psi \rangle = 1$ so

$$(c_0^x)^2 - (c_0^t)^2 + (c_1^x)^2 - (c_1^t)^2 = 1,$$

or using standard hyperboloid coordinates $(\theta, \varphi_1, \varphi_2)$ equivalently,

$$
\begin{aligned}
|\psi\rangle &= \left(\cos\left(\frac{\theta}{2}\right) e^{\gamma_0^t\gamma_0^x\varphi_0}\gamma_0^x + \sin\left(\frac{\theta}{2}\right) e^{\gamma_1^t\gamma_1^x\varphi_1}\gamma_1^x\right) \\
&= e^{\gamma_0^t\gamma_0^x\varphi_0}\left(\cos\left(\frac{\theta}{2}\right)\gamma_0^x + e^{\gamma_1^t\gamma_1^x\varphi_1-\gamma_0^t\gamma_0^x\varphi_0}\sin\left(\frac{\theta}{2}\right)\gamma_1^x\right)
\end{aligned}
\tag{5}
$$

where $e^{\gamma_0^t\gamma_0^x\varphi_0} = \cosh(\varphi_0)+\sinh(\varphi_0)\gamma_0^t\gamma_0^x$ and $e^{\gamma_1^t\gamma_1^x\varphi_1} = \cosh(\varphi_1)+\sinh(\varphi_1)\gamma_1^t\gamma_1^x$, so qubits belong to hyperboloid given by (5). Algebraically we can describe a vector ψ from (4) as an expression

$$\psi = c_0^x\gamma_0^x + c_0^t\gamma_0^t + c_1^x\gamma_1^x + c_1^t\gamma_1^t = (c_0^x + c_0^t\gamma_0^t\gamma_0^x)\gamma_0^x + (c_1^x + c_1^t\gamma_1^t\gamma_1^x)\gamma_1^x,$$

such that $(\gamma_i^t\gamma_i^x)^2 = \gamma_i^t\gamma_i^x\gamma_i^t\gamma_i^x = -\gamma_i^t\gamma_i^t\gamma_i^x\gamma_1^x = 1$, such that $i \in \{0,1\}$. So we can represent alternatively a qubit as a couple of elements from geometric algebra $\mathbb{G}_2$.

6 Lie Algebra of QBA Bivectors

QBA contains a 6-dimensional subalgebra of bivectors, denoted as $Cl_2(2,2)$, which is also a Lie algebra since it is closed under the commutator given by

$$[x,y] = xy - yx,\ \ x,y \in Cl_2(2,2),$$

where the multiplication is a standard geometric product in $Cl(2,2)$. For better readability we denote the basis elements by $E_1,\ldots,E_6$ and define them as

$$
\begin{aligned}
E_1 &= \frac{1}{2}\gamma_0^x\gamma_0^t, \quad E_2 = \frac{1}{2}\gamma_0^x\gamma_1^x, \quad E_3 = \frac{1}{2}\gamma_0^x\gamma_1^t, \\
E_4 &= \frac{1}{2}\gamma_0^t\gamma_1^x, \quad E_5 = \frac{1}{2}\gamma_0^t\gamma_1^t, \quad E_6 = \frac{1}{2}\gamma_1^x\gamma_1^t.
\end{aligned}
$$

For instance, the commutator of the elements E_1, E_2 is

$$
\begin{aligned}
[E_1,E_2] &= \frac{1}{4}\left(\gamma_0^x\gamma_0^t\gamma_0^x\gamma_1^x - \gamma_0^x\gamma_1^x\gamma_0^x\gamma_0^t\right) \\
&= \frac{1}{4}\left(-(\gamma_0^x)^2\gamma_0^t\gamma_1^x + (\gamma_0^x)^2\gamma_1^x\gamma_0^t\right) \\
&= \frac{1}{4}\left(-2\gamma_0^t\gamma_1^x\right) \\
&= -E_4.
\end{aligned}
$$

This example also clarifies why we are using the factor $\frac{1}{2}$ at each element. By the same procedure for all combinations of basis elements we obtain a multiplication table:

[,]	E_1	E_2	E_3	E_4	E_5	E_6
E_1	0	$-E_4$	$-E_5$	$-E_2$	$-E_3$	0
E_2	E_4	0	$-E_6$	$-E_1$	0	E_3
E_3	E_5	E_6	0	0	E_1	E_2
E_4	E_2	E_1	0	0	E_6	E_5
E_5	E_3	0	$-E_1$	$-E_6$	0	E_4
E_6	0	$-E_3$	$-E_2$	$-E_5$	$-E_4$	0

Clearly, elements E_4, E_5, E_6 span a Lie subalgebra. If we choose a different basis K, L, M of this subalgebra such that

$$K := E_4 + E_5, \quad L := E_6, \quad M := E_4 - E_5,$$

then commutators are

$$[K, L] = K, \quad [K, M] = -2L, \quad [L, M] = M$$

and we can represented them by multiplication table

[,]	K	L	M
K	0	K	−2L
L	−K	0	M
M	2L	−M	0

Since this is the table of commutators for special linear algebra $\mathfrak{sl}(2)$, the algebra spanned by generators E_4, E_5, E_6 is isomorphic to $\mathfrak{sl}(2)$. The adjoint representation (the representation provided by commutator operation, see [14]) is given by the following matrices:

$$ad_{E_4} = \begin{pmatrix} X_1 & 0 \\ 0 & -Z_1 \end{pmatrix}, \quad ad_{E_5} = \begin{pmatrix} Y_1 & 0 \\ 0 & -Y_1 \end{pmatrix}, \quad ad_{E_6} = \begin{pmatrix} Z_1 & 0 \\ 0 & -X_1 \end{pmatrix},$$

where

$$X_1 = \begin{pmatrix} 0 & 0 & 0 \\ 0 & 0 & 1 \\ 0 & 1 & 0 \end{pmatrix}, \quad Y_1 = \begin{pmatrix} 0 & 0 & 1 \\ 0 & 0 & 0 \\ -1 & 0 & 0 \end{pmatrix}, \quad Z_1 = \begin{pmatrix} 0 & -1 & 0 \\ -1 & 0 & 0 \\ 0 & 0 & 0 \end{pmatrix}.$$

Elements E_1, E_2, E_3 do not span a Lie algebra, however if we represent these elements in adjoint representation we obtain block matrices

$$ad_{E_1} = \begin{pmatrix} 0 & Z_2 \\ -X_2 & 0 \end{pmatrix}, \quad ad_{E_2} = \begin{pmatrix} 0 & Y_2 \\ -Y_2 & 0 \end{pmatrix}, \quad ad_{E_3} = \begin{pmatrix} 0 & X_2 \\ -Z_2 & 0 \end{pmatrix},$$

where

$$X_2 = \begin{pmatrix} 0\,1\,0 \\ 0\,0\,1 \\ 0\,0\,0 \end{pmatrix}, \quad Y_2 = \begin{pmatrix} -1\,0\,0 \\ 0\,\;0\,0 \\ 0\,\;0\,1 \end{pmatrix}, \quad Z_2 = \begin{pmatrix} 0 & 0 & 0 \\ -1 & 0 & 0 \\ 0 & -1 & 0 \end{pmatrix}.$$

Matrices X, Y, Z span a Lie algebra with the multiplication table

[,]	X	Y	Z
X	0	X	Y
Y	−X	0	Z
Z	−Y	−Z	0

which leads again to a Lie algebra isomorphic to $\mathfrak{sl}(2)$ and adjoint representation represents the splitting $\langle E_1, E_2, E_3\rangle \oplus \langle E_4, E_5, E_6\rangle$.

The classification of this Lie algebra can also be determined by the eigenvalues of adjoint representation. Since the adjoint representation of matrices X, Y, Z has 2 distinct real eigenvalues, it belongs to the class of Lie algebras isomorphic to $\mathfrak{sl}(2)$, see [1]. Thus the Lie algebra of bivectors of $Cl(2,2)$ can be written as semidirect product

$$Cl_2(2,2) \cong \mathfrak{sl}(2) \ltimes \mathfrak{sl}(2).$$

Finally, the adjoint representation of whole algebra is based on the following matrices:

$$ad_{\sum_{i=1}^{6} a_i E_i} = \begin{pmatrix} a_4X_1 + a_5Y_1 + a_6Z_1 & a_1Z_2 + a_2Y_2 + a_3X_2 \\ -a_1X_2 - a_2Y_2 - a_3Z_2 & -a_4Z_1 - a_5Y_1 - a_6X_1 \end{pmatrix}$$

To define the complex multiplication on qubits by the standard conjungation we need a reduction with respect to an almost complex structure. On the Lie algebra level we would have to include the Lie algebra

$$\mathfrak{gl}(2,\mathbb{C}) = \left\{ \begin{pmatrix} A & B \\ -B & A \end{pmatrix} : A, B \in Mat_2(\mathbb{R}) \right\}$$

into Lie algebra $\mathfrak{sl}(2) \ltimes \mathfrak{sl}(2)$ which is impossible.

7 Hyperbolic Qubits

Now, let us look closer to the Hyperbolic quantum formalism [15], we will see that the qubits here belong to the same Bloch's hyperboloid (5) as in QBA. In detail, hyperbolic numbers belong to linear algebra over $\mathbb{R}$, such that each element can be written as the formal linear combination $z = x + jy$, with respect to identity $j^2 = 1$. The hyperbolic conjugation is defined as $(x + jy)^* = x - jy$

and then Hyperbolic norm is defined as $|z| = \sqrt{zz^*} = \sqrt{x^2 - y^2}$. In the sense of Clifford algebras, the Hyperbolic numbers are just Geometric algebra $\mathbb{G}_1$. Let us note that $\mathbb{G}_1$ is not an integral domain, because there are zero divisors, for example $(1 + j)(1 - j) = 0$. Hyperbolic vector space is then a module $\mathcal{H}$ over geometric algebra $\mathbb{G}_1$.

We define Hyperbolic scalar product as a $\mathbb{G}_1$-bilinear map $(\ ,\ ) : \mathcal{H} \times \mathcal{H} \to \mathbb{G}_1$ together with properties

1. $(au + bv, w) = a(u, w) + b(v, w)$ for all $a, b \in \mathbb{R}$ and $u, v, v \in \mathcal{H}$
2. $(u, v) = (v, u)^*$ for all $u, v, v \in \mathcal{H}$
3. $(u, v) = 0$ for all $u \in \mathcal{H}$ if and only if $v = 0$

The classical Hyperbolic scalar product on $\mathcal{H}^2$ can by written by matrices as $(u, v) = (u^*)^T v$. In the hyperbolic quantum computing the qubits belong to the two dimensional hyperbolic vector space $\mathcal{H}^2$ and we can describe them by basis $|0\rangle$ and $|1\rangle$ as

$$|\psi\rangle = \alpha|0\rangle + \beta|1\rangle,$$

where $\alpha, \beta \in \mathbb{G}_1$ such that $\langle\psi|\psi\rangle = 1$, where

$$\begin{aligned}
\langle\psi|\psi\rangle &= \langle(\alpha|0\rangle + \beta|1\rangle)|(\alpha|0\rangle + \beta|1\rangle)\rangle \\
&= \langle\alpha 0|\alpha 0\rangle + \langle\alpha 0|\beta 1\rangle + \langle\beta 1|\alpha 0\rangle + \langle\beta 1|\beta 1\rangle \\
&= \alpha\alpha^*\langle 0|0\rangle + \alpha\beta^*\langle 0|1\rangle + \beta\alpha^*\langle 1|0\rangle + \beta\beta^*\langle 1|1\rangle \\
&= \alpha_0^2 - \alpha_1^2 + \beta_0^2 - \beta_1^2
\end{aligned}$$

and the hyperbolic qubits belongs to Bloch's hyperboloid (5).

8 GAALOPWeb for Qubits

For optimizing Geometric Algebra algorithms there is an online tool originally designed by Dietmar Hildebrand called GAALOPWeb[1], see [5]. With the help of this online software we can optimize a code based on geometric algebras and we obtain a code for programming languages such as C/C++, Python, Matlab, etc. GAALOP can handle various algebras and recently authors have added a possibility to work with QBA.

At first the QBA setting allows a user to set the number of qubits. Then the user can edit the GAALOPScript code and by pressing the *Run* Button the GAALOP automatically optimizes the code. A new code designed in a chosen programming language is then displayed to the user. The QBA vectors are represented as displayed in Table 1.

[1] http://www.gaalop.de/gaalopweb-for-qubits/.

Table 1. The representation of basis vectors of QBA in GAALOP.

	Basis vector	Signature	GAALOPScript
$\vert 0\rangle$ (space-like)	γ_0^x	$+1$	e0x
$\vert 0\rangle$ (time-like)	γ_0^t	-1	e0t
$\vert 1\rangle$ (space-like)	γ_1^x	$+1$	e1x
$\vert 1\rangle$ (time-like)	γ_1^t	-1	e1t

9 The Example of the Basic Quantum Gate NOT

For example the not gate in Fig. 3

$$(\alpha_0 + \alpha_1 i)|0\rangle + (\beta_0 + \beta_1 i)|1\rangle \mapsto (\beta_0 + \beta_1 i)|0\rangle + (\alpha_0 + \alpha_1 i)|1\rangle$$

can be realized in QBA through the conjugation of the element

$$\psi = \alpha_0\gamma_0^x + \alpha_1\gamma_0^t + \beta_0\gamma_1^x + \beta_1\gamma_1^t, \tag{6}$$

by the element $\frac{1}{2}(\gamma_0^x + \gamma_1^x)(\gamma_0^t + \gamma_1^t)$:

$$\begin{aligned}&(\frac{1}{2}(\gamma_0^x + \gamma_1^x)(\gamma_0^t + \gamma_1^t))(\alpha_0\gamma_0^x + \alpha_1\gamma_0^t + \beta_0\gamma_1^x + \beta_1\gamma_1^t)(\frac{1}{2}(\gamma_0^t + \gamma_1^t)(\gamma_0^x + \gamma_1^x))\\ &= (\beta_0\gamma_0^x + \beta_1\gamma_0^t + \alpha_0\gamma_1^x + \alpha_1\gamma_1^t).\end{aligned}$$

To demonstrate an optimization of the QBA code using Gaalop, we can implement this gate by trivial input code displayed in Fig. 5. By running optimization process, Gaalop produces the C++ function, where all computation is done by elementary operations multiplication and addition of real numbers (in our case there are only assignments of values without any computations), see Fig. 6. For example the not gate can be used in the circuits which takes as input a factorize state and output as entangled Bell states is in Fig. 7.

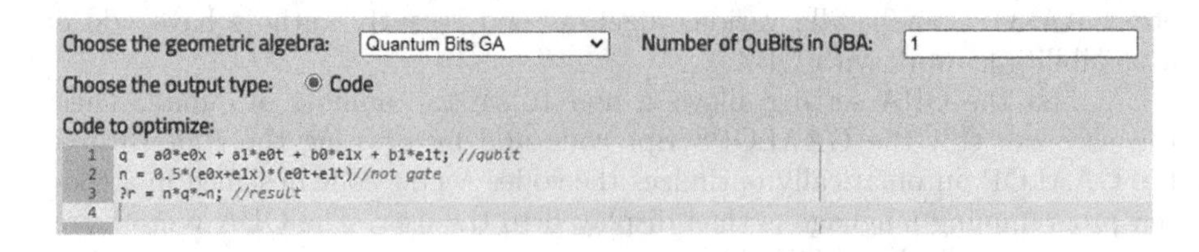

Fig. 5. NOT gate input code for Gaalop

CODE:

```
void calculate(float a0, float a1, float b0, float b1, float r[16]) {
        r[1] = b0; // e0x
        r[2] = a0; // e1x
        r[3] = b1; // e0t
        r[4] = a1; // e1t
}
```

Fig. 6. NOT gate output code from Gaalop

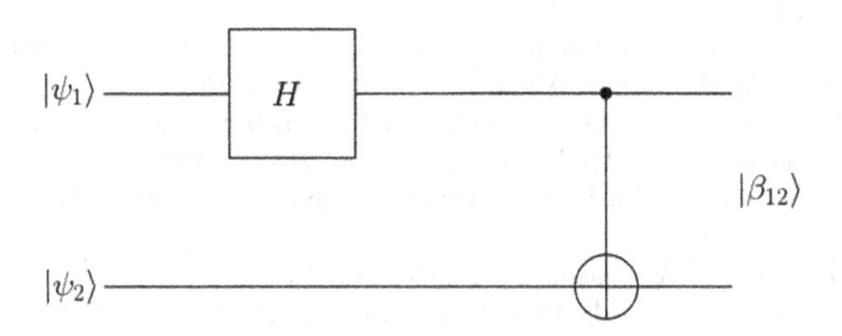

Fig. 7. Example of quantum circiuts

10 Conclusions

We introduced the concept of qantum computing in geometric algebra framework. We have showed two different representations of a qubit, one in algebra $\mathbb{G}_3$, where qubits belong to the Bloch sphere, and in algebra QBA ($\mathbb{G}_{2,2}$) where qubits lie on a hyperboloid. In Sect. 3, the subalgebra of bivectors of QBA as a Lie algebra was discussed. It has been showed that the algebra of the infinitesimal symmetries of QBA is $\mathfrak{sl}(2) \ltimes \mathfrak{sl}(2)$. It rather leads to the concept where one can define qantum gates as transformations belonging to $\mathfrak{sl}(2)$. For implementation of QBA quantum computing purposes we presented online GAALOP. This software can transform an algebra code into optimized code that computes with real numbers only.

References

1. Bowers, A.: Classification of three-dimensional real liealgebras, April 2005. http://www.math.ucsd.edu/~abowers/downloads/survey/3d_Lie_alg_classify.pdf
2. de Lima Marquezino, F., Portugal, R., Lavor, C.: A Primer on Quantum Computing. SCS. Springer, Cham (2019). https://doi.org/10.1007/978-3-030-19066-8
3. Dorst, L., Fontijne, D., Mann, S.: Geometric Algebra for Computer Science. An Object-Oriented Approach to Geometry. Morgan Kaufmann, Burlington (2007)
4. Hildenbrand, D., Steinmetz, C., Alves, R., Hrdina, J., Lavor, C.: An online calculator for Qubits based on geometric algebra. In: Magnenat-Thalmann, N., et al. (eds.) CGI 2020. LNCS, vol. 12221, pp. 526–537. Springer, Cham (2020). https://doi.org/10.1007/978-3-030-61864-3_45

5. Hildenbrand, D., Steinmetz, C., Tichý, R.: Gaalopweb for matlab: an easy to handle solution for industrial geometric algebra implementations. Adv. Appl. Clifford Algebr. **30**(4) (2020)
6. Hildenbrand, D.: Foundations of Geometric Algebra Computing. Springer, Heidelberg (2013). https://doi.org/10.1007/978-3-642-31794-1
7. Hildenbrand, D.: Introduction to Geometric Algebra Computing. Taylor & Francis Group, Boca Raton (2019)
8. Homeister, M.: Quantum Computing verstehen. Grundlagen - Anwendungen - Perspektiven. Friedrich Vieweg & Sohn Verlag (2018). https://doi.org/10.1007/978-3-658-22884-2
9. Hrdina, J., Návrat, A.: Binocular computer vision based on conformal geometric algebra. Adv. Appl. Clifford Algebras **27**, 1945–1959 (2017)
10. Hrdina, J., Návrat, A., Vašík, P.: Control of 3-link robotic snake based on conformal geometric algebra. Adv. Appl. Clifford Algebr. **26**(3) (2016)
11. Hrdina, J., Návrat, A., Vašík, P.: Geometric algebra for conics. Adv. Appl. Clifford Algebras **28**, 66 (2018)
12. Hrdina, J., Návrat, A., Vašík, P., Matoušek, R.: CGA-based robotic snake control. Adv. Appl. Clifford Algebras **27**(1), 621–632 (2016). https://doi.org/10.1007/s00006-016-0695-5
13. Hrdina, J., Vašík, P., Matoušek, R., Návrat, A.: Geometric algebras for uniform colour spaces. Math. Methods Appl. Sci. **41**, 4117–4130 (2018)
14. K. Erdmann, M.J.W.: Introduction to Lie Algebras. Springer, London (2011). https://www.ebook.de/de/product/4018299/k_erdmann_mark_j_wildon_introduction_to_lie_algebras.html
15. Khrennikov, A.: Hyperbolic quantum mechanics. Adv. Appl. Clifford Algebras **13**(1), 1–9 (2003)
16. Lounesto, P.: Clifford Algebra and Spinors. CUP, Cambridge (2006)
17. McMahon, D.: Quantum Computing Explained. Wiley, New York (2008)
18. Perwass, C.: Geometric Algebra with Applications in Engineering. Springer, Heidelberg (2009). https://doi.org/10.1007/978-3-540-89068-3
19. Stodola, M.: Monocular kinematics based on geometric algebras. In: Mazal, J. (ed.) MESAS 2018. LNCS, vol. 11472, pp. 121–129. Springer, Cham (2019). https://doi.org/10.1007/978-3-030-14984-0_10

On Building Communication Maps in Subterranean Environments

Martin Zoula, Miloš Prágr(✉), and Jan Faigl

Faculty of Electrical Engineering, Czech Technical University in Prague, Technická 2, 166 27 Prague, Czech Republic
{zoulamar,pragrmi1,faiglj}@fel.cvut.cz
https://comrob.fel.cvut.cz

Abstract. Communication is of crucial importance for coordinating a team of mobile robotic units. In environments such as underground tunnels, the propagation of wireless signals is affected by nontrivial physical phenomena. Hence, both modeling of the communication properties and the consequent task to estimate where communication is available becomes demanding. A communication map is a tool assessing the characteristic of communication between two arbitrary spatial coordinates. The existing approaches based on interpolation of a priori obtained spatial measurements do not provide precise extrapolation estimates for unvisited locations. Therefore, we propose to address the extrapolation of the signal strength by a position-independent model based on approximating the obstacle occupancy ratio between the signal source and receiver. The proposed approach is compared to the existing attenuation models based on free-space path loss and spatial projection using a natural cave dataset. Based on the reported results, the proposed approach provides more accurate predictions than the existing approaches.

Keywords: Subterranean · Communication map · Gaussian process

1 Introduction

In multi-robot scenarios, reliable wireless communication between individual units is an important feature, together with estimating the communication availability in cases where signal propagation is difficult due to natural or artificial constraints. Our research is motivated by deployments of a multi-robot team in communication denied subterranean environments such as cave systems, where most of the signal obstructing mass is static. These environments suffer from nontrivial signal propagation, which cannot be efficiently predicted using simple attenuation models. Even though robots can collaborate without mutual communication while using preset task division algorithms, on-line negotiation is the key factor when reacting to dynamic events. Hence, it is desirable to reliably predict whether and how well a communication channel can be established given the current mission state. A signal propagation model could support tasks like

J. Mazal et al. (Eds.): MESAS 2020, LNCS 12619, pp. 15–28, 2021.
https://doi.org/10.1007/978-3-030-70740-8_2

building communication infrastructure using communication beacons [19]. Also, fast and robust network repairs [4] and reconnection [16] can be maintained. Furthermore, it is possible to predict changes in the network topology [20].

Communication models simulate and approximate the complicated phenomena behind the physical media. The communication models range from statistical Bernoulli or Gilbert-Elliot [7] models to complex raytracing-based physics simulators [22]. The former class yields limited information about modeled communication such as packet delivery probability, yet they are computationally efficient. The latter class tends to involve computationally demanding operations, but yields precise and relevant results, e.g., about the total received power since it accounts for various rigorously defined physical phenomena. Probably the simplest case of physics-based communication models is the *Free-Space Path Loss* (FSPL) attenuation model based on the Friis formula [25]. Note, these approaches rely on a priori known parameters such as signal wavelength or material properties. Besides, setups using no communication model can be used [1], where the robot in need of communication tries to establish a link with another robot by a costly blind broadcast.

Given a communication model that encapsulates the underlying physical media characteristics into a prediction, a communication map is defined as a computational tool determining communication characteristics based on two arbitrary spatial coordinates of the transmitting device and the receiving device. The predicted characteristics may be various, e.g., *Received Signal Strength Indicator* (RSSI) or packet loss ratio. The communication map can be used by a robot to assess the supposed difficulty of communicating to an arbitrary location in the environment. The map can be implemented in various ways, e.g., utilizing self-organizing maps in an acoustic setup as proposed in [6]. However, recent work on communication maps is based on a Gaussian Process (GP) regressor, a soft model that relies on training samples to serve as an interpolation corpus [2,3,18]. Due to the nature of GPs, the model also yields a predictive variance together with the estimate. However, the existing spatial projection method [3] is limited to interpolating the measured data and does not provide satisfiable measurement extrapolations for the unvisited areas.

In this paper, we propose a more general approach to implementing communication maps based on a new computational layer allowing the model to comprehend basic structural properties of the environment. The key benefit of the proposed approach is the ability to extrapolate previous measurements to the so-far unvisited locations. Furthermore, the learned model for one specific environment could be utilized for different environments with similar physical properties. The proposed idea is based on a descriptor of the queried locations that comprehend the spatial information of in-vivo measurements of signal characteristics transformed into a low-dimensional vector according to the vicinity of the queried points. The computed value of the descriptor is fed as the training data to learn a GP-based regressor. In particular, the RSSI metric is utilized to model the signal characteristics. The proposed approach is compared to the baseline approaches of the FSPL [25] and spatial projection (SP) based on a GP regressor [3]. The reported results support the feasibility of the proposed approach and show significant improvements in the prediction accuracy in the extrapolation tasks.

The rest of the paper is organized as follows. A brief overview of the related work with the focus on multi-robot wireless communication is presented in Sect. 2. The proposed approach for communication map building is introduced in Sect. 3. Results of the experimental evaluation are reported and discussed in Sect. 4, and the work is concluded in Sect. 5.

2 Related Work

Communication in a group of mobile robots can be understood as an instance of the *Mobile Ad-Hoc Network* (MANET). MANETs are networks with a topology that changes over time as individual network nodes change their spatial coordinates. Due to their nature, MANETs are often implemented using wireless technology allowing for more flexible communication link management. In such networks, fast and robust routing protocols [15] are needed to ensure timely packet delivery with an overhead as low as possible. Since the network nodes are dynamic, some knowledge about future signal properties need to be known in advance; a communication map can be used to obtain it.

The effect and impact of network nodes' mobility are investigated in [12], showing the importance of the prior information about nodes' behavior in multi-hop message routing problems. The current state-of-the-art multi-robot applications use prior signal or motion models to predict future positions of individual nodes and associated signal characteristics. The impact on the message routing decision-making is forthright; nodes known to leave communication range or enter signal-denied region can be disqualified from the link establishing process.

Even though communication is crucial in multi-robotic tasks, to the authors' best knowledge, the notion of the on-line building or learning a direct communication map approach is elaborated only sparsely. Most of the existing approaches use a simple fixed signal propagation model to determine whether and how well a communication link can be established between two spatial coordinates. Although a thorough review of communication modeling methods can be found in [1], a short overview of the related works is provided in the rest of this section to provide background and context of the addressed problem.

The most straightforward approach to communication modeling is the constant range method, e.g., used in [27]. It determines whether a communication link can be established between two points by comparing Euclidean distance with a threshold. Fixed-radius communication has been utilized for Kilobots [24] that use a low-power ground-directed infrared channel characterized by a relatively short and accessible communication range under the defined environmental conditions. The line-of-sight method [26] assumes that a link can be established if and only if a straight line segment connecting two spatial positions exists such that it does not intersect any obstacle. Both the fixed radius and line-of-sight methods can be combined [17].

Signal attenuation model [25] accounts for natural signal strength decay that occurs with the growing distance from the signal source. Since obstacles are not modeled, the model is called *Free-Space Path Loss* (FSPL). Its predictions of relative signal decay L_p are computed given the distance between receiver and transmitter d, their respective antennas directivities D_r, D_t, and the signal wavelength λ:

$$L_p = 10 \log_{10} \left(D_t D_r \left(\frac{\lambda}{4\pi d} \right)^2 \right). \quad (1)$$

Unlike the constant-range and line-of-sight methods, the signal attenuation model returns a continuous value of signal strength instead of the binary predictions. An extension of the attenuation modeling by considering obstacles present in the environment is presented in [14]. However, explicit knowledge about signal and material properties must be known beforehand. Thus both explorative deployments and the ability of on-line adaption to environment change are limited.

Another research path is led by Malmirchegini et al. [13], who proposed a purely probabilistic model for estimation of signal characteristics with the notion of some spatial dependence. Their model also provides mathematical reasoning about the predictability of the communication parameters. However, since their model lacks an explicit environment model, it cannot infer the attenuation thereupon; it always needs some prior measurements in the area of interest.

More sophisticated models are based on raytracing the signal propagation from the source by casting rays through the environment [22]. These models need to comprehend phenomena such as multi-path signal propagation, waveguide effect, or reflections. Hence, the raytracing-based approaches require detailed geometric environment models that include surface normals and carry information about obstacle material properties. Further, such approaches are computationally very demanding and sensitive to inaccurate or biased data about the environment that make these approaches unsuitable for on-board deployments with limited resources and field measurement devices.

The methods mentioned above tend to yield pessimistic but guaranteed results when used under particular assumptions in an austere environment. However, in sites that suffer from waveguide effect, non-trivial reflections, or non-trivial attenuation, these basic methods can return wrong predictions even in the close vicinity of the queried points. Only one group of existing approaches for communication map building is, to the best of our knowledge, based on pure machine learning without relying on any strong prior knowledge of the physical properties of the environment. In [3], a communication map is built using a spatial GP, where the core idea is to interpolate samples of communication link quality gathered in-vivo. The GP provides the most likely regression of the initially unknown signal strength function, and it interpolates the input samples together with the variance of the predicted estimates. It can thus be used to determine the most probable link quality with the confidence estimate. Consequently, the GP-based model can be utilized in active perception tasks such as [11] because information about predictive variance can be used as a valuation of the model uncertainty for the given configurations. If the whole space of all spatial pairs is sampled with sufficiently high resolution, it is possible to predict the signal characteristics perfectly. Even though the efforts are being made to sample the whole space in [3,18], such sampling is a problem with the complexity that can be bounded by $\mathcal{O}(n^4)$ because we need to sample all (two-dimensional) coordinate pairs, which can be too costly or unfeasible. Therefore, we consider GP-based modeling to address this drawback and develop a more general model that scales better with the environment size without dense sampling before a particular deployment in the same environment.

3 Proposed Communication Map Model

The proposed approach to communication modeling follows the communication map $\mathcal{M}$ defined as a data structure with the accompanied procedure to predict a characteristic value of the communication channel for two given arbitrary positions of the transmitter and receiver placed in the environment. Thus, we formalize a generic communication map as $\mathcal{M} : \mathcal{P} \to \mathcal{R}$, where $p = (a_1, a_2) \in \mathcal{P}$ is a pair of two spatial coordinates $a = [a_x, a_y, a_z]^{\mathrm{T}} \in \mathbb{R}^3$, $r \in \mathcal{R}$ is a value characterizing the communication quality; a_1 is the position of the transmitter, a_2 of the receiver. We assume w.l.o.g. r is a scalar $\mathcal{R} \subseteq \mathbb{R}$ for the sake of simplicity in this paper; in particular, r is a value of the RSSI.

The mapping $\mathcal{M}$ is realized by a descriptor function δ chained to the GP regressor [21] $\mathcal{G}|_X$. Formally, $\mathcal{M} = \delta \circ \mathcal{G}|_X$, where X is a set of the training data. Since the GP output provides the predictions associated with predictive variance, the GP-based regression is suited for active perception scenarios, where the learner improves its model by adding new samples for configurations with high predictive variance. Hence, we follow the GP-based approach [3]. However, the key novelty is in the proposed descriptor function that provides data such as distance between the queried locations or obstacle occupancy metric. Thus, unlike the approach [3], the proposed method provides low-variance predictions also in spatially unexplored locations given the corresponding descriptor value is correlated to some previously sampled measurement's descriptor.

3.1 Gaussian Process

We use GP regressors in each model to infer the signal strength from samples transformed by respective descriptors. In this section, we briefly describe the GP regression to make the paper self-contained. First, let the function of interest $f(x)$ be observed with Gaussian noise ϵ

$$y = f(x) + \epsilon, \quad \epsilon \in \mathcal{N}(0, \sigma^2). \tag{2}$$

The GP is then a distribution over all possible functions [21]

$$f(x) \sim \mathcal{GP}(m(x), K(x, x')), \tag{3}$$

where $m(x)$ and $K(x, x')$ are the mean and covariance, respectively, defined as

$$m(x) = E[f(x)], \tag{4}$$
$$K(x, x') = E[(f(x) - m(x))(f(x') - m(x'))]. \tag{5}$$

The latent values f_* of testing data X_* are computed given training data X as

$$\begin{aligned} \mu(X_*) &= K(X, X_*)\left[K(X, X) + \sigma^2 I\right]^{-1} y, \\ (\sigma(X_*))^2 &= K(X_*, X_*) \\ &\quad - K(X, X_*)^T \left[K(X, X) + \sigma^2 I\right]^{-1} K(X, X_*), \end{aligned} \tag{6}$$

where $K(X, X')$ is the covariance function. Two distinct covariance functions were used in individual descriptors; the first is the squared exponential kernel

$$K(x, x') = \sigma^2 \exp\left(-\frac{\|x - x'\|_2}{2l^2}\right), \tag{7}$$

where σ^2 is the output variance and l is the lenghtscale. The second covariance function is the Matern $3/2$ kernel

$$K(x, x') = \sigma^2 \frac{2^{1-3/2}}{\Gamma(3/2)} \left(\sqrt{3}\frac{\|x - x'\|_2}{l}\right)^{3/2} L_{3/2}\left(\sqrt{3}\frac{\|x - x'\|_2}{l}\right), \tag{8}$$

with σ and l defined analogously as before, Γ being the gamma function and $L_{3/2}$ the modified Bessel function of the second kind.

3.2 Descriptor Functions in GP-based Communication Map

The GP can be coupled with a descriptor function characterizing the input data; three different functions are considered. The first descriptor (9) is the spatial projection utilized in [3] further denoted also as $\delta_{\text{spatial}}(p)$, where the descriptor is a four-dimensional vector created by truncating the z-axis coordinates of the pose vectors a_a, a_b

$$\delta_{\text{spatial}}(p) = \delta_{\text{spatial}}\left(\left(\begin{bmatrix} a_{1x} \\ a_{1y} \\ a_{1z} \end{bmatrix}, \begin{bmatrix} a_{2x} \\ a_{2y} \\ a_{2z} \end{bmatrix}\right)\right) = \begin{bmatrix} a_{1x} \\ a_{1y} \\ a_{2x} \\ a_{2y} \end{bmatrix}. \tag{9}$$

The second examined descriptor δ_{FSPL} is based on Euclidean distance between the two pose vectors in p. It follows the FSPL [25] signal attenuation model under the assumption that both the signal transmitter and receiver are omnidirectional and the environment is without obstacles. However, when it is combined with the GP regressor, it provides the estimate of the mean signal strength and its variance. The descriptor is defined as

$$\delta_{\text{FSPL}}(p) = [\|p\|_2], \tag{10}$$

where $\|p\|_2$ is the three-dimensional Euclidean distance

$$\|p\|_2 = \left\| \left(\begin{bmatrix} a_{1x} \\ a_{1y} \\ a_{1z} \end{bmatrix}, \begin{bmatrix} a_{2x} \\ a_{2y} \\ a_{2z} \end{bmatrix} \right) \right\|_2 = \left\| \begin{bmatrix} a_{1x} - a_{2x} \\ a_{1y} - a_{2y} \\ a_{1z} - a_{2z} \end{bmatrix} \right\|_2 . \quad (11)$$

Finally, the third proposed descriptor is called the *Projected Free Space Ratio* (PFSR) denoted δ_{PFSR}. It extends the FSPL-based descriptor by examining obstacles between the connection of two locations using the available map of the environment. The descriptor δ_{PFSR} is a two-dimensional vector, where the first dimension corresponds to δ_{FSPL}. The second dimension is defined as the ratio of the occupied and free space along the straight line connecting the queried points a_1 and a_2. However, since the exact volumetric computation is not practical due to data noise, we use a robust off-line estimation of the environment.

The environment map is based on point cloud built from depth measurements of RGB-D or LiDAR sensors localized to the same reference frame as the pose vectors a. For model learning and inference, the point cloud is discretized into an equally-sized spatial grid summed along the vertical axis. The resulting two-dimensional grid is normalized by the maximum value and thresholded to create a binary occupancy grid. After that, Canny edge detector [9] with eight-fold image erosion and dilation gap-filling technique is performed. Then, given two query points, a_1 and a_2, a set of grid cells intersected by the straight-line $\mathcal{B}(p)$ between the points' ground plane projections is generated using Bresenham's line algorithm [8].

The proposed PFSR descriptor δ_{PFSR} is defined as

$$\delta_{\text{PFSR}}(p) = \begin{bmatrix} \|p\|_2 \\ O(p) \end{bmatrix}, \quad (12)$$

with the occupied ratio $O(p)$ computed as

$$O(p) = \frac{\sum_{b \in \mathcal{B}(p)} occupied(b)}{|\mathcal{B}(p)|}, \quad (13)$$

where $occupied(b)$ returns 1 if the cell b is occupied and 0 otherwise.

4 Results

The prediction of the signal strength using all three introduced descriptors employed in GP-based regression has been examined with real experimental data. We also compare the GP-based models with pure FSPL model [25] as defined in (1). All the methods have been implemented in C++ using Limbo [10]. The GP-based regressor with the descriptor $\delta_{\text{spatial}}(p)$ has been used with the exponential kernel (7) as in [3]. The same kernel is also utilized with $\delta_{\text{FSPL}}(p)$ because of the identical dimension as for $\delta_{\text{spatial}}(p)$. However, Matern three-halves function (8) has been used with $\delta_{\text{PFSR}}(p)$ because better results have

Table 1. Utilized hyperparameters in GP-based communication models.

Descriptor	$\delta_{\text{spatial}}(p)$ [3]	$\delta_{\text{FSPL}}(p)$	$\delta_{\text{PFSR}}(p)$
Kernel function	Exponential (7)	Exponential (7)	Matern $3/2$ (8)
Prior constant mean	−45.00	−45.00	−45.00
l	1.17	0.50	10.00
σ	0.01	0.37	0.05

been achieved than with the exponential kernel. We used a grid-search method to find the individual hyperparameters; the best performing values in the cross-validation are listed in Table 1.

The training and testing dataset has been collected using measured RSSI in the *Bull Rock* cave system using communication nodes with RFM69HCW transceiver [23] with an output power of 100 mW operating at 868 MHz that exhibits better around-corner propagation in comparison, e.g., to 2.5 GHz and 5.0 GHz [28]. The module uses a quarter wavelength long whip antenna and frequency-shift keyring (FSK) modulation schema. Direct measurement of the RSSI is provided via the internal module circuitry. The RSSI value is numerically equal to the measured received signal power in units of dBm, rounded to the nearest integer. The RSSI measurements are accompanied by spatial information from the Leica TS16 total station. The used environment map for the descriptor $\delta_{\text{PFSR}}(p)$ has been built using a point cloud assembled from eight full-dome scans captured by Leica BLK360; each scan contained roughly five million points.

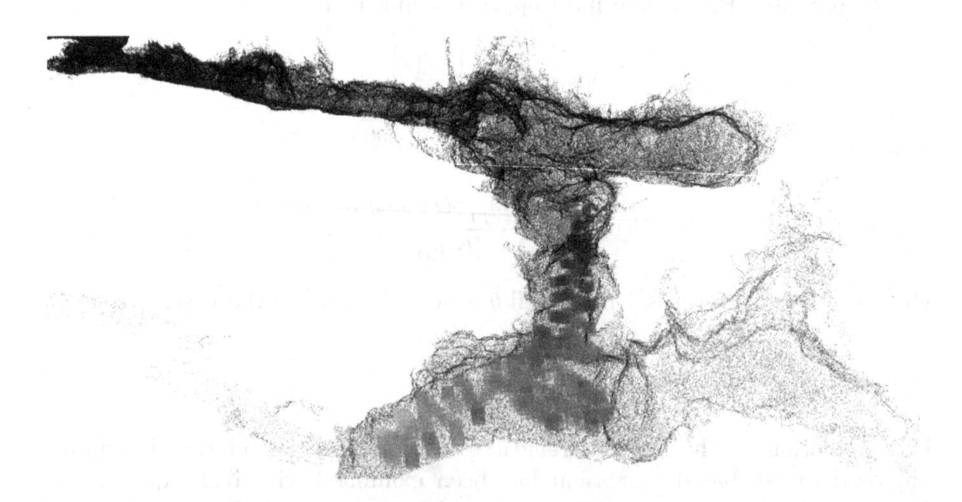

Fig. 1. Visualization of the created environment model of the testing area in the *Bull Rock* cave system. The localized point cloud is shown in black color, and a part of the localized RSSI measurements is depicted in the yellow-purple color map. The yellow color represents full received signal strength; purple samples are those with a more attenuated signal. The figure presents a situation with the transmitter position near the cave entrance (Color figure online).

The whole dataset contains about 4500 samples of the RSSI with the positions of both transmitter and receiver registered to the global reference frame. The samples have been collected by fixing the transmitter at five different locations and moving the receiver through the testing area. A visualization of the created model of the environment is depicted in Fig. 1. An example of the created grid map utilized for computing *occupied*(b) in (13) is shown in Fig. 2. The communication models are evaluated in two setups assessing interpolation and extrapolation capabilities, respectively.

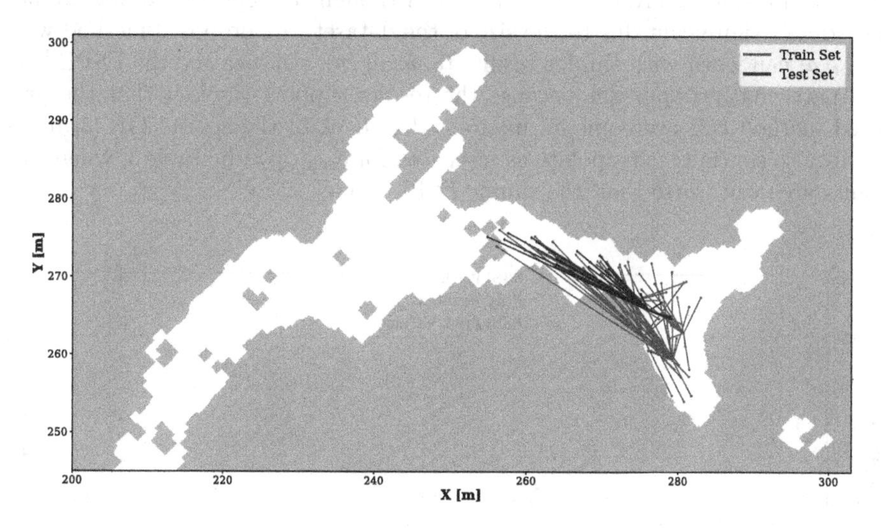

Fig. 2. A cutout of the relevant area of the occupancy grid with an instance of dataset in extrapolation setup. Training part is depicted in the red and testing dataset in the blue (Color figure online).

Table 2. Cross-validation results of RSSI prediction in the interpolation setup.

Method	FSPL [25]	$\delta_{\text{spatial}}(p)$ [3]	$\delta_{\text{FSPL}}(p)$	$\delta_{\text{PFSR}}(p)$
MAE [dBm]	8.99	4.05	8.19	6.51
RMSE [dBm]	11.31	5.46	10.01	8.93

Monte Carlo cross-validation schema has been used for evaluation in the interpolation setup. The dataset has been randomly divided into a training set and testing set 10 times with the train-to-test set size ratio $1:9$, producing mutually independent data. The results of the interpolation cross-validation are summarized in Table 2, where MAE stands for the *Mean Absolute Error* and RMSE for the *Root Mean Squared Error*.

For the evaluation of the methods in the extrapolation setup, one representative transmitter location with associated measurements has been utilized as

Table 3. Cross-validation results of RSSI prediction in the extrapolation setup.

Method	FSPL [25]	$\delta_{\text{spatial}}(p)$ [3]	$\delta_{\text{FSPL}}(p)$	$\delta_{\text{PFSR}}(p)$
MAE [dBm]	9.08	16.62	7.71	7.88
RMSE [dBm]	11.42	21.30	9.77	10.37

the test set; the rest of the dataset has been utilized as the training set. The evaluation results are reported in Table 3. Although the reported results are not statistically significant due to the size of the dataset, the proposed method with $\delta_{\text{PFSR}}(p)$ can learn with similar results to $\delta_{\text{FSPL}}(p)$, still beating the FSPL [25]. It is particularly important because the results support the fact that the proposed method can overcome an inherent drawback of the spatial GP [3] in its inability to spatially extrapolate as indicated for $\delta_{\text{spatial}}(p)$ in Table 3, where the results are even worse than the simple FSPL model.

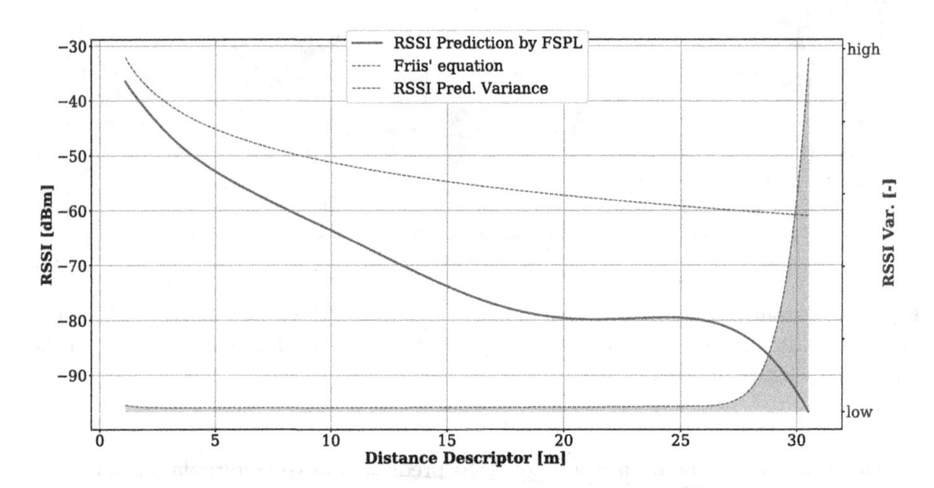

Fig. 3. RSSI predicted while utilizing $\delta_{\text{FSPL}}(p)$ as a function of the distance only and its prediction variance. The FSPL model assumes omnidirectional antennas, and it is parametrized with the nominal frequency 868 MHz.

In addition to the cross-validation, we further investigate the examined descriptor functions. The descriptor space of $\delta_{\text{FSPL}}(p)$ can be understood as regression of the FSPL [25] applied in an obstructed environment. The prediction plot is shown in Fig. 3. Analogously, the descriptor space of $\delta_{\text{PFSR}}(p)$ is depicted in Fig. 4. Expectedly, the values predicted by the GP are greatly outlying in areas where sparse or no samples have been taken as it can be seen in the top right part of the plot.

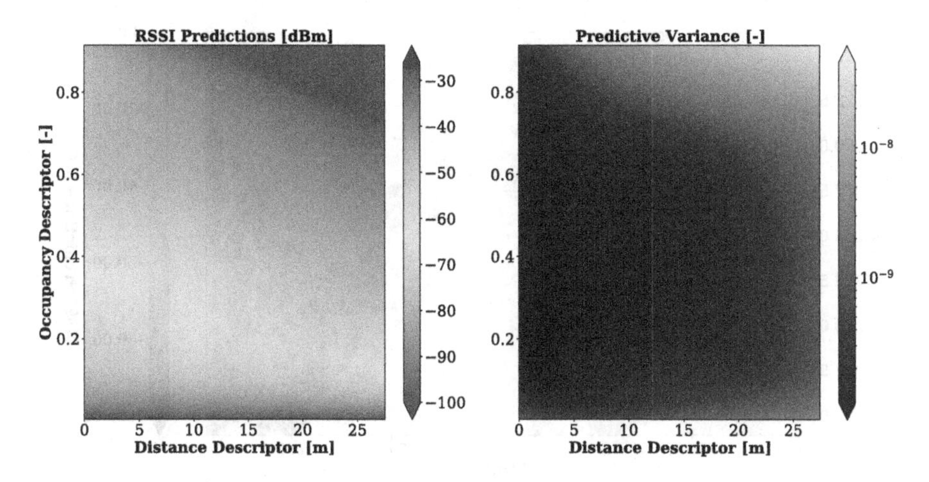

Fig. 4. Predicted RSSI mean values with variances in the descriptor space of $\delta_{\mathrm{PFSR}}(p)$. The plots are clipped to the interval $[-20, -100]$ dBm because the underlying GP predicts highly outlying values in unobserved feature-value regions with high variances.

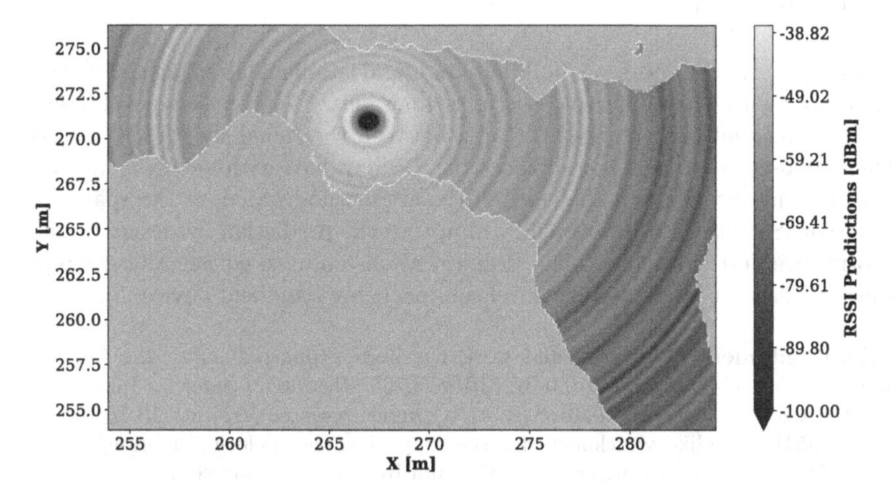

Fig. 5. Illustration of extrapolated communication map for fixed transmitter $\delta_{\mathrm{FSPL}}(p)$.

Finally, we provide examples of the extrapolated predictions for the $\delta_{\mathrm{FSPL}}(p)$ and $\delta_{\mathrm{PFSR}}(p)$ descriptors in Fig. 5 and Fig. 6, respectively. The figures illustrates that the proposed descriptor $\delta_{\mathrm{PFSR}}(p)$ is able to account for obstacles whereas $\delta_{\mathrm{FSPL}}(p)$ accounts only for the distance.

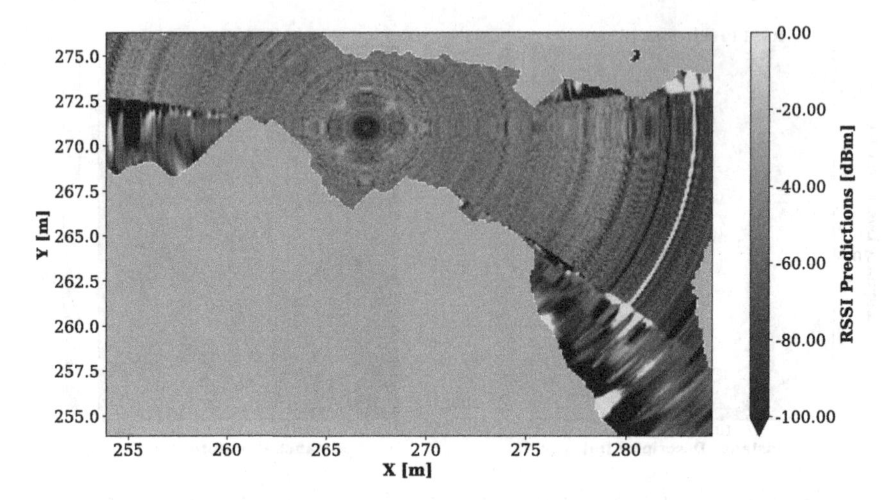

Fig. 6. Illustration of extrapolated communication map for fixed transmitter $\delta_{\mathrm{PFSR}}(p)$.

5 Conclusion

In this paper, we report on building communication maps using a GP-based framework with three different descriptor functions. We propose a novel descriptor function that characterizes the area between the signal source and receiver. The proposed approach extrapolates the RSSI values with accuracy an order of magnitude better than the state-of-the-art method based on the spatial GP regressor. For future work, we aim to improve the prediction by designing more informative descriptor functions. Besides, we also aim to collect a larger representative dataset by exploiting novel self-localizing handheld device [5].

Acknowledgment. The presented work has been supported under the OP VVV funded project CZ.02.1.01/0.0/0.0/16_019/0000765 "Research Center for Informatics" and the Czech Science Foundation (GAČR) under research Projects 18-18858S and GA20-29531S. We like to acknowledge the support of the speleologist branch organization ZO 6-01 for providing access to the Bull Rock cave testing site.

References

1. Amigoni, F., Banfi, J., Basilico, N.: Multirobot exploration of communication-restricted environments: a survey. IEEE Intell. Syst. **32**(6), 48–57 (2017). https://doi.org/10.1109/MIS.2017.4531226
2. Amigoni, F., Banfi, J., Basilico, N., Rekleitis, I., Li, A.Q.: Online update of communication maps for exploring multirobot systems under connectivity constraints. In: Distributed Autonomous Robotic Systems, pp. 513–526 (2019). https://doi.org/10.1007/978-3-030-05816-6_36

3. Banfi, J., Li, A.Q., Basilico, N., Rekleitis, I., Amigoni, F.: Multirobot online construction of communication maps. In: IEEE International Conference on Robotics and Automation (ICRA), pp. 2577–2583 (2017). https://doi.org/10.1109/ICRA.2017.7989300
4. Banfi, J.: Recent advances in multirobot exploration of communication-restricted environments. Intelligenza Artificiale **13**(2), 203–230 (2019). https://doi.org/10.3233/IA-180013
5. Bayer, J., Faigl, J.: Handheld localization device for indoor environments. In: International Conference on Automation, Control and Robots (ICACR), pp. 60–64 (2020). https://doi.org/10.1109/ICACR51161.2020.9265494
6. Berglund, E., Sitte, J., Wyeth, G.: Active audition using the parameter-less self-organising map. Auton. Robots **24**(4), 401–417 (2008). https://doi.org/10.1007/s10514-008-9084-9
7. Bildea, A., Alphand, O., Rousseau, F., Duda, A.: Link quality estimation with the Gilbert-Elliot model for wireless sensor networks. In: IEEE Annual International Symposium on Personal, Indoor, and Mobile Radio Communications (PIMRC), pp. 2049–2054 (2015). https://doi.org/10.1109/PIMRC.2015.7343635
8. Bresenham, J.E.: Algorithm for computer control of a digital plotter. IBM Syst. J. **4**(1), 25–30 (1965). https://doi.org/10.1147/sj.41.0025
9. Canny, J.: A computational approach to edge detection. IEEE Trans. Pattern Anal. Mach. Intell. PAMI **8**(6), 679–698 (1986). https://doi.org/10.1109/TPAMI.1986.4767851
10. Cully, A., Chatzilygeroudis, K., Allocati, F., Mouret, J.B.: Limbo: a flexible high-performance library for Gaussian processes modeling and data-efficient optimization. J. Open Source Softw. **3**(26), 545 (2018). https://doi.org/10.21105/joss.00545
11. Luo, W., Sycara, K.: Adaptive sampling and online learning in multi-robot sensor coverage with mixture of Gaussian Processes. In: IEEE International Conference on Robotics and Automation (ICRA), pp. 6359–6364 (2018). https://doi.org/10.1109/ICRA.2018.8460473
12. Maan, F., Mazhar, N.: MANET routing protocols vs mobility models: a performance evaluation. In: International Conference on Ubiquitous and Future Networks (ICUFN), pp. 179–184 (2011). https://doi.org/10.1109/ICUFN.2011.5949158
13. Malmirchegini, M., Mostofi, Y.: On the spatial predictability of communication channels. IEEE Trans. Wirel. Commun. **11**(3), 964–978 (2012). https://doi.org/10.1109/TWC.2012.012712.101835
14. Micheli, D., Delfini, A., Santoni, F., Volpini, F., Marchetti, M.: Measurement of electromagnetic field attenuation by building walls in the mobile phone and satellite navigation frequency bands. IEEE Antennas Wirel. Propag. Lett. **14**, 698–702 (2015). https://doi.org/10.1109/LAWP.2014.2376811
15. Mohseni, S., Hassan, R., Patel, A., Razali, R.: Comparative review study of reactive and proactive routing protocols in MANETs. In: IEEE International Conference on Digital Ecosystems and Technologies, pp. 304–309 (2010). https://doi.org/10.1109/DEST.2010.5610631
16. Muralidharan, A., Mostofi, Y.: Statistics of the distance traveled until successful connectivity for unmanned vehicles. Auton. Robots **44**(1), 25–42 (2019). https://doi.org/10.1007/s10514-019-09850-7
17. Nestmeyer, T., Robuffo Giordano, P., Bülthoff, H.H., Franchi, A.: Decentralized simultaneous multi-target exploration using a connected network of multiple robots. Auton. Robots **41**(4), 989–1011 (2016). https://doi.org/10.1007/s10514-016-9578-9

18. Penumarthi, P.K., et al.: Multirobot exploration for building communication maps with prior from communication models. In: International Symposium on Multi-Robot and Multi-Agent Systems (MRS), pp. 90–96 (2017). https://doi.org/10.1109/MRS.2017.8250936
19. Qin, J., Wei, Z., Qiu, C., Feng, Z.: Edge-prior placement algorithm for UAV-mounted base stations. In: IEEE Wireless Communications and Networking Conference (WCNC), pp. 1–6, April 2019. https://doi.org/10.1109/WCNC.2019.8885992
20. Li, A.Q., et al.: Multi-robot online sensing strategies for the construction of communication maps. Auton. Robots **44**(3), 299–319 (2020). https://doi.org/10.1007/s10514-019-09862-3
21. Rasmussen, C.E., Williams, C.K.I.: Gaussian Processes for Machine Learning. Adaptive Computation and Machine Learning. MIT Press, Cambridge (2006). http://www.gaussianprocess.org/gpml
22. Remley, K.A., Anderson, H.R., Weisshar, A.: Improving the accuracy of ray-tracing techniques for indoor propagation modeling. IEEE Trans. Veh. Technol. **49**(6), 2350–2358 (2000). https://doi.org/10.1109/25.901903
23. Rouček, T., et al.: Darpa subterranean challenge: multi-robotic exploration of underground environments. In: 2019 Modelling and Simulation for Autonomous Systems (MESAS), pp. 274–290 (2020). https://doi.org/10.1007/978-3-030-43890-6_22
24. Rubenstein, M., Ahler, C., Hoff, N., Cabrera, A., Nagpal, R.: Kilobot: a low cost robot with scalable operations designed for collective behaviors. Robot. Auton. Syst. **62**(7), 966–975 (2014). https://doi.org/10.1016/j.robot.2013.08.006
25. Shaw, J.A.: Radiometry and the Friis transmission equation. Am. J. Phys. **81**(1), 33–37 (2013). https://doi.org/10.1119/1.4755780
26. Stump, E., Michael, N., Kumar, V., Isler, V.: Visibility-based deployment of robot formations for communication maintenance. In: IEEE International Conference on Robotics and Automation (ICRA), pp. 4498–4505 (2011). https://doi.org/10.1109/ICRA.2011.5980179
27. Sheng, W., Yang, Q., Ci, S., Xi, N.: Multi-robot area exploration with limited-range communications. In: IEEE/RSJ International Conference on Intelligent Robots and Systems (IROS), vol. 2, pp. 1414–1419 (2004). https://doi.org/10.1109/IROS.2004.1389594
28. Zhou, C., Plass, T., Jacksha, R., Waynert, J.A.: RF propagation in mines and tunnels: extensive measurements for vertically, horizontally, and cross-polarized signals in mines and tunnels. IEEE Antennas Propag. Mag. **57**(4), 88–102 (2015). https://doi.org/10.1109/MAP.2015.2453881

Combat UGV Support of Company Task Force Operations

Jan Nohel(✉), Petr Stodola, and Zdeněk Flasar

University of Defense, Brno, Czech Republic
{jan.nohel,petr.stodola,zdenek.flasar}@unob.cz

Abstract. Effective deployment of military forces and equipment in diverse operations requires the widest possible support of modern technical means. Robotic, semi-autonomous or autonomous means using artificial intelligence can also be important for reconnaissance and orientation in the operational area, as well as for identifying and destroying the enemy, and saving soldiers' lives. The paper describes the research into the requirements for these unmanned systems, with an emphasis on their abilities to bypass or pass obstacles, to ascertain a wide range of information from the theatre of operations and to transmit it to the commander in real time. The tasks can be fulfilled either semi-autonomously, under the control of the operator or autonomously, including diverse offensive activities. With the use of the Maneuver Control System CZ the possibilities of maneuver planning and the use of the unmanned ground vehicle group to support the combat action of the company task force are described here. A case study was conducted as a basis for the development of the article. The scenarios of three tactical situations describe the possibilities of the effective tactical use of a group of autonomous means in a wooded and open terrain, as well as in attacking the enemy.

Keywords: Unmanned ground vehicle · Decision-making process · Offensive maneuver · Multiple maneuver model · Target enemy

Military operations of NATO forces are conducted in diverse terrain and environment of the whole world in 21st century. Such operations frequently focus on urban, i.e. densely populated areas. The enemy represents an asymmetric threat in the form of armed rebel gangs, or members of organized crime groups fully defined in [1, pp. 39]. Their fight is not fair and open. They press home their current time advantage over the allied forces and the advantage of a detailed knowledge of the area. They set improvised booby traps, spread religious propaganda and ideology, and carry out suicide attacks. Members of armed gangs are impossible to distinguish from the civilian population that frequently supports them, voluntarily or under duress.

The tactics of troops serving in these operations reflect efforts to adapt to these asymmetric threats. Instead of deploying large forces in the entire operational area, only smaller but well-armed units are put in particular areas. Their success lies in up-to-minute information on the situation in the operational area. Sensors acquiring information as well as decision-making elements and maneuver and fire support units are interconnected with a communications network, see [2, 3]. The facts ascertained or the presence of the

J. Mazal et al. (Eds.): MESAS 2020, LNCS 12619, pp. 29–42, 2021.
https://doi.org/10.1007/978-3-030-70740-8_3

enemy revealed by sensors are shared as real-time information with the commanders and the staff who can make prompt and informed decisions about how the forces and equipment should react. The units charged and operators of weapon systems more easily comprehend the situation on the battlefield thanks to information directly transferred by sensors – sources of information. Sharing and processing information, assigning tasks via a computer network of sensors, decision-making elements and maneuver and fire support units form the Digitized Battlefield. Thanks to the Digitized Battlefield the process of commanding and controlling can be decentralized, commanders of small units can be more independent, and more favorable conditions are created for gaining dominance on the battlefield. For further information on the Digitized Battlefield, battlefield situation monitoring, and the development of the above in the U.S. Army, see [4]. Common Operation Picture (COP) displaying available information based on geographical context and requiring authorized access is a characteristic element of the Digitized Battlefield, see [5, 6].

Company Task Force (TF) is one of the smallest units formed in order to fulfil specific tasks of a particular operation independently. During military operations, the company task force is usually formed on the basis of an organic infantry, motorized, or mechanized company supported by intelligence, combat, and logistic systems. The company TF can fulfil a range of tasks, e.g. the defense of a detached building of interest, an attack on a detached poorly protected building, an escort of a convoy, searching, reconnaissance of an area of interest, etc. Scientific literature deals with the issues concerning the use of the company TF [7]. The company TF can be supported by engineer, artillery, anti-aircraft, medical, and intelligence systems with respect to the size and required independence when fulfilling tasks. The combat UGV (Unmanned Ground Vehicle) can represent one of such support systems which can be used as forward scouting sensors, or weapon systems destroying the enemy in endangered areas.

At present the weapon systems of armies include a wide range of combat UGVs having various sensory systems, some of them are described in [8]. Combat UGVs usually have wheel or caterpillar undercarriages, diesel or hybrid electric drive reaching a speed of up to 77 km.h^{-1}, and driving range from 20 to 300 km. They usually carry 7.62 or 12.7 mm caliber machine guns supported e.g. by a 30 mm caliber cannon, anti-tank guided missiles, or grenade launchers. Tests of humanoid robots are conducted too; they are intended for future use in reconnaissance and combat missions in built-up areas. However, combat missions of humanoid robots are not feasible at present especially due to their considerable instability when moving across uneven terrain; based e.g. on test results of The DARPA Robotics Challenge 2015 - Running Robots Competition [9].

1 Literature Review

The use of robotic and autonomous systems, artificial intelligence and their implementation on the battlefield are described in [10–12]. A strategy of robotic and autonomous systems integration into future U.S. Army operations and psychological, humanitarian, and legal consequences of deploying autonomous robotic systems in military operations can be found in [13–16]. Possibilities of automated ground units maneuver planning, effectively supported by firepower, while performing tasks during military operations are described and checked in scenarios in [17]. The development of capabilities of automated robots when supporting squad-level soldiers, including priorities, functions, and responsibilities when supporting squad-level soldiers are contained in [18]. The ability of UGVs to cross difficult terrain and its evaluation is dealt with in [19, 20]. Optimal maneuver planning for two cooperative military elements under an uncertain enemy threat is described in [21]. The model suggested in this article minimizes endangering of elements and optimizes the length of a maneuver route. Autonomous navigation for legged robots in challenging terrain not passable for wheel and caterpillar undercarriages of vehicles is dealt with in [22]. It examines the potential of autonomous navigation of legged robots. The navigation system has been integrated and tested on a four-legged robot.

Modelling and simulation of the movement of forces and equipment in the urbanized area of the 21st century battlefields and the autonomy level of unmanned systems are presented in [23–25]. These articles further describe the potential use of autonomous systems on the battlefield, including the levels of autonomy and human control of unmanned systems. Practical examples illustrate the capabilities of UGVs and levels of their autonomy. The increasing efficiency of units by deploying unmanned systems and integrating them into military operations is dealt with in [26]. Specific manners of use of combat UGVs, including their technical parameters and types of carried arms, are described in [27].

The interoperability, autonomy, and security of communications network and UGV types are examined in [16]. The mathematical algorithmic model of maneuver route planning that can be adopted during military operations and the UGV deployment is specified in [28–30]. Besides the description of algorithmic correlations and evaluation criteria, the article includes the evaluation of various types of maneuvers in the terrain, including time characteristics.

2 Requirements for Company Combat UGVs

As part of the research the author carried out controlled semi-structured interviews with 10 current and former company commanders of mechanized or motorized infantry. Most of these commanders gained experience in operations abroad on the territories of Iraq and Afghanistan. Each of the commanders was asked to consider possible manners of use of reconnaissance/combat is to support their company if they could take advantage of these vehicles during a military operation.

They suggested using UGVs primarily for surveillance and search of more distant or risky areas focusing on the enemy activities, terrain, weather, and CBRN threats. During combat missions, the company commanders would use combat UGVs to directly conduct the activities of units, as protective forces in retreat, or as a means of destruction in built-up areas. Other suggested uses of preferably a single UGV included: transfer of wounded persons, providing supplies to separated units, and a relay station.

The manner of use of 3–4 UGVs of the company task force should be well-planned, focused on endangered areas where the UGVs would acquire intelligence information from fixed observation posts. The movement to the area where the task should be performed should be mostly autonomous following a route planned and programed in advance. The UGVs would use GPS navigation during movement. Barriers and avoiding them should be identified by LIDAR – laser radar capable of a high resolution, together with software for planning the maneuver route and geographical location. Identified targets could be destroyed automatically or semi-automatically. Automatic target destruction means that the target is identified, compared with pre-defined targets in the database and if they fit, the target is destroyed. Semi-automatic target destruction means that movement or a target is identified automatically but the instruction to destroy the target is given by an UGV operator. The maintenance and transport would be performed by an UGV operation squad/group.

It was the commanders' opinion that the UGV undercarriage should ensure maximum passability in rough terrain, which can be provided by the caterpillar type. Two company commanders suggested that the caterpillar and retractable wheel undercarriages be combined to increase the speed on the pavement. The vehicle should be hybrid combining an electromotor and gasoline generator recharging batteries. The generator should guarantee a minimum 50-km radius of action, 12-h uninterrupted operation, and a 3-day emergency. The UGV weapon system should include a small-caliber machine gun, 3–4 antitank guided missiles, smoke grenades, and possibly a mortar or mines. The vehicle could also contain a guided self-destruct mechanism.

The sensory equipment should enable the vehicle to identify a moving target within a 10-km range by day and night, and also in adverse weather conditions. To ensure the above, the UGV should use optoelectronic observation equipment, night-vision equipment, a thermal imaging camera, and laser radar. A laser rangefinder would determine the distance to the target. An electromagnetic spectrum jammer would protect the vehicle from RCIED and radio communication of the enemy in the vicinity. A radiation indicator would warn of being localized by the weapon systems of the enemy. A robotic hand would increase the UGV general usefulness by enabling it to survey small objects, remove barriers, and carry out acts of sabotage.

Tasks should be assigned to an UGV group by a company commander in accordance with key requirements for information about the situation and activities of friendly units in the area. The person responsible for the fulfillment of tasks or requirements for information of the company commander would be the company ISR element commander. Some of the company commanders suggested that the ISR element commander of company TF propose activities for the UGV group for the company commander who would approve and specify them. However, this passive approach may fail to utilize the UGV group as a means of destroying targets, taking active measures, or providing support to the company. The UGV group – sensors can also provide an overview on the situation in the area where the UGV operates which would be crucial for the decision-making process.

Information on the activity results of the GV group and information acquired flows from the UGV operator through the ISR element commander of the company TF to the company commander. The ISR element commander records general information about the situation in the area in COP (Common Operational Picture). The ISR element commander gives information to the company commander according to the company commander's requirements for information in order of priority.

2.1 Other Manners of the Combat UGV Group Utilization

Combat UGVs having a caterpillar undercarriage, armed with machine guns and antitank guided missiles, equipped with optoelectronic systems, GPS, LIDAR, and a robotic hand are able to perform a broad range of tasks. Examples of supporting the company task force activities:

- removing explosives in the operational area or line of advance,
- allowing passage through entanglements, obstacles and barriers,
- making a reconnaissance of dangerous areas or buildings possibly occupied by the enemy,
- observing areas beyond the visual range of the units (concealed area),
- destroying the identified enemy including lightly-armored forces,
- securing maneuvers across open terrain,
- attacking protected firing positions of the enemy.

Combat UGVs could perform individual tasks automatically or semi-automatically, using a database of targets and maneuver planning software. They are able to carry out FIND AND IDENTIFY or MOVE AND OBSERVE tasks completely automatically. They will produce reports (warnings) on identified/observed objects. If UGVs use the ATR (Automatic Target Recognition) system, see [31], and an updated (enemy) target database, they can also provide tactical and technical data of the object identified. UGVs can also report exclusively weapon systems and vehicles of the enemy defined by the commander. Any subsequent activity, such as changing the observation post, using a robotic hand, or the weapon systems, would have to be ordered by the UGV group commander or company commander. FIND AND DESTROY tasks could also be carried out automatically provided that the weapon systems and vehicles used by the enemy would be precisely defined in the target database. However, the risk of mistaken identification cannot be eliminated.

The movement route can be planned before the fulfilment of tasks begins by using a digital model of the territory, digital elevation model, and multiple maneuver route planning software. Direct navigation in the terrain, identification, and avoidance of barriers will be provided by LIDAR in cooperation with GPS and multiple maneuver route planning software.

The MCS CZ can be used to execute a group synchronized maneuver using position data of the on-board GPS receiver navigating the UGV.

3 The MCS CZ Multiple Maneuver Route Planning

The calculation of a multiple maneuver route by the MCS CZ is based on a model of the optimal maneuver route described in [28]. It runs using Tactical Decision Support System (TDSS), a platform that has been developed at University of Defense in the Czech Republic, by associate professor Ing. Jan Mazal, PhD and professor Ing. Petr Stodola, PhD since 2006, see [32]. The model of optimal maneuver route represents the output of MCS CZ operations. It uses a raster digital data model, map algebra algorithms, and associated criterion assessments of the effects of the situation to process the effects of the situation on the battlefield. The raster format of the network graph allows the layers of individual situation variables to be flexibly updated and mathematically combined. The layers of the individual situation variables affect the process of creating the movement route. An important feature of each raster cell is its value (attribute), which is specified by a particular or continuous character of the represented terrain area. The attribute represents the time of covering the raster cell in hundredths of a second depending on the importance or the character of information.

The model of the optimal maneuver route assesses whether the area is passable taking into consideration a combination of important factors of the situation on the battlefield which are expressed as layers, see Fig. 1. The basis for calculating the optimal movement route is the ground surface layer which characterizes the passability in various types of soil and plant cover. The influences of individual situation factors on general passability are represented by the layers of hypsometry, weather, and the activities of the enemy and friendly forces and equipment providing support in the military operation. The influence of every factor including their combined influence on passability is calculated by means of algorithms and specific criterion correlations among individual layers. The resulting calculation of the combined influence of all layers on the passability of the area has the form of a cumulative layer (matrix) of the cost surface of passability. On the basis of this numerical assessment matrix of passability, a particular maneuver (movement) route is calculated by means of an algorithm seeking the shortest route. The result of mathematical model calculations is a time-optimized and safe movement route.

The algorithm of the optimal maneuver route model then searches for the path between the two selected points with the lowest total sum of attribute values on the route. This allows the estimated total time of covering the route to be obtained. The TDSS uses the combination of Floyd-Warshall and Dijkstra's principle, described in [33, 34].

To calculate the movement routes of individual maneuver elements, the MCS CZ multiple maneuver uses the same cumulative matrix of the cost surface of passability as the optimal maneuver route model. The space specification of maneuver routes is defined by the "invisible" layer of space limitations having a fundamental effect on individual maneuvers; see Fig. 1. Individual maneuvers are described in detail in [31]. In the event of encirclement, the impassable area is in the form of a circle with the diameter equal to the distance D between the maneuver element and the target, but not more than 1 km. The impassable maneuver area is defined by two semi-ellipsoids intersecting in the distance $D \leq 1$ km, and having the length 1.3D and width 0.5D. The attack by fire maneuver calculates the shortest safe route to the nearest edge of the direct target visibility area. The calculation of a frontal attack is made using exclusively the optimal maneuver route model. The separation in space of individual maneuver routes is executed through a conflict-free area along each maneuver route not passable for the other group elements and having an adjustable length (minimum 100 m). Two elements, therefore, cannot move along the same route within a distance ≥ 1 km. The entire group moves with minimum tactical distances according to its size.

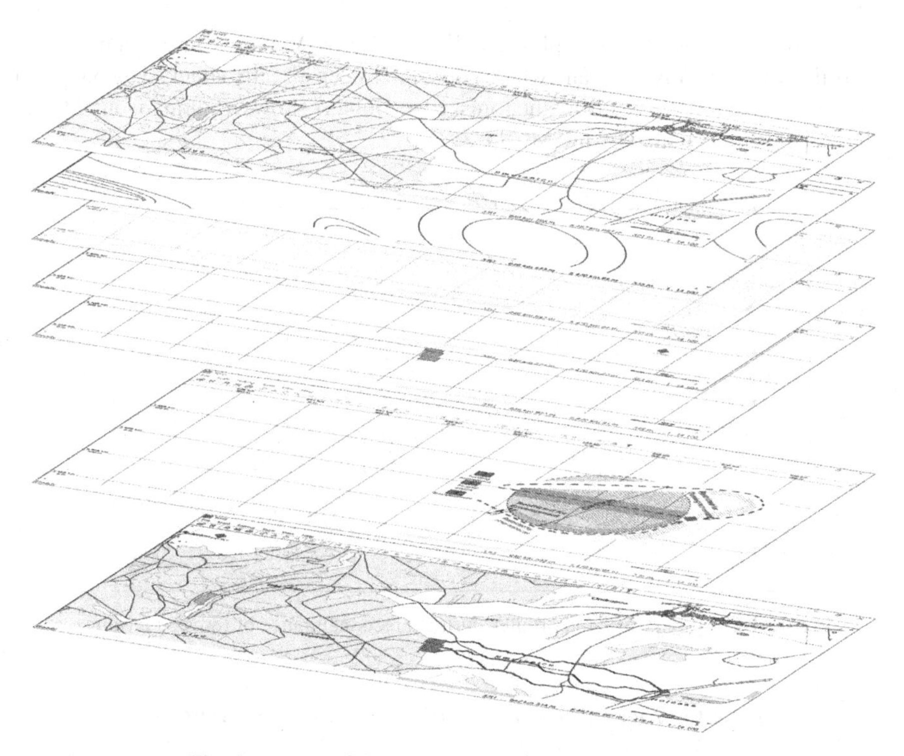

Fig. 1. Layers of the MCS CZ multiple maneuver model

4 Case Studies of the Combat Use of the Company Task Force UGV Group

A case study concerning company task force maneuvers on the battlefield was conducted which examined the possibilities of maneuver planning and the UGV combat use. The case study included several simulated scenarios of tactical situations dependent on a particular type of terrain, identified enemy, and the target of friendly unit activities.

The company task force was charged with a task to capture a transmitter tower (COMM) situated approximately 3500 m to the north-west of the company task force position. During the movement to the area of the transmitter tower, it was necessary to cross 2000 m of forested land with an extensive network of paved roads. Subsequently, the company task force had to proceed approximately 2000 m through open terrain to the transmitter tower. A preliminary reconnaissance conducted before the operation revealed 4 motorized squads moving in the area to the north of the company position.

The scenario of the first tactical situation focused on crossing forested terrain behind the opposite edge of which one of the enemy squads was moving. The company task force commander decided to conduct a reconnaissance of the forested terrain when a group of three UGVs would advance in a coordinated manner in front of the company. After having received the specification of the element type forming the subject-matter of the task, the identified posts of the enemy, and the weather, the MCS CZ calculated the maneuver route for each UGV displayed in Fig. 2. The UGV maneuver target point was the position of the enemy company which could have been directly attacked by on-board weapon systems, or observed from the forest edge. Table 1 shows a comparison of times of the maneuver calculated by the MCS CZ.

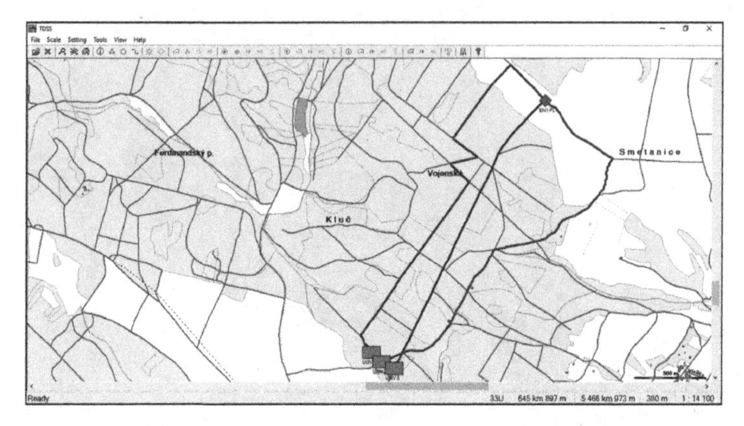

Fig. 2. UGV group maneuver routes during a reconnaissance of forested terrain

Table 1. Characteristics of the UGV maneuver in forested terrain

Vehicle/route	Time calculated	Distance calculated
UGV1 (middle)	5:31	2187 m
UGV2 (left)	6:42	2973 m
UGV3 (right)	7:03	2904 m

The middle UGV1 would get in contact with the enemy company as the first vehicle; executing the maneuver took 4:01 from the maneuver beginning. Covered at the edge of the forest from where it was possible to directly observe the enemy company, the UGV1 observed the enemy activities in the area. It checked the position of the enemy company, transferred the information to the company commander and secured the advancement of other two UGVs and the entire company. The maneuver of the UGV1 in this tactical situation could also be dealt with as an attack by fire when the target of the vehicle would be the nearest edge of the area from where the enemy could be directly observed, see Fig. 3.

The scenario of the second tactical situation focused on the advancement of the company task force through open terrain after having crossed the forest. Observation and direct fire was possible from here at a maximum distance of 1500 m, especially in the east – west direction which was the direction of the company maneuver. The recently revealed enemy forces and equipment were unchanged. The company commander decided to use an UGV group as a frontal protection of the company movement again. Its task was to

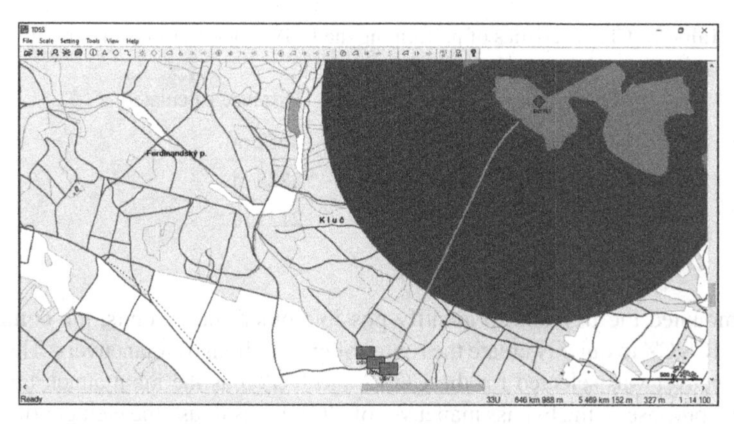

Fig. 3. The maneuver route of the UGV1 to be able to observe the enemy

search for and warn the commander in a timely manner of the enemy identified, ground barriers, and other terrain passability limitations. Before getting closer to the transmitter tower, the UGV should have obtained a visual picture of the area on the basis of which the possible enemy secured positions could have been identified. The maneuver routes through open terrain calculated by the MCS CZ can be seen in Fig. 4 and times needed for the performance of the maneuver can be compared in Table 2.

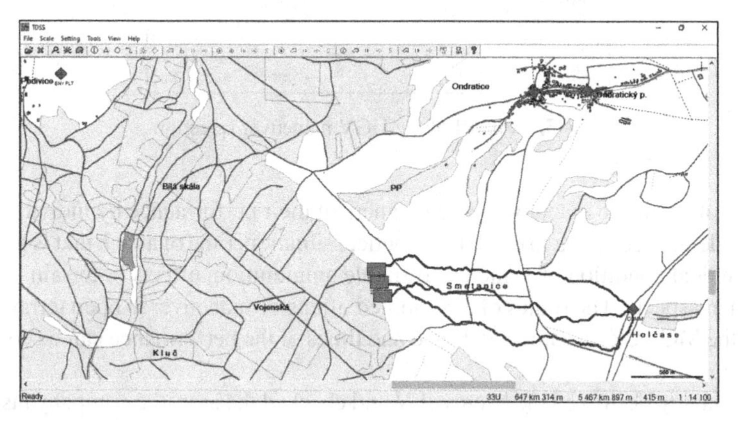

Fig. 4. Maneuver routes of the UGV group during a reconnaissance in open terrain

The scenario of the third tactical situation focused on a coordinated offensive maneuver of an UGV group against an enemy IFV (infantry fighting vehicle). An IFV was unexpectedly identified approximately in two thirds of the distance to the transmitter tower which had to be covered by the company task force in open terrain. The commander decided to use the closest UGV group to destroy it. The distance between the middle vehicle of the UGV group and the IFV was 665 m. Thanks to this distance it was possible to use an antitank guided missile carried by the UGV. To destroy the infantry even if they

Table 2. Characteristics of performing the UGV maneuver in open terrain

Vehicle/route	Time calculated	Distance calculated
UGV1 (middle)	10:46	2052 m
UGV2 (left)	11:22	2225 m
UGV3 (right)	10:31	2131 m

have dismounted the IFV and to identify possible other enemy units, the commander used the MCS CZ to quickly create the routes of group offensive maneuvers. The frontal attack maneuver was selected for the UGV2 and UGV3 as the main attack target; see Fig. 5. The purpose of the bypass maneuver of UGV1 was to use the element of surprise for a flanking maneuver, intensify the pressure on the enemy from more directions, and partially divert the attention of the enemy from the UGV frontal attack.

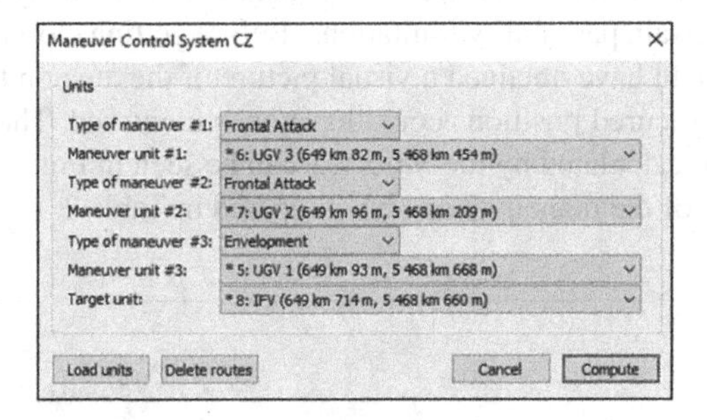

Fig. 5. A panel of the UGV maneuver selection

Types of offensive maneuvers and manners of their performance by individual units can be modelled according to the current tactical situation, time of attack that is given priority, technical condition of vehicles, available ammunition, effects of terrain, weather, and civilian aspects. The routes of a combined offensive maneuver in open terrain calculated by the MCS CZ are shown in Fig. 6 and times of the performance can be compared in Table 3.

After having destroyed the enemy IFV and checked the area, the company task force proceeded towards the transmitter tower with the UGV group as the frontal protection.

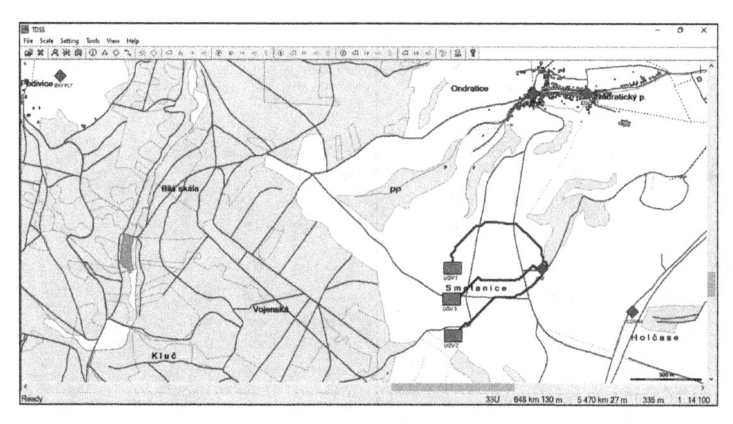

Fig. 6. Routes of the UGV offensive maneuvers

Table 3. Characteristics of the UGV group offensive maneuver

Vehicle/route	Time calculated	Distance calculated
UGV1 (left)	8:32	1052 m
UGV2 (right)	3:58	814 m
UGV3 (middle)	3:43	736 m

4.1 Setting Speed of the MCS CZ in Tactical Situations

As mentioned above, the speeds of maneuver elements on various types of surface can be set in accordance with tactical, technical, geographical, and meteorological factors and experience or physical readiness of operators or foot soldiers. Different speeds of the same UGV group were used to calculate the above tactical situations. The speeds for the calculation of the first tactical situation were set differently from the second and third situations; see Fig. 7. When calculating the maneuver route for the first tactical situation, standard UGV speeds were taken into consideration which had been checked as maximum speeds when moving in the terrain without limitations. The speeds on paved roads in the second and third tactical situations were adjusted to reflect the speed in fields. It was required by the geographical features of the area of the second and third tactical situations characterized by fields with a limited network of paved roads. The maneuver direction and characteristics were not favorable for a faster movement of UGVs on roads. It was necessary to scout the entire width of the maneuver area in front of the transmitter tower. Similarly, faster movement on roads failed to give any advantage during a spatially synchronized offensive maneuver of UGVs against the IFV. Due to the above reasons the movement speeds were equalized to the speed in a field as the prevailing type of surface.

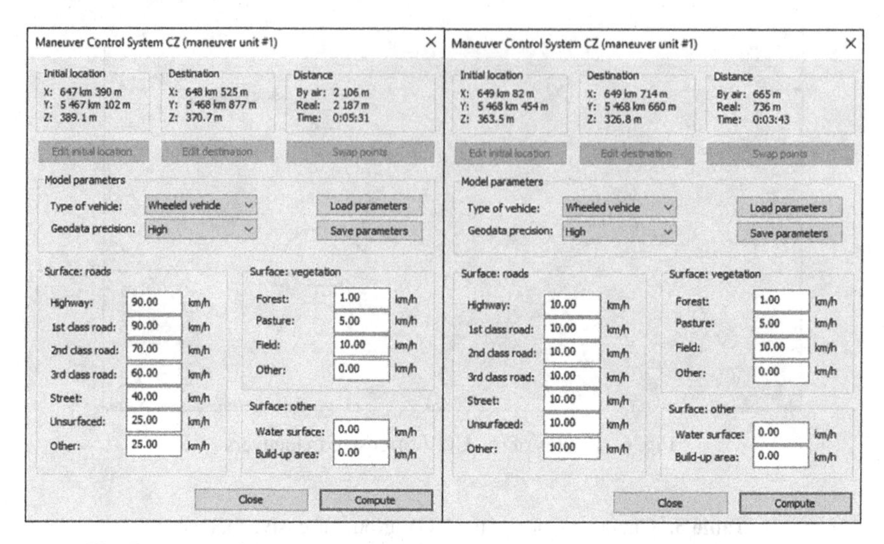

Fig. 7. Panels of the UGV speeds under the first and second (third) scenarios

5 Conclusion

Tasks and tactical situations for the performance of which it is suitable to use an UGV group can be characterized as risky, having a high probability of contact with the enemy, taking place in challenging terrain, or areas separated from the friendly units. These tasks and situations include reconnaissance in front of friendly units, observation of concealed places, and destruction of armored vehicles and fortified positions of the enemy.

However, the manners of the UGV deployment will always depend on technical parameters, sensory systems, weapon systems carried, time for which an UGV can stay operational, and cooperating technical equipment and units. An UGV providing support to a company task force can be defined as a vehicle having hybrid drive, caterpillar undercarriage, satellite navigation and communication system, weapon system consisting of a 7.62 or 12.7 mm caliber machine gun, antitank guided missiles, and smoke grenades. A radar and optoelectronic sensors are equally important because of direct navigation in the microrelief, obstacle avoidance, and target identification.

The practical usability of the MCS CZ in planning an UGV group maneuver as a company task force support can be confirmed on the basis of the combat possibilities of current UGVs, requirements of company commanders, and results of the case study. The MCS CZ can speed up the decision-making process provided that topical information from the battlefield is available. It is possible to model a tactical situation by giving variable input parameters of the maneuver calculation to suggest and find the most effective alternative of solution. When combined with a GPS receiver the MCS CZ can give the UGV considerable autonomy and speed during maneuver planning.

References

1. AAP-6 (2018): NATO glossary of terms and definitions (English and French). NATO Standardization Office, Brussels (2018)

2. Federated Mission Networking: NATO Allied Command Transformation. Brussels (2015). https://web.archive.org/web/20190128083216/https://www.act.nato.int/fmn
3. Network Centric Warfare Capabilities & Indian Armed Forces. Defence 360. Next Generation Weapons Technology (2016). https://defence360officials.blogspot.com/2016/12/network-centric-warfare-capabilities.html
4. Niel, G. Siegel, Madni, A.M.: The digital battlefield: a behind the scenes look from the systems perspective. Procedia Comput. Sci. **28**, 799–808 (2014). https://doi.org/10.1016/j.procs.2014.03.095
5. Spak, U.: The common operational picture: a powerful enabler or a cause of severe misunderstanding? In: 22st International Command and Control Research and Technology Symposium (ICCRTS): Frontiers of C2, vol. Topic 4 (2017). ISSN 2577-1604
6. Sophronides, P., Papadopoulou, C., Giaoutzi, M., Scholten, H.: A common operational picture in support of situational awareness for efficient emergency response operations. J. Future Internet **2**, 10–35 (2017). https://doi.org/10.18488/journal.102.2017.21.10.35
7. Infantry Company Operations: US Marine Corps, Department of the Navy, Headquarters United States Marine Corps, p. 387. Washington, D.C. (2014). MCWP 3–11.1, PCN 143 000117 00
8. The Buzz: Top 10 Unmanned Ground Combat Vehicles (UGCVs), Top Military Robots in the World (2019). In: YouTube. 23 November 2019
9. DARPA Robotic Challenge -The Wins and Fails. In: YouTube. Channel of Robot Time Machine with Gray Bright. Tomorrow Show (2015). https://www.youtube.com/watch?v=wX0KagJ1du8
10. Feickert, A., Kapp, L., Elsea, J.K., Harris, L.A.: U.S. Ground Forces Robotics and Autonomous Systems (RAS) and Artificial Intelligence (AI): Considerations for Congress. Washington D.C. (2018). https://digital.library.unt.edu/ark:/67531/metadc1442984/m1/.
11. Digital Infantry Battlefield Solution, Research and Innovation, Part III, p. 120. Milrem robotics, Tallinn (2019). ISBN 978-9934-567-37-7
12. Digital Infantry Battlefield Solution, Introduction to Ground Robotics, Part I, p. 128. Milrem robotics, Tallinn (2016). ISBN 978-9984-583-92-1
13. Galliott, J.: The soldier's tolerance for autonomous systems. J. Behav. Robot. 124–136 (2018). https://doi.org/10.1515/pjbr-2018-0008
14. European Group on Ethics in Science and New Technologies, Artificial Intelligence, Robotics and 'Autonomous', Systems, Brussels, p. 24. (2018). https://doi.org/10.2777/786515.ISBN 978-92-79-80328-4
15. The U.S. Army Robotic and Autonomous Systems Strategy, U.S. Army Training and Doctrine Command, p. 43. Fort Eustis (2017)
16. Unmanned Systems Integrated Roadmap FY2017–2042, p. 58 (2017). https://www.defensedaily.com/wp-content/uploads/post_attachment/206477.pdf
17. Harder, Byron R.: Automated battle planning for combat models with maneuver and fire support, theses and dissertations, Monterey. Naval Postgraduate School, California, p. 477 (2017)
18. Swiecicki, C., Elliott, L.R, Wooldridge R.: Squad-level soldier-robot dynamics: exploring future concepts involving intelligent autonomous robots, army research laboratory, Aberdeen Proving Ground, p. 152 (2015). https://doi.org/10.13140/2.1.3575.6326.
19. Pokonieczny, K., Rybanský, M.: Method of developing the maps of passability for unmanned ground vehicles. In: 9th IGRSM International Conference and Exhibition on Geospatial & Remote Sensing (IGRSM 2018) Kuala Lumpur, Malaysia: IOP Conference Series: Earth and Environmental Science, vol. 169 (2018). https://doi.org/10.1088/1755-1315/169/1/012027. ISSN 1755-1307.

20. Rybanský, M.: Trafficability analysis through vegetation. In: Conference Proceedings of ICMT 2017. Institute of Electrical and Electronics Engineers Inc., Piscataway, pp. 207–210 (2017). https://doi.org/10.1109/MILTECHS.2017.7988757. ISBN 978-1-5386-1988-9.
21. Hrabec, D., Mazal, J., Stodola, P.: Optimal maneuver for two cooperative military elements under uncertain enemy threat. Int. J. Oper. Res. **35**(2), 263–277 (2019). https://doi.org/10.1504/IJOR.2019.10022439.ISSN1745-7645
22. Wermelinger, M., Fankhauser, P., Diethelm, R., Krüsi, P., Siegwart, R., Hutter, M.: Navigation planning for legged robots in challenging terrain, Daejeon, South Korea, pp. 1184–1189 (2016). https://doi.org/10.1109/IROS.2016.7759199.
23. Hodický, J., Castrogiovanni, R., Lo Presti, A.: Modelling and simulation challenges in the urbanized area. In: 2016 Proceedings of the 17th International Conference on Mechatronics - Mechatronika (ME), pp. 429–432 (2016). Czech Technical University in Prague, Prague. ISBN 978-80-01-05882-4
24. Hodický, J., Procházka, D.: Challenges in the implementation of autonomous systems into the battlefield. In: Proceedings of the 2017 International Conference on Military Technologies (ICMT), pp. 743–747. Institute of Electrical and Electronics Engineers Inc., Piscataway (2017). https://doi.org/10.1109/MILTECHS.2017.7988855. ISBN 978-1-5386-1988-9.
25. Bostelman, R., Messina, E.: A-UGV capabilities, Naples, Italy, pp. 1–7. (2019). https://doi.org/10.1109/IRC.2019.00130.
26. Braun, W. G., Nossal, K. R., Hlatky, S.: Robotics and military operations. In.: Kingston Conference on International Security. U.S. Army War College, New York, p. 77 (2018). https://doi.org/10.1117/12.720422. ISBN 1-58487-780-4
27. Tilenni, G.: Unmanned ground vehicles for combat support. Eur. Secur. Defence 74–77 (2019). https://euro-sd.com/2019/11/articles/15191/unmanned-ground-vehicles-for-combat-support/ ISSN 1617-7983
28. Nohel, J.: In: Mazal, J. (ed.) MESAS 2018. LNCS, vol. 11472, pp. 553–565. Springer, Cham (2019). https://doi.org/10.1007/978-3-030-14984-0_41
29. Nohel, J., Stodola, P., Flasar, Z.: Model of the optimal Maneuver route, IntechOpen, London, pp. 79–100 (2019). https://doi.org/10.5772/intechopen.85566. https://www.intechopen.com/online-first/model-of-the-optimal-maneuver-route
30. Nohel, J., Flasar, Z.: Maneuver control system CZ. In: Mazal, J., Fagiolini, A., Vasik, P. (eds.) MESAS 2019. LNCS, vol. 11995, pp. 379–388. Springer, Cham (2020). https://doi.org/10.1007/978-3-030-43890-6_31
31. Schachter, B. J.: Automatic Target Recognition, 3rd edn, p. 330. SPIE Press, Bellingham (2018). https://doi.org/10.1117/3.2315926. ISBN 978-1-510-61857-2
32. Stodola, P., Mazal, J.: Tactical decision support system to aid commanders in their decision-making. In: Hodicky, J. (ed.) MESAS 2016. LNCS, vol. 9991, pp. 396–406. Springer, Cham (2016). https://doi.org/10.1007/978-3-319-47605-6_32
33. Risald, R., Mirino, A., Suyoto, S.: Best routes selection using Dijkstra and floyd-warshall algorithm, Surabaya, Indonesia, pp. 155–158 (2017). https://doi.org/10.1109/ICTS.2017.8265662
34. Pradhan, A., Kumar, M.G.: Finding all-pairs shortest path for a large-scale transportation network using parallel Floyd-Warshall and parallel Dijkstra algorithms. J. Comput. Civ. Eng. **27**(3), 263–273 (2013). https://doi.org/10.1061/(ASCE)CP.1943-5487.0000220

M&S Driven Experimentation for Concept Development on Combined Manned and Unmanned Vehicles in Urban Environment: Simulation Design

Jan Hodicky[1], Pilar Caamano[2], Alberto Tremori[2](✉), and Giovanni Luca Maglione[2]

[1] ACT JFD MSTT, NATO Supreme Allied Command for Transformation, Norfolk, VA 23455, USA
Jan.hodicky@act.nato.int
[2] NATO STO Centre for Maritime Research and Experimentation (CMRE), Viale San Bartolomeo 400, 19126 La Spezia, Italy
{pilar.caamano,alberto.tremori,giovanni.maglione}@cmre.nato.int

Abstract. In the paper we demonstrate the role of modelling and simulation in NATO Concept development and experimentation process both from individual training and analysis perspective. In the use case a simulation prototype is designed to support a proposed concept validation through experimentation. The stochastic simulation prototype design and development followed modified Distributed Simulation Engineering and Execution Process (DSEEP). The use case concept is framed around future operation of combined manned and unmanned vehicles in urban environment. There are two objectives in the use case. The first one is to design and develop a simulation environment replicating experimental trials like the experiment is happening in the field. The role of M&S is here limited to the generation of data based on the user design of the field experiment. The second objective is to support the experiment by M&S in its full potential employing different experimental designs, like parameters variation, optimization, compare runs, Monte Carlo and sensitivity analysis. The paper demonstrates the importance of a design phase of a simulation that can be used to support logistic experimentation in the urbanized environment and serves to the other researches as a basic building block for their own specific experiment being supported by simulation.

Keywords: Concept development · Modelling and simulation · Experimentation · Unmanned ground vehicle

1 Introduction

Concept Development and Experimentation (CD&E) is one the principle of transformation within NATO and National environment. CD&E provides a basis for developing credible solutions to identified capability shortfalls by capturing the best ideas and enabling potential solutions to be explored via experimentation and validated [1]. Moreover, CD&E is an integral part of the development of required capabilities and it

J. Mazal et al. (Eds.): MESAS 2020, LNCS 12619, pp. 43–61, 2021.
https://doi.org/10.1007/978-3-030-70740-8_4

provides valuable input to the NATO Capability Development Process (CDP). The concept the concept is simply a proposed solution that, even after approval, could be subject to testing and evaluation. However, every concept is different, varying in complexity, timescales and resources and consequently each one may require a unique application of the CD&E approach. Therefore a generic guidance and methodology is needed that must be tailored to a concept's specific needs and an organization's processes. Figure 1 describes the overall CD&E methodology. CD&E is applied in three main spheres:

- Transformation oriented. It supports the identification of required capabilities and needs. Transformation-orientated CD&E is likely to generate insights in the problem domain.
- Solution oriented. It finds and develops solutions to specific capability shortfalls/gaps or the exploitation of new opportunities to address capability requirements not yet identified.
- Test and Validation oriented. It focuses on specific solutions and works to confirm its suitability and completeness.

Modeling and Simulation (M&S) is well established in NATO, through the policy level document, NATO M&S Master Plan version 2 [2]. The second part of the document identifies application areas where M&S can be applied. Basically, it can be divided into two main categories. The first one, Training and Education is the most widespread M&S application area, almost all NATO nations have their own Training and Simulation Centre. However, the second category, the M&S to support analysis, is somehow cornered and needs to prove its value in NATO. Support the operations, capability development, mission rehearsal and procurement are the examples of the second category.

In the context of CD&E, M&S can play the most important role in the experimentation. Experimentation is a tool to get insight, to test specific hypothesis or validate the solution [3]. It is perfectly coherent with the previously identified spheres of CD&E applicability. Military experimentation being supported by simulation has been applied at different level of command and in different application areas. Examples of very tactical experiments in command and control domain [4, 5] and in the medical treatment domain [6] and strategic level experiments in defense planning [7] can be taken.

Figure 1 shows a proposed way to use modelling and simulation to support phases of the CD&E methodology. In the Pre-initiation phase there is no specific role of M&S. It uses mainly tools like NATO Defense Planning Process and Strategic Foresight Analysis to identify gaps in current NATO capabilities needed to be addressed.

In the Initiation phase the main product is the Concept Proposal. Because it contains a statement of the challenge the concept is to address, a clear outline of the problem and the scope of the concept, there is a role of M&S mainly in the form of informal and formal modeling of a problem domain. It helps to articulate the problem within the stakeholders.

In the Research phase, the initial concept is developed containing the refined problem statement and the identified compared solutions even with Measures of Performances (MoPs) that serves for the future experimentation as a part of the exercises. In this phase the modeling is still playing the most important role in the problem domain refinement, however in the context of MoPs development, a simulation can be consider to engineer the most relevant ones to the problem domain.

In the Development phase, the concept become mature with the number of limited experimentation that depends heavily on a simulation whether in the form of a wargame or human in the loop constructive simulation. Due the experimentation supported by simulation bringing the qualitative and quantitative dada, the individual solutions can be compared and if deemed disregard from the concept.

In the Refinement and Validation phase, the concept is being tested and validated in the operational environment. It can happen in NATO exercises where a simulation is already used or in the synthetic environment is created on demand composed of a federation of simulations and functional services. This phase should generate more quantitative then qualitative data available for the analysis. The Approval phase is mainly about staffing the concept by the stakeholders with final approval of the concept by NATO Military Committee. There is no role of M&S in this phase.

All in all, M&S can be used to support CD&E phases to reach the required common understanding and credibility of a proposed concept.

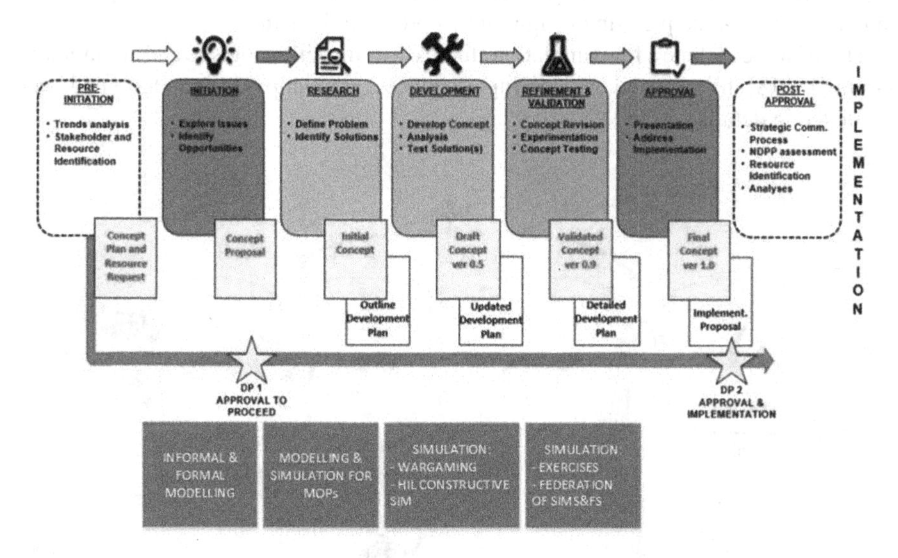

Fig. 1. CD&E methodology and role of M&S.

In NATO nations, there is a permanent effort to make the autonomous system fully operational, not only including the technological observation [8, 9], but the legal aspects as well [10–12]. In NATO Allied Command for Transformation (ACT), there has been an idea of creating a concept proposal of manned and unmanned vehicles deployed to sustain the mission form the logistic point of view in the urbanized area when applying the CD&E methodology. In [13, 14], the authors reviewed processes and techniques from industry, academia and NATO Nations which use M&S methods to enhance the NATO CD&E process when developing concepts involving autonomous systems.

Case Study: Resupply in the "Last Mile" in Hostile Urban Environments
This paper shows the use of M&S methodologies and techniques in support of the CD&E NATO. The case study chosen by the authors is described in the following.

The operation relevant to the developing concept is the last part of a logistic support operation described in [15], also known as "last mile resupply" involving the delivery of combat supplies (mainly ammunition, fuel and lubricant, rations, water, medical supplies and spare parts) from the forward-most logistic base to personnel engaged in combat operations [16].

As depicted in Fig. 2, the logistic operation considers four echelons or levels. The concept of operation to be developed during the CD&E course supports the 2nd and 1st echelon, where trucks and logistic units experience the majority of the problems due to the characteristics of the urban environment and the opposition of local bandits groups.

The main problem is that manned trucks with trailer get stuck when fighting in cities. There turning around time take too long and it endangered the expected mission objective. Their mobility and speed in small streets are very limited. Therefore there is a proposal to find a new concept of operations when employing Unmanned Ground Vehicles (UGVs) that could mitigate this risk and there is a proposal to implement a simulation that can help getting insight in this problem domain.

There has been done a few simulation studies dealing within logistic and autonomous system [17, 18], however no engineering aspect of a simulation design phase had been articulated in details.

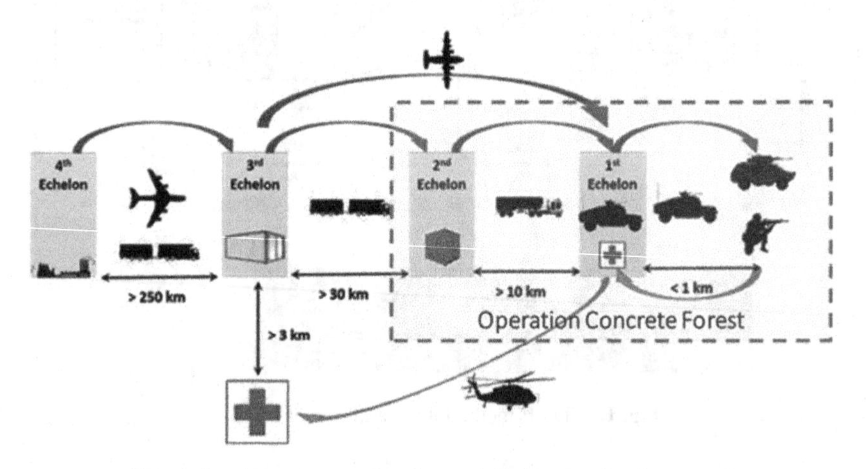

Fig. 2. Logistics Approach to be modelled in the simulation tool.

2 Problem Formulation–Study Objectives

The development of a new operation concept of UGVs in the urbanized environment comes from the need of the students of the NATO CD&E Course to learn how to design and conduct experiments in the context of the NATO CD&E process. In the current situation, the students do not have means to test the concept under development and to generate quantitative data for its validations, so they discuss about possible outcomes based on made up results.

For these students, the Operation Concrete Forest (OCF) simulation is an education tool allowing them to investigate the potential of experimentation and the use of simulation-based experiments to collect and analyze quantitative data.

To fulfil the identified user's need, the following list of objectives for the simulation tool has been identified.

- The OCF simulation tool should allow students to replicate experimental trials like the ones carried out in the field. That is, the simulation tool will provide the students the capability of simulating field experiments.
- The OCF simulation tool should support students to run experimentation by M&S in its full potential, employing different experimental designs, like Parameter Variation, Optimization, Monte Carlo Simulation, Sensitivity Analysis and Calibration.

The following list includes a minimum and initial set of functionalities that the project team currently envisioned as required for fulfilling the objectives.

- The OCF simulation tool should allow students to run stochastic experiments varying a set of factors to assess the impact of these factors on the experiment results. An initial set of factors include:
  - The level of complexity of the environment in terms of density and size of obstacles, or width of streets and roads. These parameters will affect to the mobility of the vehicles in the urban environment in terms of, for instance, speed or turning time. More characteristics could be added as the project evolve.
  - Number, type, size and capacities of the vehicles. The simulation tool is meant to allow students to compare the performance of Unmanned Ground Vehicles (UGVs) and Manned Trucks of different sizes and with different mobility, sensing, communication and transportation capacities.
- The OCF simulation tool should allow to run stochastic simulation and to collect metrics to compute Key Performance Indicators (KPIs) to assess the performance of the simulated system providing graphical representations of these KPIs and allowing the comparison of the KPIs of the simulated system for different experimental configurations.

This list of functionalities created the foundations for the design of the Simulation Environment Requirements (DSEEP Activity 2.3) further described in the methodology section.

3 Methodology

For the development of the simulation tool, the project team will apply the IEEE 1730 Std. – Distributed Simulation Engineering and Execution Process (DSEEP) [19] and

IEEE 1516.4 Std. - Recommended Practices for Verification, Validation and Accreditation (VV&A) of a Federation [20]. To help on the communication between the stakeholders involved in the projects (i.e., Sponsor, Customer and Technical Team), the products developed at each DSEEP stages, see Fig. 3, are linked with the NATO Architecture Framework [21] views. SCRUM is used as an agile framework for the development and delivering of the Simulation SW products [22].

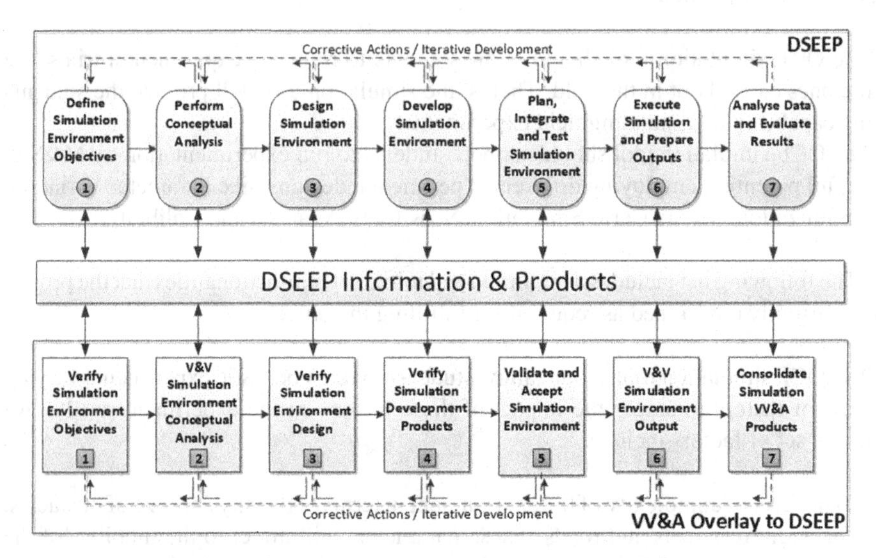

Fig. 3. Top-level view of the VV&A overlay for the DSEEP process

The work here presented covers the second phase of DSEEP shown at Fig. 4, i.e., Perform Conceptual Analysis. The purpose of the following sections is to provide a conceptual representation of the intended problem space and the real world domain based on the needs statement and objectives already identified. The conceptual model provides an implementation-independent representation that serves as a vehicle for transforming objectives into functional and behavioural descriptions. It covers the second step of DSEEP and describes activities taken by the team to design OCF simulation, as shown at the Fig. 4.

Therefore, methodologically the very first activity is to develop a scenario or set of scenarios that will create the boundaries of the problem domain and creates the full picture of operational environment allowing the specification of the KPIs for future experimentation and all the operational environment entities and their attributes. The second activity is to develop the conceptual model describing in the formal way the model of urbanized environment where manned and unmanned vehicles are operating in the logistic domain. The very last activity brings engineered detailed requirements that simulation must reflected.

As stated before, the whole conceptual model has been developed using NATO Architectural Framework (NAF) views. NAF is an enterprise architecture framework [20] that provides guidelines on how to describe and document an architecture or a system.

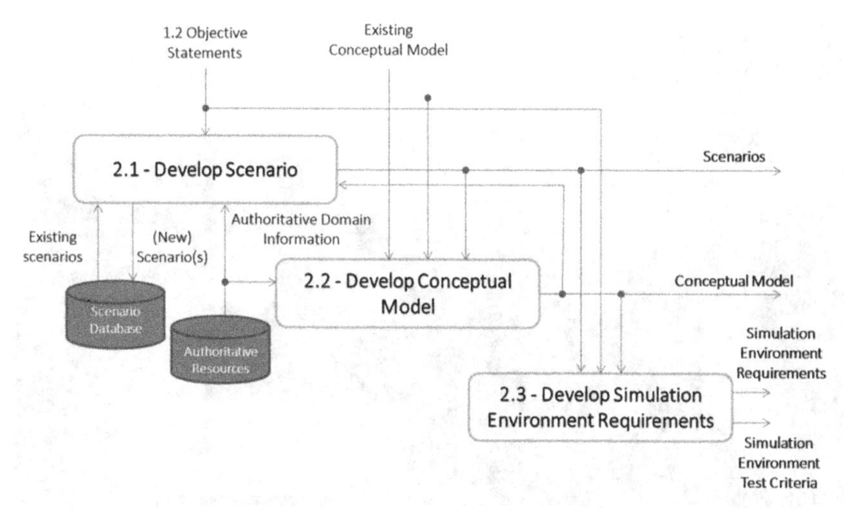

Fig. 4. Main activities carried out at DSEEP Stage 2 – Perform Conceptual Analysis

It consists of rules, guidance, and products for developing and presenting architecture or systems descriptions that ensure a common denominator for their understanding, comparison, and integration.

4 Results

The following section describes activities under the second step of DSEEP, the conceptual model design.

4.1 Development of Scenario

From [14] we can describe general scenario for the use case:

"The approaches to the city of Lyonesse, where the 2nd and 1st echelon operations take place, are limited to a few narrows. Groups of bandits have infiltrated the city attacking the front and rear of convoys, slowing down and even stopping the supply to the 2nd echelon. The trucks with trailers have a hard time to turn and pass the stricken trucks. Even if this situation itself does not cause much damage, it creates shortages up the chain and it hampers the supply operations and draws some manpower away to serve as organic force protection. They also have problems to find their way in the city. They need the bigger streets to move around, but sometimes these streets are blocked. Therefore, the transfer times are longer. Problems in the 1st echelon gets even worse, crews sometimes got disoriented and, in some cases, isolated and surrounded. In some other cases, miscommunications lead to deliver the wrong cargo to the wrong unit."

Figure 5 provides a simplified description and representation of the scenario to be simulated.

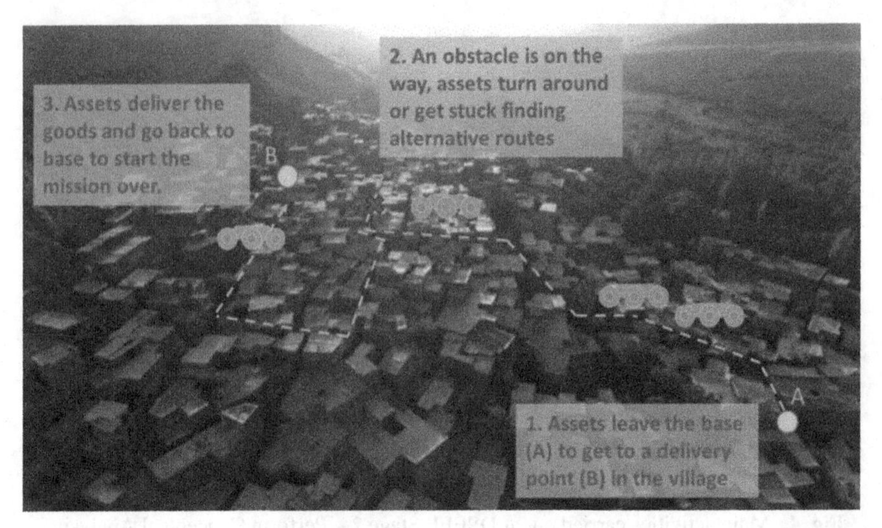

Fig. 5. Assets are used to shuttle between to fixed points through a complex hostile urban environment. A manned truck with trailer gets stuck when fighting in cities. Truck turning around time is too long since the mobility and speed in small streets are limited. There is a suggestion that using UGVs could solve this problem.

The scenario depicted in Fig. 5 will be used to compare the performance of systems with different characteristics. The users should be able to run this scenario using Unmanned Ground Vehicles (UGVs) or Manned Trucks, and within these macro categories of vehicles the users will be able to choose from a list of configurations.

To assess the performance of the system, in the described scenario, data will be collected during the simulation runs to compute the KPIs. The following KPIs have been identified:

- *Supplies delivered per hour*. Weight of goods that successfully reached the destination point per hour.
- *Operational Cost per Ton*. Estimated operational cost of the assets required in current operations per ton of material deployed.

Urban Environment Model Description and Environmental Conditions
The urban environment to be represented in the simulation tool covers an area of 6.4 km by 4.6 km, i.e., a total area of approximately 30 km^2. Streets in this urban environment range from streets seven meters wide and relatively easy to transit, streets five meters wide imposing more difficulties to large vehicles, to narrow streets of three meters wide passable only with small vehicles. The urban environment will be populated with obstacles which can hamper and limit the mobility capacity of the vehicles. Obstacles will be characterized by size, depending on the characteristics of the vehicles and size of the obstacles, these can be overcome or not. Environmental conditions such as type of terrain, day/night operations, or climate are not considered in this version of the simulation environment.

The following list provides a description of the elements to be considered in the modelling and implementation of the urban environment:

- Street -Walkable path. The simulation environment should allow representing paths that can be transit by vehicles representing streets in the urban environment. Obstacles can also appear in the streets.
- Building. Non-walkable area. There are areas in the urban environment that cannot be transit by vehicles because they are occupied by buildings.
- Forward-most logistic base. Entry point. One or more entry points to the urban environment representing the forward-most logistic base where the "last mile resupply" operation starts.
- Delivery point. One or more locations in the urban environment where the supplies have to be delivered.
- Obstacles. Elements in the urban environment that can limit or hamper the mobility of the logistics units.

Vehicles

Vehicles, the only entities being modeled, represent delivery assets or platforms which have to complete the mission of delivering a certain amount of goods to the delivery points in the urban environment. Vehicles can be unmanned ground vehicles (UGVs) or manned trucks. Figure 6 shows two pictures of a UGV developed by QinetiQ and Milrem Robotics to support operations.

Fig. 6. TITAN is a QinetiQ and Milrem Robotics system developed to modeled in the simulation.

Both types of vehicles have sensing capabilities to collect imagery information from the environment and to process it to take decision on how to proceed in the mission. In order to exploit this information, the vehicles, both UGVs and Manned Trucks, keep an internal representation of the environment in the form of map.

The delivery mission is executed by two or more vehicles of the same type which have the capacity to communicate to other members of the team the position of obstacles so the assets can reconsider the paths followed to optimize their performance.

The goal of the vehicles is to deliver an assigned amount of cargo from an entry point to a delivery point as quickly as possible. The performance of the single vehicles and the aggregated performance of the whole system are going to be evaluated in terms of operating cost and cargo delivered by hour.

Vehicles are characterized by the parameters summarized in the following list. The difference between the two types of vehicles are given for different capabilities in terms of mobility, sensing and communication capabilities, and processing and quality of information stored in the internal map.

- Type of vehicle, i.e., Unmanned Ground Vehicle (UGV) or Manned Truck. In a simulation run all the vehicles are of the same type.
- Speed [km/h]. Average speed of a vehicle.
- Turning Time [s]. Average time to turn or modify the direction of the path followed.
- Payload capacity [Ton]. Capacity of cargo of the vehicle.
- Load/Unload time [min]. Average time required for loading and unloading operations.
- Mean time between failure [h]. Mean time between two failures of the vehicle.
- Detection distance [m]. Maximum distance at which a vehicle can detect an obstacle placed in the urban environment.
- Communication range [km]. Average distance the vehicle can shared information about obstacles found in the environment.
- Off-road capacity level. Capacity of the vehicle to overcome obstacles of a given size.

Interaction Among Vehicles

Vehicles can sense obstacles and share position and characteristics of the obstacles with other vehicles in the operation. This will provide information to those vehicles to generate more efficient routes.

Obstacles are placed in the urban environment to hamper or impede the movement of vehicles. However, in some situations, depending on the characteristics of both vehicles and obstacles, the later can be overcome and the vehicles can continue its pre-defined path. For instance, if a big vehicle with a high off-road capacity finds a small obstacle in a wide street, the vehicle can overcome it. When the situation is such that the obstacle cannot be overcome, the vehicle will have to turn and modify the path is following to reach the destination.

Initial and Termination Conditions of Simulation

The following list contains all possible initial conditions:

- A complexity level for the environment is selected, and
- At least one obstacle is placed, and
- Vehicle type and number is selected, and
- All vehicles has an entry point (A) assigned, and
- At least one delivery point (B) is set up, and
- A defined amount of goods have to be delivered, and
- All vehicles receive a mission.

The following list contains all possible termination conditions:

- All the goods have been delivered, or
- All the vehicles are stuck, or
- 24 h has passed.

4.2 Development of the Conceptual Model

The conceptual model is a conceptual representation of the intended problem space based on the user needs, sponsor objective and the description of the scenario provided in the previous sections. The conceptual model provides an implementation-independent representation that serves as a vehicle for transforming objectives into functional and behavioural descriptions for system and software engineers.

The goal of the conceptual model is to identify relevant entities within the domain of interest, to identify static and dynamic relationships between entities, and to identify the behavioural ad transformational (algorithmic) aspects of each entity [18].

Static Relationship

Figure 7 depicts the NAF L7 View – Logical Data Model of the simulation tool to be implemented. This Logical Data Model has been developed after the analysis done in the previous section on the scenario to be simulated. The Logical Data Model is concerned with identifying information elements, and describing their relationships. The Logical Data Model identifies information elements relevant for the architecture, relationships between information elements, attributes of information elements and associate attributes with data entities.

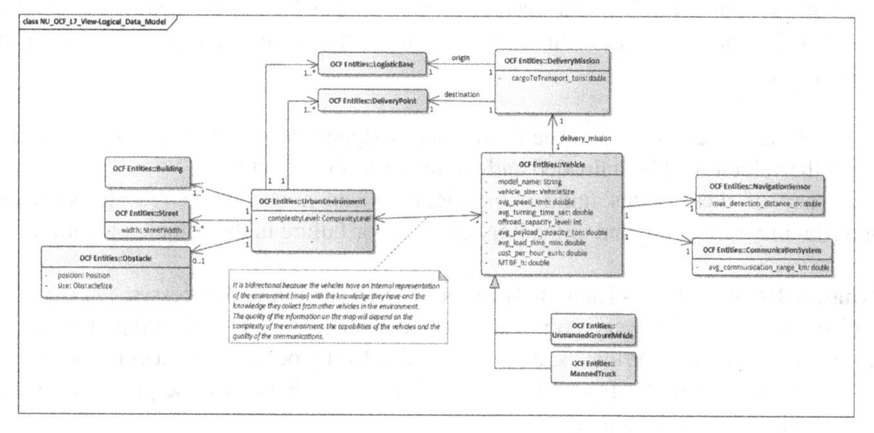

Fig. 7. NAF v4.0 L7 View – Operation Concrete Forest Simulation Tool - Logical Data Model.

Dynamic Relationships – Vehicle Behavior

Figure 8 shows the NAF v4.0 L5 – Logical States view for the entity Vehicle when executing a delivery mission, i.e. delivery an amount of goods from an entry point to ad delivery point. Logical States viewpoint is concerned with the identification and definition of the possible states a node (entity in our case) may have, and the possible transitions between those states.

At the beginning of the operation vehicles are assigned with a logistic base (entry point) in which the vehicles are going to be load with the cargo assigned to them.

If the loading operation does not fail, the vehicle is assigned with a delivery point and starts the execution of its mission, which goal is to reach the assigned delivery point and deliver the goods assigned. During the delivery mission, the vehicles loop through four states:

- Collect Information: in this state, the vehicles using their available sensors collect information of the surrounding environment.
- Process Information: in this state, the vehicles compile the information collected in the previous state and process it with information collected in previous steps or received from other vehicles that are also in operation. Their internal map of the environment is updated with the processed information and this information is shared with the rest of the team using a broadcast (not direct message) strategy.
- Make Decision: with the information compiled in the internal map, the vehicles evaluate the alternative movements available and based on the strategy implemented for path planning (i.e., shortest path, avoid obstacles, etc.) the alternatives will be weighted and one will be chosen to be executed.
- Execute action: vehicles' actions are mainly turn around, if needed, and move in the chosen direction. The kinematic model of the vehicle which governs the turning and moving actions depends on the average turning time and average speed of the vehicle, and on the characteristics of the transiting street. It will be further described later in this document.

Once the vehicle arrives to the destination assigned, it will enter in a new state in which the vehicle will be unloaded and the mission is completed.

In any of these states failures of operation can arise due to reasons such as unavailability of loading and unloading resources or failure in the vehicle operation.

Dynamic Relationships - Logistic Bases and Delivery Points Processes

Logistic Bases and Delivery points are also represented as entities in the simulation model to be developed. At this location, load and unload processes are executed. Figure 9 and Fig. 10 shows the NAF v4.0 L4 View – Logical Activities for the processes to be implemented at these locations.

At the Logistic Bases, the vehicles are load with the assigned cargo; while at the delivery points the vehicles are unloaded once reached the assigned destination. In both cases, a queue will be used as loading/unloading services (i.e., loading/unloading machinery or soldiers itself) are limited in number and several assets can arrive at the same time not being able to be served in parallel.

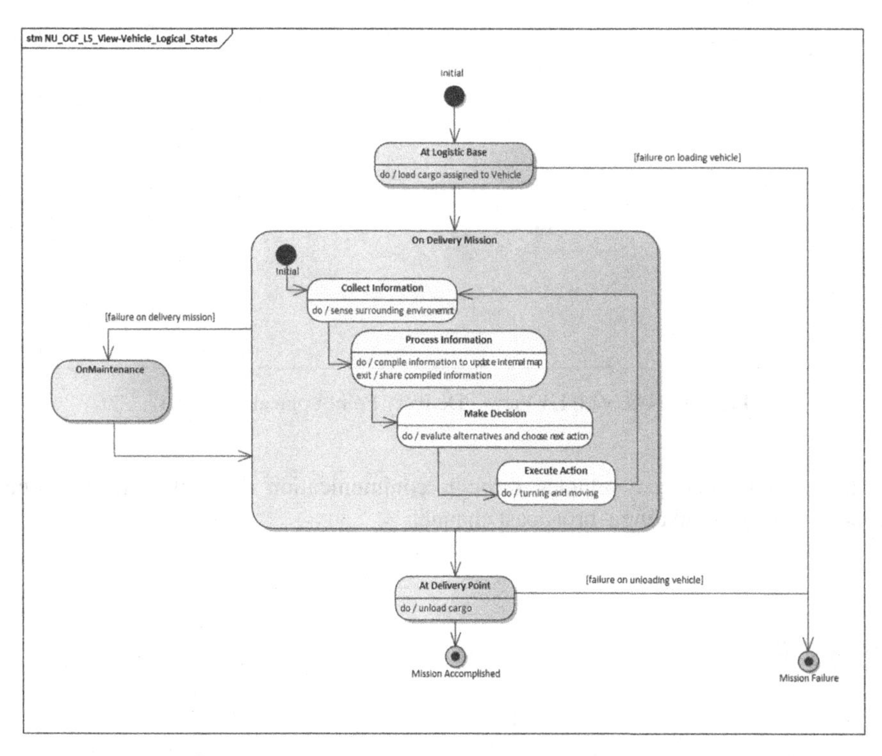

Fig. 8. NAF v4.0 L5 View - Vehicle Behaviour Logical States.

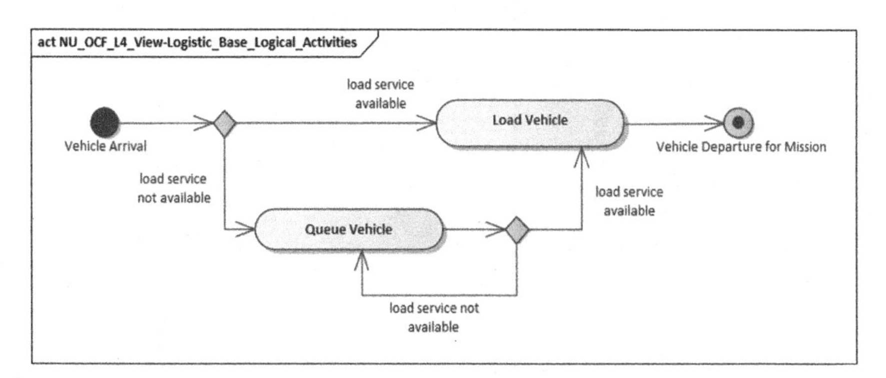

Fig. 9. NAF v4.0 L4 View - Logistic Base Logical Activities.

Dynamic Relationship: Entities Interactions and Sequence of Activities in the Scenario

Figure 11 shows the NAF v4.0 L6 View – Logical Sequence of the scenario to be developed in the simulation tool. The sequence of activities and logical flows described in this diagram are those needed for the execution of on delivery mission by one vehicle.

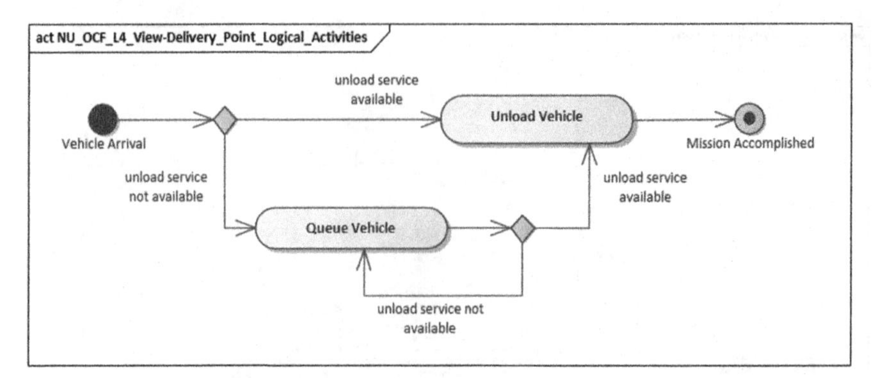

Fig. 10. NAF v4.0 L4 View – Delivery Point Logical Activities.

Interactions amongst vehicles through communication channels will be done asynchronously simulating a broadcast channel.

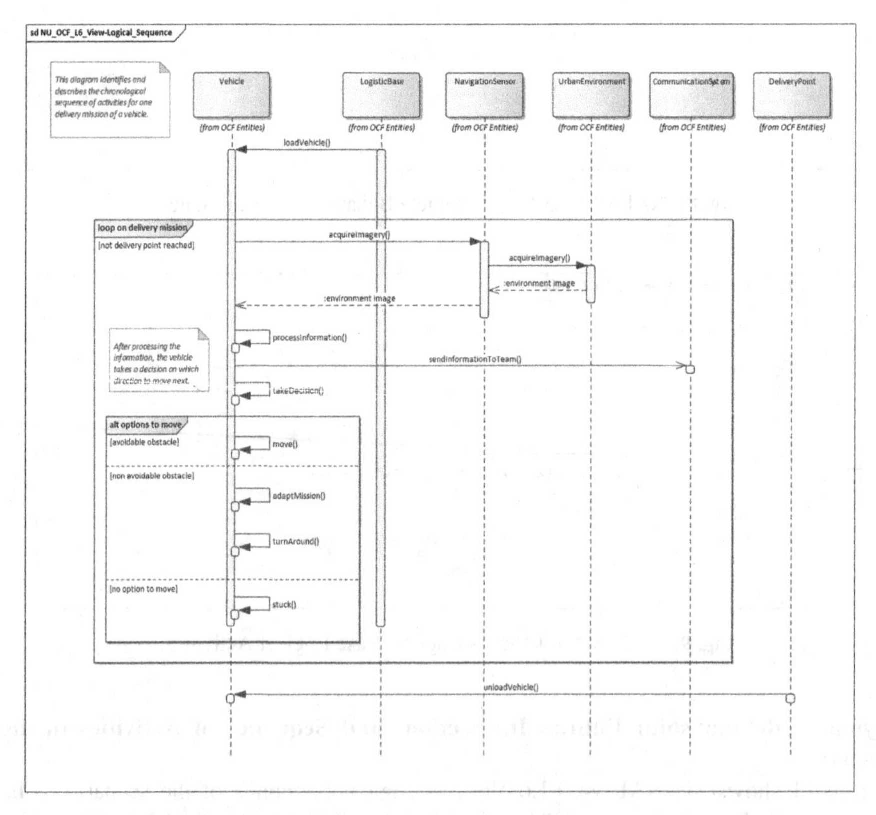

Fig. 11. NAF v4.0 L6 View – Logical Sequence of the event in the scenario.

4.3 Develop Simulation Environment Requirements

This subsection describes the requirements for the simulation environment that are derived from the objectives and from the scenario and conceptual model developed in previous sections.

OCF Students and Instructors Requirements

The first set of requirements that have been derived from the analysis of the user needs statements and the simulation environment objectives are those related with the functionalities that are going to be provided to the two main users (actors) of the simulation tool: the OCF Student and the OCF Instructor. Figure 12 shows the requirements to meet the needs and fulfil Objective 1, i.e. providing a tool to provide the capability of simulating field experiments.

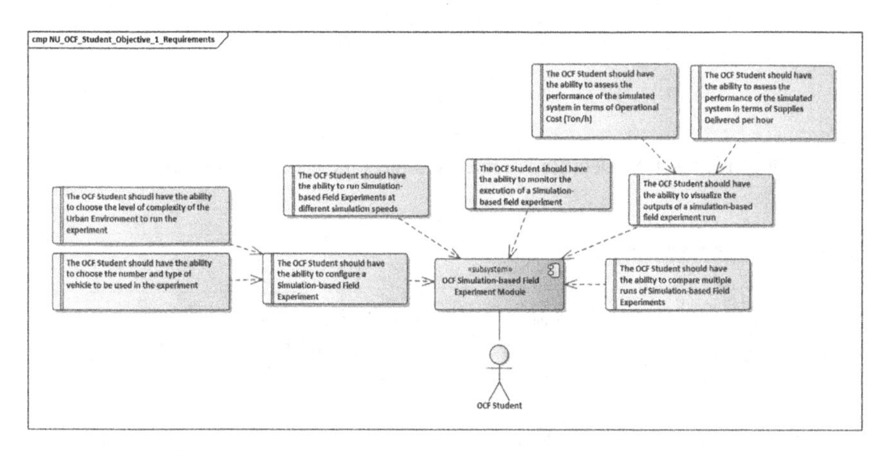

Fig. 12. OCF Student Requirements to meet the needs of Objective 1.

Figure 13 shows the requirements to meet the needs and fulfil Objective 2, i.e. to run experimentation by M&S to its full potential using methods such as Parameter Variation, Optimization, Monte Carlo Simulation, Sensitivity Analysis and Calibration.

For a sake simplified visualization, Fig. 13 do not elaborate in detail on the requirements related to inputs variations or outputs visualization. For those requirements concerning variation ranges or distribution of input parameters, the parameters that will be considered are number of vehicles, type of vehicles and level of complexity. For those requirements related to visualization and comparison of results, the outputs to be considered are those related with the identified KPIs: Operational Cost and Supplies delivered per hour.

From the analysis of needs and objectives, and after conversations with the project customers, it has also arisen the need of providing CD&E Instructors with a module within the OCF Suite of Tools to configure and set-up the exercises that the students are going to simulate. Figure 14 shows the requirements for this module.

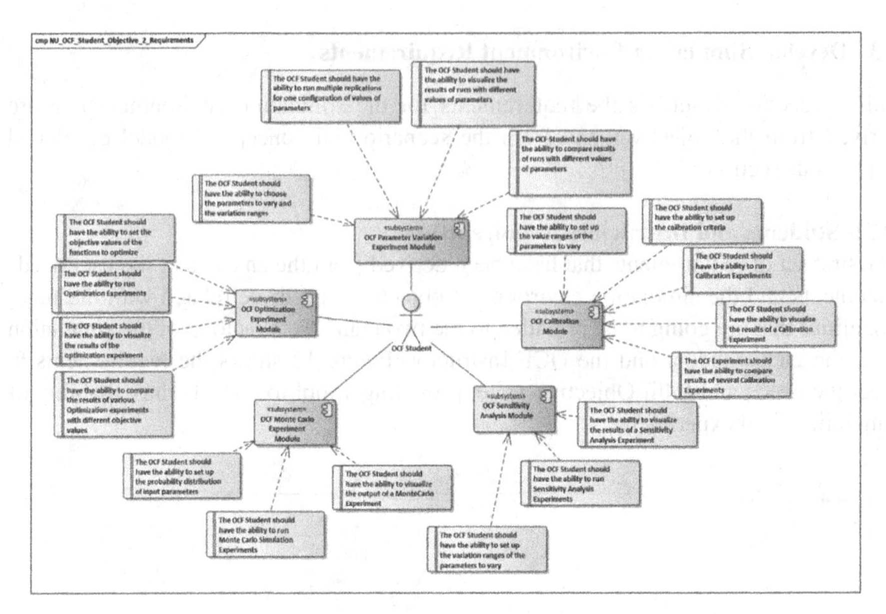

Fig. 13. OCF Student Requirements to meet the needs of Objective 2.

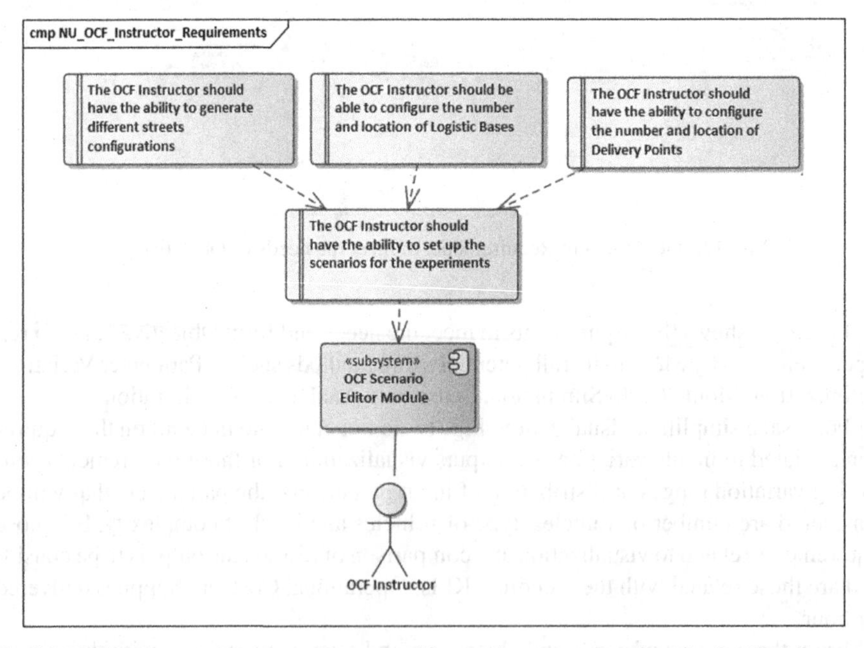

Fig. 14. OCF Instructor Requirements to provide a tool to set up exercises for the CD&E course.

5 Conclusion

The paper discusses the need to have an engineering approach to design and develop a simulation that serves as an education and experimentation tool to support the design of a new concept of operations in the urbanized area.

The authors recommend, following DSEEP steps together with the agile development technique, to deliver a final product as soon as possible to the customer. DSEEP standard needs to be tailored to better fit the unique project requirements and it is strongly recommended to use the NATO Architectural Framework as the formalized way do develop the conceptual model. The conceptual model is the most important tool to achieve a common agreement on the project with the customer.

The main outcome described in this paper is the designed conceptual model, that could serve to other researchers as on a building block for their own simulation development. This would clearly facilitate to focus on the specific research objectives by adopting the described approach and models. In such approach, conceptual model can serve as a template for the future simulation studies in the urbanized environment. Further work will be focused on the implementation of the simulation. At time being Fig. 15 shows the current user interface. A simulation prototype is being developed with AnyLogic SW where entity behavior is modeled by Agent Based Modeling and discrete Event Simulation paradigms.

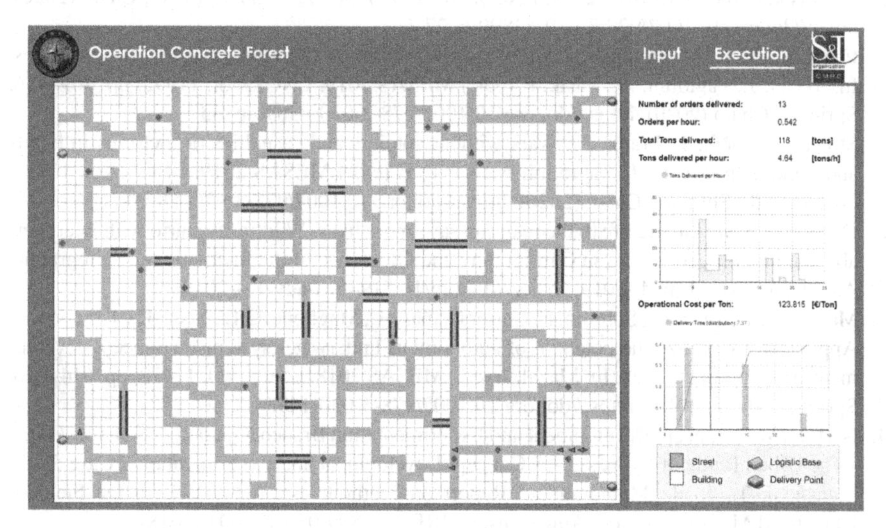

Fig. 15. Graphical User Interface for monitoring the execution and evolution of KPIs of Operation Concrete Forest M&S-based field experiments.

References

1. Concept Development and Experimentation Handbook. Allied Command Transformation, Norfolk, USA (2018)

2. NATO ACT. NATO Modelling and Simulation Master Plan. Edition 2. 2021. AC/323/NMSG(2012)-015
3. NATO ACT. Bi-Sc 75-4. Experimentaiton Directive. Norfolk. US: HQ SACT (2010)
4. Nohel, J., Flasar, Z.: Maneuver control system CZ. In: Mazal, J., Fagiolini, A., Vasik, P. (eds.) MESAS 2019. LNCS, vol. 11995, pp. 379–388. Springer, Cham (2020). https://doi.org/10.1007/978-3-030-43890-6_31
5. Stodola, P., Drozd, J., Nohel, J., Michenka, K.: Model of observation posts deployment in tactical decision support system. In: Mazal, J., Fagiolini, A., Vasik, P. (eds.) MESAS 2019. LNCS, vol. 11995, pp. 231–243. Springer, Cham (2020). https://doi.org/10.1007/978-3-030-43890-6_18
6. Hodický, J., Procházka, D., Jersák, R., Stodola, P., Drozd, J.: Optimization of the casualties' treatment process: blended military experiment. Entropy **22**, 706 (2020). https://doi.org/10.3390/e22060706
7. Hodický, J., et al.: Computer assisted wargame for military capability-based planning. Entropy **22**, 861 (2020). https://doi.org/10.3390/e22080861
8. Fučík, J., Melichar, J., Kolkus, J., Procházka, J.: Military technology evolution assessment under growing uncertainty and complexity: methodological framework for alternative futures. In: Proceedings of the 2017 International Conference on Military Technologies. Piscataway, NJ 08854-4141 USA: Institute of Electrical and Electronics Engineers Inc., pp. 682–689 (2017). ISBN 978–1–5386–1988–9.
9. David, W., Pappalepore, P., Stefanova, A., Sarbu, B.A.: AI-powered lethal autonomous weapon systems in defence transformation. Impact and challenges. In: Mazal, J., Fagiolini, A., Vasik, P. (eds.) MESAS 2019. LNCS, vol. 11995, pp. 337–350. Springer, Cham (2020). https://doi.org/10.1007/978-3-030-43890-6_27
10. Fučík, J., Frank, L., Stojar, R.: Legality and legitimacy of the autonomous weapon systems. In: Mazal, J., Fagiolini, A., Vasik, P. (eds.) MESAS 2019. LNCS, vol. 11995, pp. 409–416. Springer, Cham (2020). https://doi.org/10.1007/978-3-030-43890-6_33
11. Stojar, R., Fučík, J., Frank, L.: Wars without soldiers and casualties or victories in hearts and minds? In: Mazal, J., Fagiolini, A., Vasik, P. (eds.) MESAS 2019. LNCS, vol. 11995, pp. 372–378. Springer, Cham (2020). https://doi.org/10.1007/978-3-030-43890-6_30
12. David, W., et al.: Giving life to the map can save more lives. Wildfire scenario with interoperable simulations, Advances in Cartography and GIScience of the International Cartographic Association, vol. 1, p. 4 (2019). https://doi.org/10.5194/ica-adv-1-4-2019
13. Mansfield, T., Caamaño Sobrino, P., Carrera Viñas, A., Maglione, G.L., Been, R., Tremori, A.: Approaches to realize the potential of autonomous underwater systems in concept development and experimentation. In: Mazal, J. (ed.) MESAS 2018. LNCS, vol. 11472, pp. 614–626. Springer, Cham (2019). https://doi.org/10.1007/978-3-030-14984-0_46
14. NATO SACT Concept Development and Experimentation Branch, Operation Concrete Forest - The Urban Battle of Lyonesse - A Concept Development and Experimentation Case Study
15. Mansfield, T., et al.: Memorandum Report on Methodology for CD&E in the Field of Unmanned Maritime System Autonomy, CMRE-NU-809-03-0106-Q4 (2018)
16. Defence Science and Technology Laboratory (DSTL), Competition document: autonomous last mile resupply, 29 June 2017. https://www.gov.uk/government/publications/accelerator-competition-autonomous-last-mile-supply/accelerator-competition-autonomous-last-mile-resupply. Accessed 29 July 2020
17. Foltin, P., Vlkovský, M., Mazal, J., Husák, J., Brunclík, M.: Discrete event simulation in future military logistics applications and aspects. In: Mazal, J. (ed.) MESAS 2017. LNCS, vol. 10756, pp. 410–421. Springer, Cham (2018). https://doi.org/10.1007/978-3-319-76072-8_30

18. Tulach, P., Foltin, P.: Research Methods In Humanitarian Logistics – Current Approaches And Future Trends. Business Logistics in Modern Management, Josip Juraj Strossmayer University of Osijek, Faculty of Economics, Croatia, vol. 19, pp. 459–474 (2019)
19. IEEE: IEEE Recommended Practice for Distributed Simulation Engineering and Execution Process (DSEEP) (2011)
20. IEEE: IEEE Recommended Practices for Verification, Validation and Accreditation of a Federation - An Overlay to the High Level Architecture Federation Development and Execution Process (2007)
21. Architecture Capability Team Consultation, Command & Control Board, NATO Architecture Framework Version (2018)
22. Schwaber, K., Sutherland, J.: The Scrum Guide: The Definitive Guide to Scrum: The Rules of the Game (2017)

UAV Based Vehicle Detection with Synthetic Training: Identification of Performance Factors Using Image Descriptors and Machine Learning

Michael Krump(✉) and Peter Stütz

Institute of Flight Systems, Bundeswehr University Munich, Neubiberg, Germany
{michael.krump,peter.stuetz}@unibw.de

Abstract. Vehicle detection on aerial imagery plays an important role in autonomous systems especially on board of UAVs (Unmanned Aerial Vehicles). Deep learning based object detectors are often used to overcome the resulting detection challenges. To achieve the necessary variance, real training data are extended or completely replaced by synthetic data sets from virtual simulation environments. Differences between these two domains can lead to performance differences in the later real application, which are called reality gap. Our current research interests are focused on the identification of image properties that on the one hand allow a distinction between real and synthetic domains and on the other hand are related to detection performance. We present a method to rank the most relevant image descriptors using a machine learning based classification chain. In the first part of the investigations real and synthetic image pairs from own UAV flights and from the publicly available KITTI/Virtual KITTI data set will be analyzed. Finally, it will be shown which image descriptors are responsible for the false detections of a purely synthetic trained YOLOv3 vehicle detector and which conclusions can be drawn for future training data generation.

Keywords: UAV · Vehicle detection · Synthetic training dataset · Virtual environment · YOLO · Deep learning · CNN · Image descriptor · MPEG7 · Classification · Decision tree · Performance factor

1 Introduction

Sensor payload systems are an important component of unmanned flight systems to enable more autonomous behavior in the field of flight safety (e.g. Sense & Avoid), navigation (Feature-based navigation) or mission conduction (acquisition of reconnaissance and surveillance information). In an attempt to enhance the sensor performance of such systems on mission level, we chose the application field of vehicle detection from an airborne position. It plays a role in missions related to traffic monitoring, surveillance and rescue.

Striving for a higher level of autonomy immediately requires the integration of automated data processing in such systems. However, the performance of modern, deep learning based detectors is influenced to a large extent by the availability and quantity of

J. Mazal et al. (Eds.): MESAS 2020, LNCS 12619, pp. 62–85, 2021.
https://doi.org/10.1007/978-3-030-70740-8_5

suitable test and training data. A promising approach to avoid complex flight missions and at the same time increase the recorded scenario variance is the use of virtual simulation environments to generate a database with synthetic sensor images [1–4]. These can efficiently consider a multitude of atmospheric and sensory effects and thus represent a possibility to extend existing real training data or to replace it completely. However, this raises the question of how virtual simulation environments must be designed and which image properties are mainly responsible for the differences between real and synthetic sensor images.

2 Object of Research

In [5], we used real and synthetic data sets to investigate the training behavior and the reality gap between these two domains using the common object detector YOLOv3 [6] for UAV based vehicle detection. It was analyzed to what extent algorithms trained with real data can be evaluated in the simulation, which detection performance can be achieved with exclusively synthetic training and how the performance can be improved by synthetic extension of real training data. The UAVDT data set [7] serves as the basis for the real image data and the synthetic image material is generated with the Presagis M&S Suite [8]. The training of the YOLOv3 framework on the real UAVDT data set led to an Average Precision (AP) of 69.9% at a Intersection over Union (IoU) threshold of 0.3 when using the corresponding test set. By extending the real training set with 20% synthetic data, the generalization capability of the detector could be improved and the AP increased by more than two percentage points. However, only a relatively low detection performance with AP values around 15% could be achieved by purely synthetic training.

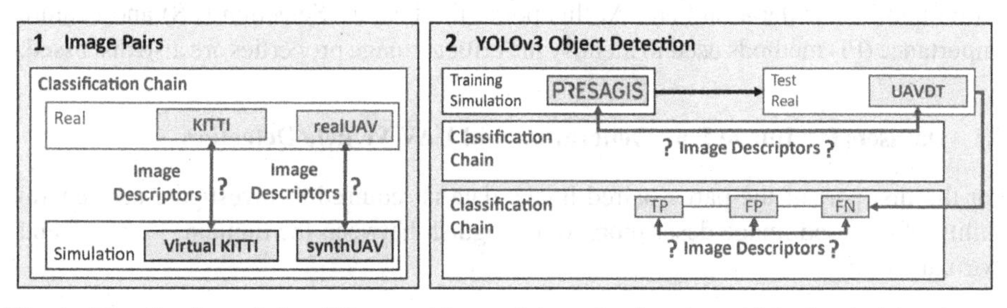

Fig. 1. Visualization of the different objects of investigation for which the decisive image properties should be determined in this paper.

Especially the object detectors relying on deep learning paradigms are largely black box models. The extracted features are abstract and cannot be directly assigned to individual image properties. In this paper we want to present an approach that uses a classification chain to determine image descriptors that cause the reality gap between real and synthetic data and that are related to correct (TP: True Positive) and incorrect (FP: False Positive, FN: False Negative) detections in purely synthetic training. This could be a way to indirectly assign the abstract features to certain image properties which affect the current detection performance.

Figure 1 shows the different objects of investigation, which are analyzed to answer the following research questions:

a) Is it possible to classify **image pairs** into real and synthetic domain with selected image descriptors as features and which image properties are decisive for the classification? This serves as a basis and legitimation for the following investigations. Image pairs from own UAV flights and from the publicly available KITTI/Virtual KITTI [1, 9] data set will be analyzed.
b) Is the approach appropriate to also assign **independent datasets** used for training and testing a vehicle detector to the correct domain? The synthetic training data set described in [5] and created with Presagis M&S Suite and the real UAVDT test set is used.
c) What is the performance in classifying the detection results of a synthetic trained deep-learning based vehicle detector into **correct and incorrect detections**. Which image characteristics are decisive, favor incorrect detections and should therefore be considered more in future training?

3 State of the Art: Datasets, Image Descriptors, Classifiers

An overview of the datasets used in the investigations is given below. Data sets with corresponding real and synthetic image pairs are presented as well as training and test datasets for UAV based vehicle detection on aerial images. In the second part the different groups of image descriptors are considered, which serve as features for the classification chain used and help to identify the relevant image properties. Finally, various classifiers are listed and it is explained why we chose the Decision Tree Classifier for the application under consideration here. At this point, the Feature Selection (FS) and Feature Importance (FI) methods used to identify the critical image properties are also discussed.

3.1 Datasets for Image Pair Generation and UAV Vehicle Detection

For the first part of the paper paired image data is required to investigate the general ability of the used image descriptors to distinguish between the domains "reality" and "virtual world".

Fig. 2. Real/synthetic image pair from our own aerial photographs taken by a UAV, which were recreated in the synthetic environment

realUAV/synthUAV. Due to the limited availability of synthetic duplicates of real aerial images for the application of UAV-based vehicle detection, we conducted own flight missions on the test area of the Bundeswehr University Munich. The quadcopter used was equipped with a gimbal with integrated camera system. A built-in GPS receiver with Real Time Kinetics (RTK) correction data enabled the precise acquisition of the current position. The resulting telemetry and sensor data are stored synchronously during the flight. To render synthetic duplicates of the real sensor data the Presagis [8] virtual simulation environments was used which provides a physically based sensor simulation. The test flight area was modelled in Common Database (CDB) format. This is based on a georeferenced satellite image with a resolution of 20 cm per pixel, which is superimposed on the elevation data of the terrain. The satellite image is improved by overlaying the corresponding terrain with fine, semi-transparent structures, e.g. for asphalt or grass areas. In addition, the CDB contains layers for light points, surface features, material classification and three-dimensional modelled buildings and trees. In this virtual world, the position and orientation of the virtual camera is now updated at each discrete time step according to the previously recorded telemetry data. The rendered images are stored and represent synthetic duplicates of the real sensor images. The observed static vehicle corresponds to the Point of Interest (POI) and is replaced by a 3D model in the simulation based on the acquired coordinates. Due to different flight altitudes (30, 60, 90 m) and different distances to the POI, the selected flight path allows different angles of view on the vehicle and different background scenarios. An example image pair from the data set is shown in Fig. 2. Visible differences in coloration and reflectivity were deliberately not corrected for the investigations presented here in order to be able to determine the factors influencing the reality gap independently of human perception. To correct camera distortions in the outer image areas the real taken aerial images were calibrated afterwards. Finally, the data set includes 823 image pairs with a resolution of 2000 $\times$ 1500 pixels. This version of the data set is not intended for the training of deep-learning detectors due to the low variance in vehicle models and scenarios. However, the synthetic data is taken from the same virtual database as the synthetic training data set used in the following and was generated with the same rendering engine. It can therefore be investigated whether the image properties responsible for the reality gap between these image pairs also have an influence on the detection performance.

KITTI/Virtual KITTI. Geiger et al. [9] presented an autonomous driving dataset called KITTI, which was captured from a car driving in Karlsruhe. It offers a wide range of synchronized sensor data with variations in terms of the objects and scenarios encountered. Due to the annotations provided, it is a commonly used public standard benchmark set for different types of computer vision tasks, such as stereo, optical flow or object detection and tracking.

Gaidon et al. [1] took up the established KITTI data set and presented a real-to-virtual world cloning method to transform it into a fully labelled, dynamic and nearly photo-realistic proxy virtual world. The Unity Game Engine was used to create the synthetic duplicates called Virtual KITTI. The use of a virtual simulation environment offers the advantage of simultaneously generating highly accurate automatic ground truth. The aim was to investigate the influence of the reality gap and different rendered variations on the detection performance for the use case of multi-object tracking. The corresponding duplicates of the original real KITTI video sequences resulted in a data set with 2131 image pairs with a resolution of 1242 × 375 pixels. Figure 3 shows a corresponding image pair. Although the data set contains ground-based image data instead of the aerial photographs examined, it is nevertheless very well suited as a publicly available benchmark. It can be analyzed to what extent the presented classification chain is also suitable for this application and which image properties are used in this case to distinguish between the image pairs, also with respect to the different underlying render engine.

Fig. 3. Real/synthetic image pair from the publicly available KITTI/Virtual KITTI data set

For the second part of the paper annotated aerial images of vehicles with high variance are needed for training and testing of the deep-learning based YOLOv3 framework.

Synthetic Training Dataset. In [5] an automated approach is described to generate a completely annotated synthetic training data set with associated ground truth in the form of bounding boxes via the programming interface of the Presagis M&S Suite. In order to cover as many later vehicle states as possible during training, the vehicle model, the scenery, the object orientation, the flight altitude and the camera radius are individually changed in predefined values in nested loops. To further increase the variation, each image also contains randomly selected values for the parameters time of day, visibility and noise. The training data set contains six different scenarios and 38 different 3D vehicle models (extended to 80 models by re-coloring them according to the worldwide car color distribution). In total this results in over 93000 automatically labelled training images with over 86000 bounding boxes and a resolution of 1024 × 540 pixels.

Fig. 4. Annotated sample images for vehicle detection from our own synthetic training dataset

Figure 4 shows sample images with the corresponding annotated vehicles for detector training. For further examples, a detailed description of the generation process and the analysis of the parameter distribution, please refer to [5]. This data set serves as a starting point for the analysis of which image properties play a role for the correctness of later detections when using purely synthetic training. To get as close as possible to the later use case, we need a real test data set, which is described in the following section.

UAVDT. The UAVDT data set [7] (see Fig. 5) was introduced in 2018 and is currently one of the most challenging large-scale drone-based data sets for object detection, single- and multi-object tracking applications. It is divided into training and test data set and contains significant variations in flight altitude, angle of view and environmental parameters. The test set contains over 16000 images with about 350000 vehicle objects at a resolution of 1024×540 pixels. It is used to test the performance of the purely synthetic trained detector on real image data.

Fig. 5. Annotated sample images for vehicle detection from the publicly available UAVDT set

3.2 Image Descriptors as Classification Features

The aim is to classify between real/synthetic data and TP/FP/FN detections solely on the basis of the image material provided. Image descriptors are used to extract the image properties that are important for classification and serve as features for the classifier algorithm. A differentiation can be made based on various criteria. On the top level, a distinction is often drawn between metrics that mainly reflect the image quality and those that rather describe the image content. The former are further distinguished between full-reference, reduced-reference and no-reference methods [10]. Only no-reference methods play a role in our considerations since they can be calculated even without a distortion-free reference image. Furthermore, we distinguish between local and global calculations, between metrics that are more based on human perception and metrics that represent more technical properties and between different mathematical calculation methods in the time and frequency domain. An attempt was made to select suitable metrics from all these categories and then assign them to specific groups of image properties: *Color, Texture, Shape, Brightness/Luminance/Contrast, Edges, Distortion/Blur/Noise, Image Quality* and *Environmental Conditions*. These are to be used to provide guidelines for the design of synthetic simulation environments by improving the image properties responsible for the reality gap and to find image properties that are related to false detections and therefore need to be considered more in the design of synthetic training data sets. Table 1 provides an overview of the metrics described below and their assignment.

MPEG-7. The MPEG-7 standard [11–13], often referred to as Multimedia Content Description Interface, provides a standardized set of descriptors for color, texture, shape, motion, audio and face description. The goal is to describe multimedia content using visual criteria to generate metadata to efficiently filter, identify, categorize, and browse images and videos based on a non-text visual content description. Fields of application are content-based image retrieval (CBIR), video analysis and keyframe extraction [14, 15].

(a) **Color:** These descriptors play a major role in human perception, are relatively robust to changes in background colors and independent of image size and orientation [12].

SCD (Scalable Color Descriptor) measures the color distribution globally over the entire image using a color histogram in the HSV color range. The histogram values are normalized and non-linearly mapped to a 4-bit representation. Coding using the hair transformation is done for dimension reduction and allows scalability to balance accuracy and descriptor size. It is used for image-to-image matching of color photographs, is hardly discriminative for synthetic content and performs badly on monochrome inputs [16].

CSD (Color Structure Descriptor) captures both the global color distribution and the local information about the spatial structure and arrangement of the colors and is therefore able to distinguish between images with the same color histogram. The calculation is done in the perceptual HMMD (Hue-Max-Min-Diff) color space. A structuring block scans the image in a sliding window approach and records the number of times a particular color is contained. The descriptor is scalable and independent of the image size. This approach corresponds to a rudimentary form recognition and improves the image retrieval performance for natural images, except monochrome inputs [13, 17].

CLD (Color Layout Descriptor) is a compact and resolution-independent method for describing the spatial distribution of the dominant colors in the YCbCr color space [14, 18].To ensure the independence of resolution and scaling, a thumbnail (8×8) of the image is calculated in the first step by partitioning and then the representative color is determined for each block by averaging. These are encoded by a discrete cosine transform (DCT) and a certain adjustable number of low frequency coefficients are selected and quantized by zigzag scanning. The CLD is particularly suitable for fast sketch-based retrieval and content filtering.

DCD (Dominant Color Descriptor) is suitable for compact and efficient global description of a small number of representative colors independent of the color space. By clustering with the Generalized Lloyd algorithm, depending on the image up to eight dominant colors are selected and additionally the percentage share and an optional variance factor are determined. A single global spatial coherence factor represents the spatial homogeneity and allows a distinction between large color blobs versus colors that are spread all over the image [11]. In [16] was shown that the descriptor is well suited for any kind of content, even though it is partially sensitive to brightness.

(b) **Texture:** Greyscale images are usually used as a basis, as the texture features are independent of the color scheme. Textures thereby reflect visual patterns and have various properties that describe the structural nature of a continuous surface.

EHD (Edge Histogram Descriptor) captures the spatial distribution of edges and proved to be a very good and discriminative texture descriptor, especially for describing images with clear edges and strong contrasts [16]. The input image is divided into 16 non-overlapping blocks of equal size. For each block an edge histogram is generated by categorizing the edges into five fixed types according to their direction: vertical, horizontal, 45° diagonal, 135° diagonal and non-directional. This results in a fixed descriptor size of 80 bins.

HTD (Homogeneous Texture Descriptor) is based on the human visual system and is suitable for the quantitative characterization of repetitive structures and textures by describing the directionality, coarseness and regularity of patterns [12, 19]. By means of a filter bank with scaling and orientation sensitive Gabor filters, the input image is divided into 30 frequency channels. The corresponding feature vector is formed by calculating the energy and the standard deviation of the energy of each channel in the frequency domain. The sorting and retrieval of aerial photographs and satellite images is often mentioned as an application field [13].

(c) **Shape:** The aim is to describe the spatial arrangement of points belonging to a certain object or region [20]. A distinction is made between contour-based descriptors, which use only the boundary information, and region-based descriptors, which are useful when objects have a similar spatial distribution of pixels.

RSD (Region Shape Descriptor) expresses pixel distribution within a 2-D object region and can describe complex objects with multiple disconnected regions as well as simple objects with or without holes [20]. The calculation is based on the determination of moments from a set of ART (Angular Radial Transform) coefficients, is able to retrieve objects consisting of disconnected sub-regions and is relatively stable against noise [13, 20].

In order to capture the whole range of image properties, it makes sense to extend the content-based MPEG features with further image descriptors [16], e.g. from the group of image quality metrics. Ke et al. [21] investigated the perceptual factors used to distinguish between professional photos and snapshots and, based on these results, developed several high-level features for different image characteristics. Metrics from this set are used frequently in the following groups and described in detail in [21].

(d) **Brightness/Luminance/Contrast (BLC):** This feature vector contains several content- and quality-based methods for brightness calculation in the LAB and HSV color range and an estimation of the existing contrast using an Improved Adaptive Gamma Correction (IAGC, [22]). The more precise details are shown in Table 1.
(e) **Color (Col):** Since color is a very versatile image property, some quality-related color metrics beyond the content-related MPEG descriptors are added. These are mainly based on color perception, are assigned to image properties that can be directly interpreted and adjusted and contain calculation rules for colorfulness, color cast, but also for the global color temperature.
(f) **Image Quality Metrics (IQM):** This group contains different approaches that are intended to numerically capture the quality of photos. Among other things, a neural network is considered, which provides a separate technical and aesthetic quality

value. The BRISQUE method [23] combines both and is a measure of naturalness, mixed with an assessment of disturbing factors such as noise.

(g) **Distortion/Blur/Noise (DBN):** This group includes different metrics for determining sharpness or blur, taking into account different types such as motion blur and out-of-focus blur, and for determining noise in the image. Although all contribute to the quality-related features, each is based on a technical calculation method.

(h) **Shape (Sha):** To compare and describe objects and shapes in segmented binary masks, scaling and rotationally invariant Hu-moments [24] are used. Different approaches were considered to create the segmentation mask:

  - Foreground/background segmentation with Otsu threshold and subsequent smoothing by morphological operations
  - Object/vehicle segmentation with a Pytorch implementation of DeepLabv3 [25] with a ResNet-101 backbone trained on the COCO train2017 data set with 20 PascalVOC categories
  - Semantic segmentation with a Tensorflow implementation of DeepLabv3 with a MobileNetV2 backbone trained on the Cityscapes Dataset. A mask is calculated for each of the classes "road", "object", "building", "vegetation" and "sky" and thus the proportion and form of the scenic conditions occurring in the image is recorded numerically. This allows conclusions to be drawn as to whether, for example, the proportion of roads or vegetation in the image influences the detection performance.

(i) **Environmental Conditions (Env):** Various ready trained neural networks are used to capture the environmental conditions from a single image. Thereby the shadow portion is determined and a division into the classes "Cloudy", "Foggy", "Rainy", "Snowy", "Sunny", "Sunrise" is made to determine whether certain weather phenomena have an influence.

(j) **Edges/Textures (ET):** An interesting approach is the calculation of a histogram of co-occurring greyscale values at a given offset over an image, called Grey Level Co-Occurrence Matrix (GLCM, [26]). From this matrix, features such as contrast, dissimilarity, homogeneity, angular second moment (ASM), energy and correlation can be calculated, which can be used to describe and classify textures. In addition, measures for the number of edges and smoothness of the image are included in this group, as well as a metric for calculating the spatial edge distribution, which can be used to distinguish between cluttered backgrounds and focused objects.

3.3 Classification Algorithms

Classification belongs to the supervised learning methods and means the division of samples into certain classes based on learned classification rules, which in contrast to clustering are already known a priori. The feature matrix X serves as the input variable during the learning process. The samples form the rows of the matrix and the individual features, which in this case correspond to the image descriptions just described, form the columns. The associated labels are listed in the target vector y. There are several types

Table 1. Overview of the image descriptors used and the number and positions of feature bins they contribute to the data matrix. The remaining columns contain the represented image property, use cases and the descriptor properties abbreviated as follows: G/L: Global/Local Descriptors, Percept.: Human Perception Based, C/Q: Image Content/Quality Based, Interpret.: Interpretable Values, T/A: Technical/Aesthetical, Distr.: Distribution, Hist.: Histogram, GLCM: Grey Level Co-Occurrence Matrix, Sem.: Semantic, NIMA: Neural Image Assessment, BRISQUE: Blind/Referenceless Image Spatial Quality Evaluator

Group	Name/ Bin Pos.	Size	G/L	Image Property	Percept.	C/Q	Application	Interpret.	T/A
MPEG Color	CLD	22	L	Dominant Color	○	C	Sketches	✗	T
	CSD	32	L	Color Distribution Hist.	✓	C	Shape, Natural Photos	✗	T
	DCD	33	G	Dominant Color	○	C	Logos, Flags	✓	T
	SCD	32	G	Color Distribution Hist.	○	C	Color Photo Retrieval	✗	T
MPEG Texture	EHD	80	L	Spatial Edge Distribution	○	C	Cliparts, Sketches	✓	T
	HTD	32	G	Spatial Frequencies	✓	C	Satellite Imagery	○	T
MPEG Shape	RSD	35	G	Zernike Moments	○	C	Tracking Shapes	✗	T
BLC	0-5	1	G	Brightness	✓	C	HSL, RGB, HSP	✓	T
	6-16	11	G	Contrast, Gamma [22]	○	Q	Enhance Contrast	✓	T
	17	1	G	Brightness [21]	✓	Q	Photo Assessment	✓	A
	18	1	G	Contrast, Hist. [21]	✓	Q	Photo Assessment	✓	A
	19	1	G	Lightness [27]	✓	Q	Photo Assessment	✓	A
Col	0	1	G	Colorfulness [28]	✓	Q	Color Quality	✓	A
	1-2	2	G	Color cast [29]	✓	Q	Detect Color Cast	✓	A
	3-4	2	G	Color Temp. [30, 31]	○	Q	Image Processing	✓	T
	5	1	G	Hue count [21]	✓	Q	Photo Assessment	✓	A
IQM	0-1	2*1	G	NIMA [32]	✓	Q	Photo Assessment	✗	A/T
	2	1	G	BRISQUE [23]	✓	Q	Image Quality	✗	A/T
	3-27	25	L	Rule of Thirds [33]	✓	Q	Photo Assessment	✓	A
	28-29	2*1	G	Depth of Field [27]	✓	Q	Photo Assessment	✓	A
DBN	0-1	1	G	Sharpness [34, 35]	✓	Q	Blur Detection	✓	T
	2-5	4*1	G	Blur [21, 36, 37]	✓	Q	Blur Detection	✓	T
	6-8	2, 1	L	Noise [38, 39]	○	Q	Noise Detection	✓	T
Sha	0-2	3	G	Width, Height, Area	✗	C	Image Size	✓	T
	3-11	7+2	L	Foreground Shape [24]	✗	C	Segmentation	○	T
	12-19	7+1	L	Object Shape [40]	✗	C	Segmentation	○	T
	20-59	8*5	L	Scenery Shape [40]	✗	C	Sem. Segmentation	○	T
Env	0	1	G	Shadow Map [41]	○	C	Shadow Detection	✓	T
	1-6	6	G	Weather Class [42]	○	C	Classify Weather	✓	T
ET	0-53	54	G	GLCM [26]	○	C	Descript Texture	✓	T
	54	1	L	Spatial Edge Distr. [21]	✓	Q	Photo Assessment	✓	A
	55	1	G	Edge Count	○	C	Edge Detection	✓	T
	56-57	2*1	G	Smoothness [27]	✓	Q	Photo Assessment	✓	A

of classifiers. Stochastic Gradient Descent (SGD) and Logistic Regression classifiers support only linear decision boundaries. Support Vector Machines (SVM) can solve nonlinear problems by specifying a kernel function, but they are not very efficient with a large number of samples and the model found is difficult to interpret. Bayes, Nearest Neighbor and Multilayer Perceptron classifiers are universally applicable and provide high classification accuracy but are largely black box models and do not allow the identification of influential features, which is crucial for our investigations.

Decision trees (DT) are therefore used in the classification chain described in Sect. 4. These are non-parametric, white box models, which are also suitable for multi-label classification, can capture non-linear patterns and are very easy to interpret. The set of samples is iteratively divided into two or more subsets based on the most significant differentiator in the feature variables using a splitting rule. After building the tree in this way, each internal node represents a decision based on a certain feature and each branch represents the result of the decision. The leaves contain the respective classes associated with the path traversed and should preferably contain only samples of that class.

3.4 Feature Selection Methods

Feature Selection (FS) methods are used to reduce correlation, training time and the risk of overfitting. The interpretability is preserved because only a subset of features containing the relevant information is selected [43, 44]. Most common is the division into Filter, Wrapper and Embedded methods (see Fig. 6). Filter methods are computationally inexpensive and arrange and select the features according to a certain criterion independently of the later classification algorithm. In addition to univariate methods based on simple statistical metrics, multivariate methods are also used, which consider multiple dependencies between several features. Wrapper methods combine a specific search strategy that selects a subset of features with a machine learning algorithm to evaluate the performance of this subset. The subset is adjusted according to the respective strategy until a termination criterion is met. Only search strategies based on a forward search are considered (marked in bold in Fig. 6), since they iteratively determine a certain number of the most influential features. Embedded methods perform the selection by their own methods integrated in the classification algorithm during the training of the model and are therefore fast, accurate and not susceptible to overfitting. The L1 regularization of

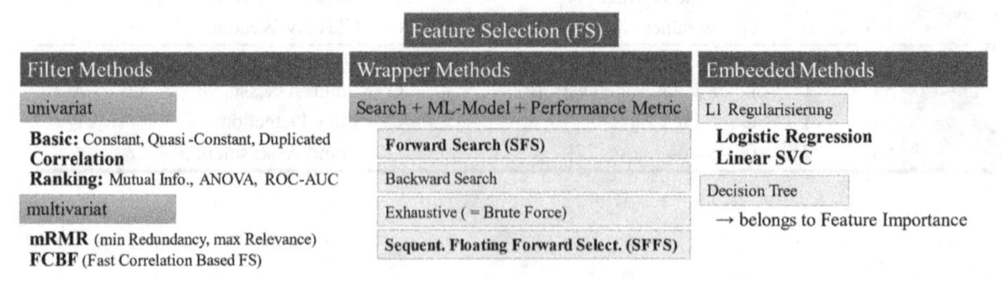

Fig. 6. Overview and categorization of common Feature Selections methods. Bold marked methods were used in the evaluation. ROC: Receiver Operator Characteristic, AUC: Area Under Curve Metric, ML: Machine Learning, SVC: Support Vector Classifier

the Logistic Regression and Linear Support Vector Classifier (SVC) algorithms introduces a dimensional reduction penalty term that shrinks non-relevant noisy coefficients to zero. This can be applied as FS regardless of the classifier used later. Decision trees already calculate the importance of a feature during the creation of the tree based on the decrease of impurity in the node. Features that tend to split nodes closer to the root have a larger importance value. Since this process takes place within the classifier used here, it is counted among the Feature Importance methods described below.

3.5 Feature Importance Methods

Feature Importance (FI) Methods are used to interpret an already trained classifier model by determining which features provide the decisive contribution to the explanation of the target variable. A distinction is made between methods that are dependent and methods that are independent of the classifier algorithm, model-specific and model-agnostic, respectively. At the same time, a distinction is also made as to whether the method analyses only a specific prediction or the entire test set, local and global, respectively. Decision Tree FI (DT-FI) mentioned in the previous chapter belongs to the local model-specific methods. Permutation FI is a global model-agnostic method and uses the decrease in a model score when a single feature is randomly shuffled as a measure. Drop-Out FI uses a similar approach, with the difference that the corresponding feature is completely removed from the data set. SHAP (SHapley Additive exPlanations) [45] is a model-agnostic, game theory-based approach that supports both local and global model interpretation. It decomposes the effect of each feature on the model output by trying all the combinations of features and allows deconstructing how the output has been constructed by each of the input feature values.

4 Statistical Evaluation Based on a Classification Chain

In the following the individual steps of the classification chain are described, which is used to assign the input images to the domains real/synth or TP/FP/FN. Figure 7 shows a typical workflow of this process. Of particular interest is the identification of the most influential features respectively image properties by FS methods before and FI methods after training the classifier. The feature extraction was already described in detail in Sect. 3.2 and provides a feature matrix X with 496 columns. For very large data sets, the number of samples is limited to 10000 randomly selected samples, otherwise the calculation time of the image descriptors would be too long. Empty entries due to variable descriptor sizes or invalid calculations are replaced by the mean values of the feature columns in the preprocessing step in order to influence the classification result as little as possible. An oversampling algorithm described in [46], called SMOTE, compensates for unbalanced data sets, since classes not represented in a similar number have a negative effect on DTs and lead to biased trees. This step is avoided as far as possible by carefully selecting the training data. To avoid the influence of different scales of features, they are standardized before training. Methods for dimension reduction are not used, because then conclusions about individual features are no longer possible.

Multicollinearity occurs when two or more feature vectors are strongly correlated, i.e. a particular feature vector can be predicted by the linear combination of two or more

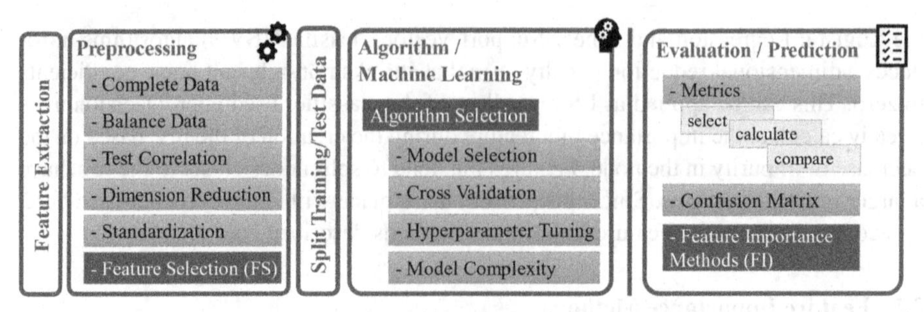

Fig. 7. Block diagram of a typical classification process

other feature vectors. While the correlation matrix only identifies pairwise correlations, the Variance Inflation Factor (VIF) also detects multiple dependencies. Figure 8 shows both units of measurement for the real/synth UAV Dataset as an example. During preprocessing, FS methods remove constant, quasi-constant and double features. Nevertheless, a high multicollinearity is still present in the data. Further methods e.g. the removal of correlated features is not advisable for the investigations presented here, because the later interpretation then becomes dependent on the order of the feature bins. However, nonlinear classification based on DTs is relatively insensitive to multicollinearity in the data.

The other FS methods described in Sect. 3.4 are now used to create a ranking of the most meaningful image descriptors based solely on the data and independent of the classification algorithm. This step serves exclusively for later analysis. The data matrix X is not reduced further to ensure comparability with the FI methods calculated later. Then a random division into training (70%) and test data (30%) is conducted. A cross validation (CV) using the parameters tree depth and splitting rule (gini/entropy) is used for model selection and hyperparameter tuning. To avoid overfitting, the model with the smallest tree depth was selected, whose score was still within the first standard deviation of the best possible score. The F1-Score with micro-averaging was used as the evaluation metric, as it represents the harmonic mean of precision and recall and is also suitable for multi-class problems. A confusion matrix is used for the detailed evaluation of the

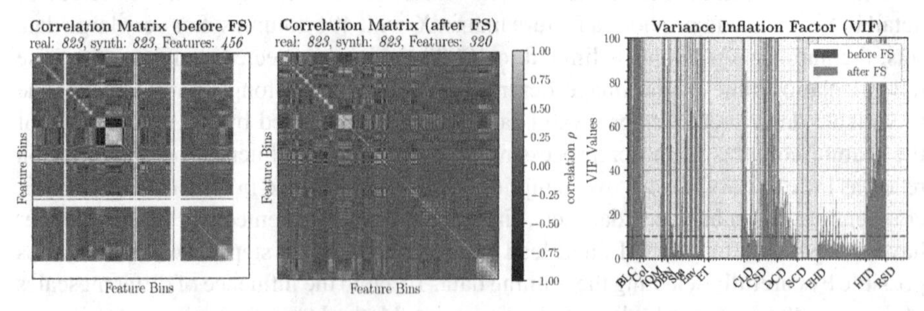

Fig. 8. Visualization of multicollinearity using the correlation matrix and VIF values before and after feature selection. White columns in the correlation matrix appear if the features do not show any variance (Data set: real/synthUAV).

detector performance for the individual classes. In the last step, the FI methods described in Sect. 3.5 are calculated. In contrast to the FS methods, these methods determine the relevant features using the trained classifier. By comparing and combining the results of the FS and FI methods, a stable and meaningful identification of the image properties that are important for the classification task should be possible.

5 Evaluation and Results

In the following the results and conclusions of the evaluation of the classification chains for the data sets described in Sect. 3.1 are presented.

5.1 Classification of Real and Synthetic Image Pairs

In the first part, it will be investigated to what extent real and synthetic image pairs can be assigned to the correct domain with the help of the calculated image descriptors and the described classification chain and which image properties are decisive for the assignment. The image pairs consist of real images of a scene, which are reconstructed in a virtual simulation environment.

First, the image pairs from the KITTI/VKITTI dataset are analyzed, which serves as a benchmark. After removal of duplicated and invariant features, the X matrix contains 325 feature bins. Comparatively high VIF values indicate multicollinearity in the data, i.e. the relevant information is contained in several features or combinations of features simultaneously. The classifier behaves ideally with both training and test data and no misclassifications occur. Figure 9 visualizes the selection process of the most influential features. The y-axis contains the FS and FI methods used for the selection. The feature bins on the x-axis are ordered according to the number of selections and represent certain image properties. This assignment can be seen in Table 1. To obtain most reliable results, feature bins selected by less than three methods are not included in the overview. The multiple available information, the ideal classification performance and the low tree depth of one suggest that even individual feature bins are sufficient to distinguish between the domains. The FI methods only analyze the current classifier and therefore focus on only one used feature bin. The FS methods analyze the complete data and show that the relevant information is present in several features. It turns out that by calculating only one metric for noise (DBN8), it is possible to distinguish clearly between the domains, since the rendered ideal synthetic images contain almost no noise. The analysis further identifies the histogram based MPEG7 descriptors SCD and CSD as influential. In [16] it is mentioned that SCD is strongly discriminative for synthetic content. CSD is a measure for the local color structure. It is assumed that the lack of fine structures on the homogeneous synthetic materials (see Fig. 3 street or sky area as an example) and the higher level of detail in reality is responsible for this influence. Furthermore, brightness (BLC17) and number of colors (Col5) are also listed. Figure 3 (sky area) again shows that this corresponds to the example images in a certain way.

Next, in comparison, we consider the real and synthetic UAV images described in Sect. 3.1. The conditions are similar. The X matrix contains 320 features, there is relative high multicollinearity in the data, the classifier with tree depth one uses only one bin

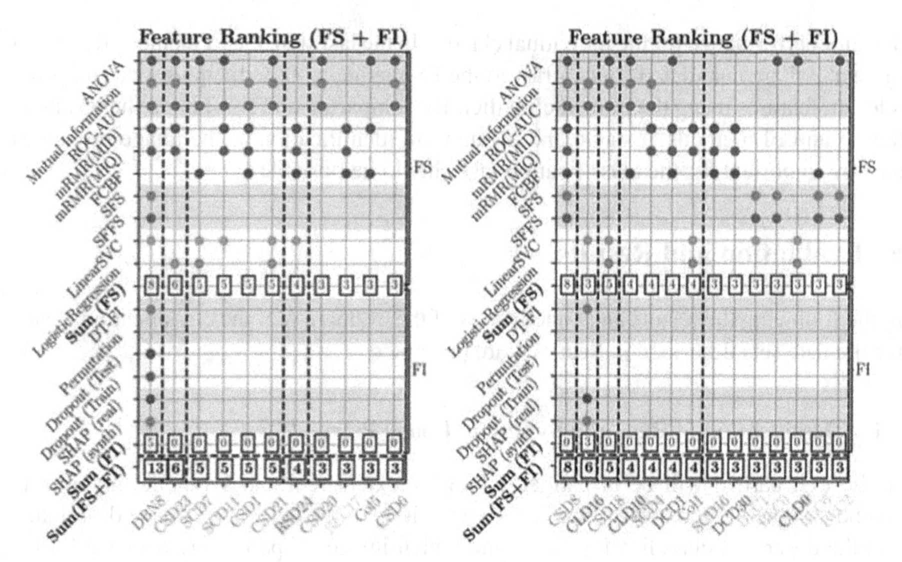

Fig. 9. Overview and global ranking of the selected image descriptors or feature bins. The y-axis lists the FS/FI methods that were used for the selection. The points represent the feature bins selected by the respective method. The feature bins are listed in descending order on the x-axis according to the number of times they have been selected and their underlying image property is described in Table 1. The first part of the abbreviation contains the group (e.g. DBN) and the number indicates the bin position (e.g. DBN8: Bin Pos. 8 in group DBN => noise). Left: KITTI/VKITTI, Right: real/synthUAV.

and achieves again an ideal assignment to the correct domain without misclassifications. Again, for the reasons mentioned above, the FS methods are more meaningful (see Fig. 9). Dropout and Permutation FI could not detect any drop in performance, because features with redundant information are available. Figure 10 shows an overview of the value distribution and illustrates that the domains differ in a multitude of image properties. The features (BLC) selected by the wrapper methods (SFS, SFFS) are not meaningful since ideal classification is already possible with one feature. The evaluation shows that all four MPEG7 color descriptors are considered relevant. This means that there are differences in the local and global histogram based color distribution as well as in the local and global dominant colors. This is clearly visible in Fig. 2. As with the KITTI/VKITTI data set, the CSD is emphasized. This can be attributed here additionally to the missing fine structures of the less high resolution satellite imagery used for the synthetic simulation. Also in [47] it was shown that ground texture with high quality is of great importance when comparing real and synthetic images based on feature detector performance. Particularly interesting is also the fact that the metric for color casts (Col1), which under certain circumstances results from the different light impression, is listed.

In summary, it can be stated that even with great variance and multiple scenarios in the data sets, the image pairs can be classified into the domains real and synthetic almost without errors. The image descriptors were chosen appropriately and contain several image properties that are responsible for the reality gap, which is emphasized

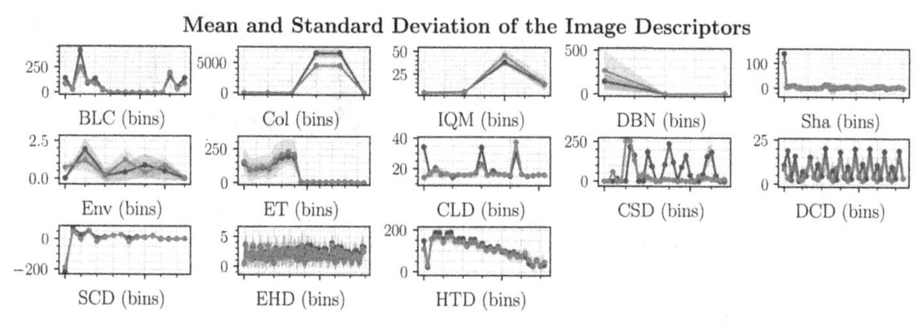

Fig. 10. Overview of the distribution of feature bins of the individual image descriptor groups for the classes real (blue) and synthetic (orange). Dataset: real/synthUAV (Color figure online)

by the high redundancy between the features. This is in agreement with the results of Hummel et al. [48], who, independent of a classification, showed that there is a disparity between real and synthetic images by comparing the distances between the respective MPEG-7 image descriptors. A comparison with the benchmark KITTI/VKITTI dataset shows that the presented method works independently from the render engine, although slightly different image properties are responsible for the reality gap. Overall, the general coloration, lighting conditions and the lack of noise and fine structures in the simulation have the greatest influence on the classification. The identified differences are also visible in example images. This forms the basis and legitimation for further investigations.

5.2 Classification of Independent Real and Synthetic Training and Test Data

In the second part it will be investigated whether the described approach is also suitable to classify independent real and synthetic sensor data into the correct domain, even if they originate from different data sets. The own synthetic training data set and the real UAVDT test data set will be used (see Sect. 3.1).

Figure 11 shows the model selection process, which prefers a compromise between maximum score and tree depth. Even with unequal images, an almost ideal classification with an F1 score of over 0.999 could still be achieved. In contrast to Sect. 5.1 the classifier model is more complex due to the higher classification difficulty and the DT has a tree depth of six. Thus, in this case the FI methods also provide information about relevant feature bins. Figure 12 shows that each method was able to select several relevant feature bins, which allows a reliable evaluation. CSD is the most influential image property in terms of both weighting and number of bins. A visual comparison of the images suggests that this difference in the structure of the color distribution, in contrast to Sect. 5.1, is due to the larger homogeneous contiguous areas (see Fig. 4 and Fig. 5: grass/asphalt areas without small parts) in the synthetic images. The CSD bins show a high correlation with themselves. IQM0 contains a metric for the aesthetic assessment of the photo quality, in which the synthetic training images have significantly lower scores. This cannot be assigned to individual image properties but corresponds to the overall impression as can be seen in the comparison with the examples in [32]. Furthermore, several DCD bins are listed. The exact cause is unclear, it is assumed that the different sceneries (real data set: chinese inner-city highways, synth. Data set: more vegetation, industrial area) could

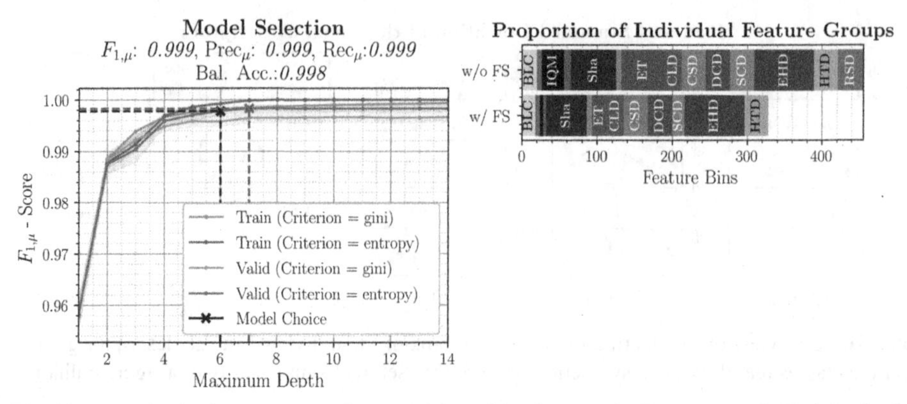

Fig. 11. Model selection process and composition of the data matrix. Dataset: synth. Training/real UAVDT test

contribute to this. Figure 11 visualizes the composition of the data matrix, which is similar for all the studies carried out and shows that CSD and DCD are not disproportionately represented. Especially interesting is that when using the same render engine as in the synthUAV set, the color cast metric (Col1) is again used as a differentiation criterion to real images. EHD70 and ET56 (smoothness) describe the spatial distribution and type of edges in the image. However, features from the groups EHD and ET are strongly correlated with the CSD values according to the correlation matrix and therefore most likely describe the same underlying cause. Lastly, the shadow proportion (Env0) is listed, which is higher in synthetic images. The cause could be a misclassification of dark areas or parts of buildings as shadows, which could be attributed to lighting effects, similar to the color cast. However, this must be confirmed in further investigations using sample images.

Overall, it has been shown that even with independent data sets, the presented approach can be used to classify between real and synthetic domains with very high reliability. It seems that image properties that mainly depend on the render engine, such as color cast, lighting effects, dominant colors, play a role, but also image properties that generally differ from reality to synthetic images, such as the structure of the local color distribution (CSD). The selected image descriptors cover all differences very well and are universally applicable.

5.3 Classification of Correct and Incorrect Bounding Boxes

Finally, in the last part it will be investigated to what extent the presented classification chain is suitable for the identification of image properties that influence the performance of deep-learning based vehicle detectors. The YOLOv3 framework was trained with the described purely synthetic training data set and evaluated on the real UAVDT test data. For the exact parameters and data set generation process please refer to [5]. The detection performance in this case is significantly lower compared to real training. Therefore, the question arises whether it is possible to distinguish between TP, FP, FN detections based on the image information in the detected bounding boxes (BB) and which of the image

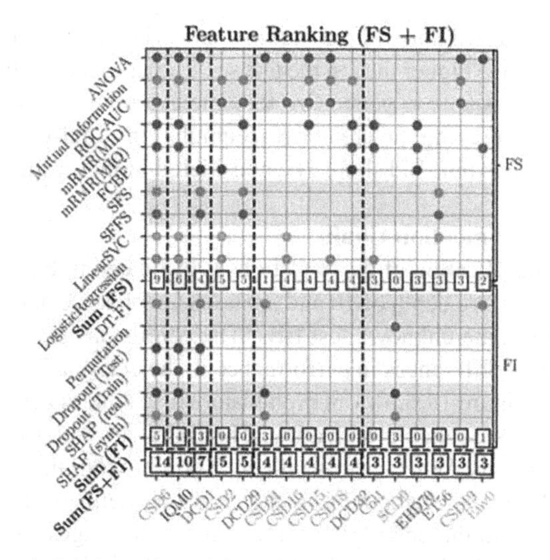

Fig. 12. Overview and global ranking of the selected feature bins. Dataset: synth. Training/real UAVDT test

properties identified for the reality gap are now relevant for this purpose and should therefore also be considered when designing synthetic training environments.

Geometric Analysis of the Bounding Boxes

At the beginning, the distribution of some geometric parameters of all BBs will be analyzed in detail. These are separated into the groups training and test set, but also into the groups TP, FP and FN, in order to analyze whether incorrect detections are related to geometric conditions, i.e. whether, for example, only those vehicles are not correctly detected that fall below or exceed a certain size. Figure 13 shows the results for the properties height, width, area and aspect ratio. The drawn anchor boxes are a kind of template bounding boxes for the YOLOv3 framework, which serve as a starting point

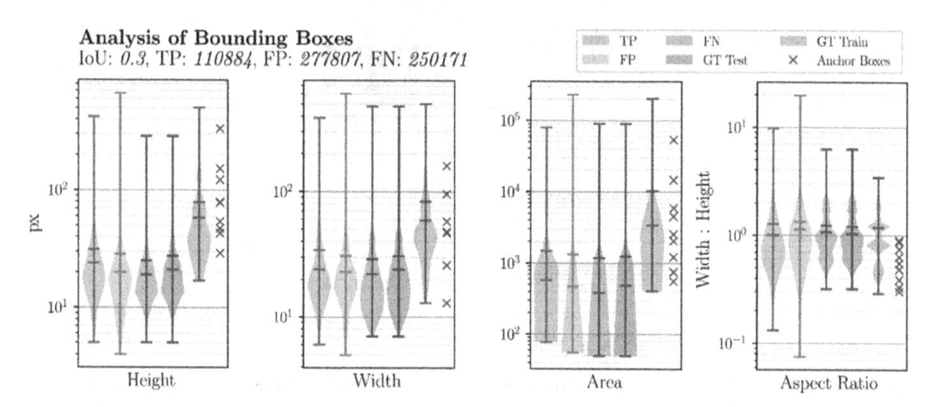

Fig. 13. Violin plot of the distribution of some geometric properties of the BBs for training and test set, but also for the assignment TP, FP and FN.

and are adapted to the distribution of the training data. The object sizes occurring in the training set are on average higher than in the test set. The influence of this is rather small, because the detector was able to generalize to some extend to the occurring BB sizes. Nevertheless, when comparing the areas, an accumulation of false detections for small sizes is visible to some extent and should be considered in the further evaluation.

Evaluation of the Classification Chain

The quality of the classification of the detections is evaluated in Fig. 14. The relatively complex task requires a tree depth of 10 and achieves an F1 score of 0.91 in training and an F1 score of 0.855 in the unknown test BBs. The confusions matrix also reveals that there are no significant differences in performance in the classification of the individual classes (TP, FP, FN). Figure 15 shows an even feature distribution between the FS and FI methods, which ensures a reliable evaluation. The DCD descriptor is again considered relevant for the detections. Whether this influence comes from the dominant colors of the background or those of the car model cannot be clearly defined. However, since DCD was also already listed in the underlying training and test images, at least a significant proportion comes from the background. Also, the aesthetic image quality (IQM0) is again included in the list. EHD71, CSD25 and ET21 (GLCM, homogeneity) describe the structure and nature of the textures and edges and, as already mentioned in the analysis of the training and test images, are strongly correlated. The analysis of the tree structure reveals that they are often used for splitting into the sub-nodes TP/FN and FP, speaks vehicle in BB and no vehicle in BB. This is obvious because vehicles have a characteristic structure of edge distribution. Since BBs magnify a certain image section, JPEG blocking artifacts can be clearly seen, especially at transitions between edges and homogeneous surfaces, which could also have effects on this. RSD1, Sha1 (height) and Sha2 (area) show the influence of smaller BBs on the detection result as observed in Fig. 13. A large part of the other shape descriptors (Sha84, Sha88, Sha78) is derived from the semantic segmentation of the complete image from which the respective BB

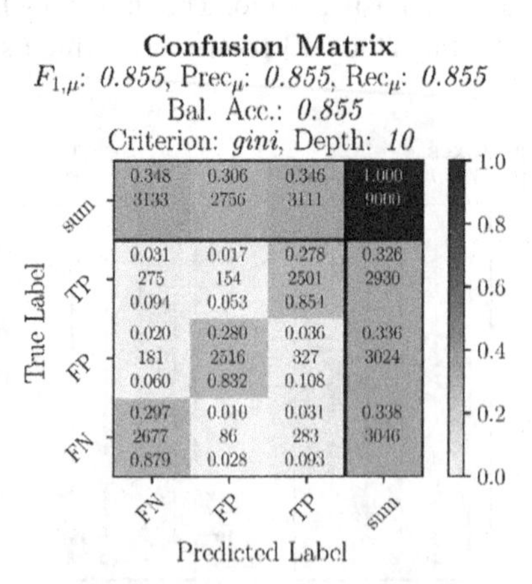

Fig. 14. Confusion Matrix for the specific evaluation of the classification quality for the individual classes. Top line: relative values, middle: absolute values, bottom: relative values, normalized to class.

originates. They describe the proportion of vegetation (Sha84) and a Hu moment from the class "vegetation" (Sha88) and "buildings" (Sha78). This indicates that the scenery or the proportion of vegetation in the complete test images has an influence on the detection results and must be considered when creating the training data set. Finally, the bins BLC1 and BLC 17 are listed, which describe brightness. This can be explained by the data, because the real UAVDT set contains test images at night, but the synthetic training set does not.

In summary, it can be stated that the classifier is able to reliable predict the detection result in advance based on the image information in the BBs. This is confirmed by the visualization of the shape weights (FI method) in Fig. 16, which weight the model influence of certain image descriptors for correct (TP) and incorrect (FP, FN) detections with different signs. The high classification score proves, that the selected image descriptors were also suitable for this task. The different signs of the weights of the LogisticRegression (FS method) for the BBs with vehicles (TP, FN) and the BBs without vehicles (FP) show that it is possible to distinguish between the BBs with and without vehicles in advance of the classification only on the basis of the data of the image descriptors (s. Fig. 17). Some image characteristics, such as the composition of the dominant colors (DCD) or the aesthetic image quality (IQM0) were already responsible for the reality gap and now also have an additional influence on the detection result. Furthermore, it has been found that image properties independent of the reality gap are also added, such as geometric sizes, brightness, or the results of the semantic segmentation of the test image. However, according to this evaluation, the considered vehicle detector is relatively insensitive to image properties such as noise, distortion, color casts and the local distribution of dominant colors (CLD). Some of these results can again be reproduced by directly visible image properties of the sample images.

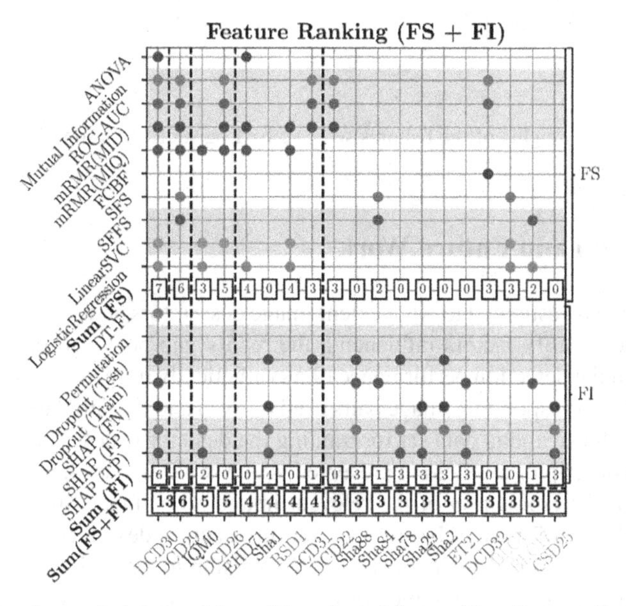

Fig. 15. Overview and global ranking of the selected feature bins. Dataset: Detection BBs

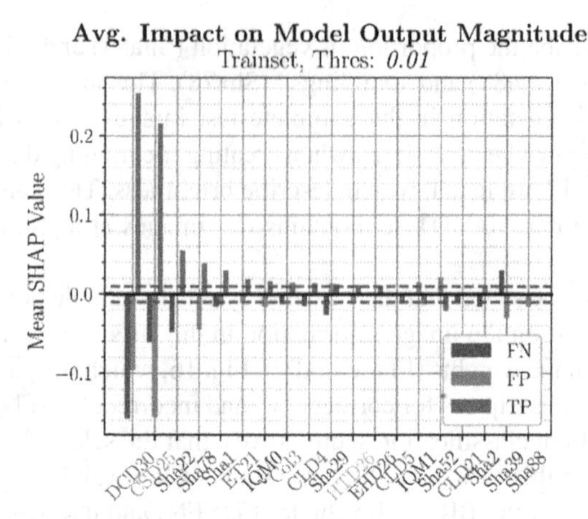

Fig. 16. Overview of the Shap values (FI method) for each class as a measure of the impact on the model output.

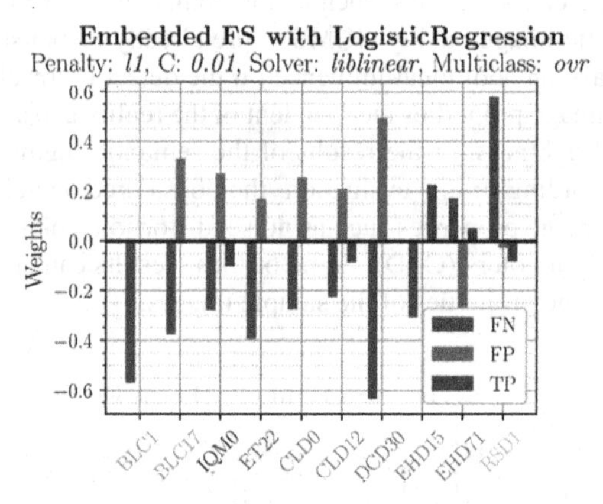

Fig. 17. Plot of the weights of the features selected by LogisticRegression (FS method) for the respective classes.

6 Conclusion and Future Work

In this paper, we investigated to what extent image descriptors and a classification chain can be used to identify factors influencing the reality gap between real and synthetic image data and the detection performance of a synthetically trained vehicle detector. An overview of the corresponding image pairs used is given and the simulation environment used to create the synthetic data set for training the detector is described. Furthermore, several image descriptor metrics are presented which serve as classification features and should allow an interpretation of the relevant image properties. Among these are the standardized MPEG-7 image descriptors. An approach is described to determine the features that have the greatest influence on the decision tree classifier by using feature importance and feature selection methods. The investigation of corresponding real and synthetic image pairs showed that the presented method was able to distinguish between the real and synthetic domain almost without misclassifications. The reality gap was largely attributed to the general coloration and the absence of noise and fine structures in the simulation. Even with independent data sets, it was possible to distinguish between real and synthetic images almost without error. Based on this, the second part of the paper analyzed the classification quality that could be achieved when differentiating between detection results. It was demonstrated that true positive (TP), false positive (FP) and false negative (FN) bounding boxes could be assigned to the correct group exclusively on the basis of the contained image information with an F1 score of 0.855. The image descriptors identified as responsible were, for example, the composition of the dominant colors in the bounding box, geometric properties and, based on semantic segmentation, also the composition of the scenery in the test image. It turned out that the described analyses are quite plausible, since some identified image properties could be verified visually using sample images or the known data set composition.

Future work will investigate to what extent the reality gap can be measurably reduced by specifically changing the simulation parameters and whether consideration of the identified image properties in the synthetic training set leads to an improvement in vehicle detection performance.

References

1. Gaidon, A., Wang, Q., Cabon, Y., Vig, E.: Virtual worlds as proxy for multi-object tracking analysis. In: Proceedings of the IEEE Computer Society Conference on Computer Vision and Pattern Recognition, pp. 4340–4349 (2016)
2. Johnson-Roberson, M., Barto, C., Mehta, R., Sridhar, S.N., Vasudevan, R.: Driving in the matrix: can virtual worlds replace human-generated annotations for real world tasks? In: IEEE International Conference on Robotics and Automation (ICRA), pp. 746–753 (2017)
3. Shafaei, A., Little, J.J., Schmidt, M.: Play and learn: using video games to train computer vision models. In: British Machine Vision Conference (2016)
4. Hummel, G., Smirnov, D., Kronenberg, A., Stütz, P.: Prototyping and training of computer vision algorithms in a synthetic UAV mission test bed. In: AIAA SciTech 2014, pp. 1–10 (2014)
5. Krump, M., Ruß, M., Stütz, P.: Deep learning algorithms for vehicle detection on UAV platforms: first investigations on the effects of synthetic training. In: Mazal, J., Fagiolini, A., Vasik, P. (eds.) MESAS 2019. LNCS, vol. 11995, pp. 50–70. Springer, Cham (2020). https://doi.org/10.1007/978-3-030-43890-6_5
6. Redmon, J., Farhadi, A.: YOLOv3: an incremental improvement (2018)
7. Du, D., et al.: The unmanned aerial vehicle benchmark: object detection and tracking. In: Ferrari, V., Hebert, M., Sminchisescu, C., Weiss, Y. (eds.) ECCV 2018. LNCS, vol. 11214, pp. 375–391. Springer, Cham (2018). https://doi.org/10.1007/978-3-030-01249-6_23
8. Presagis - COTS Modelling and Simulation Software. https://www.presagis.com/en/, https://www.presagis.com/en/page/academic-programs/
9. Geiger, A., Lenz, P., Stiller, C., Urtasun, R.: Vision meets robotics: the KITTI dataset. Int. J. Robot. Res. **32**, 1231–1237 (2013)
10. Oelbaum, T.: Design and Verification of Video Quality Metrics (2008)
11. Manjunath, B.S., Salembier, P., Sikora, T.: Introduction to MPEG-7: Multimedia Content Description Interface. Wiley, Hoboken (2002)

12. Sikora, T.: The MPEG-7 visual standard for content description - an overview. IEEE Trans. Circuits Syst. Video Technol. **11**, 696–702 (2001)
13. ISO/IEC JTC1/SC29/WG11N6828: MPEG-7 Overview (version 10) (2004)
14. Spyrou, E., Tolias, G., Mylonas, P., Avrithis, Y.: Concept detection and keyframe extraction using a visual thesaurus. Multimed. Tools Appl. **41**, 337–373 (2009)
15. Royo, C.V.: Image-based query by example using MPEG-7 visual descriptors (2010)
16. Eidenberger, H.: Statistical analysis of content-based MPEG-7 descriptors for image retrieval. Multimed. Syst. **10**, 84–97 (2004)
17. Messing, D.S., van Beek, P., Errico, J.H.: The MPEG-7 color structure descriptor: image description using color and local spatial information. In: IEEE International Conference on Image Processing (2001)
18. Kasutani, E., Yamada, A.: The MPEG7 color layout descriptor: a compact image feature description for high-speed image/video segment retrieval. IEEE International Conference on Image Processing (2001).
19. Ro, Y.M., Kim, M., Kang, H.K., Manjunath, B.S., Kim, J.: MPEG-7 homogeneous texture descriptor. ETRI J. **23**, 41–51 (2001)
20. Bober, M.: MPEG-7 visual shape descriptors. IEEE Trans. Circuits Syst. Video Technol. **11**, 716–719 (2001)
21. Ke, Y., Tang, X., Jing, F.: The design of high-level features for photo quality assessment. In: Proceedings of IEEE Computer Society Conference on Computer Vision and Pattern Recognition, vol. 1, pp. 419–426 (2006) . https://github.com/szakrewsky/quality-feature-extraction
22. Cao, G., Huang, L., Tian, H., Huang, X., Wang, Y., Zhi, R.: Contrast enhancement of brightness-distorted images by improved adaptive gamma correction. Comput. Electr. Eng. **66**, 569–582 (2018). https://github.com/leowang7/iagcwd
23. Mittal, A., Moorthy, A.K., Bovik, A.C.: No-reference image quality assessment in the spatial domain. IEEE Trans. Image Process. **21**, 4695–4708 (2012). https://github.com/bukalapak/pybrisque
24. Ming-Kuei, H.: Visual pattern recognition by moment invariants. IEEE Trans. Inf. Theory. **8**, 179–187 (1962)
25. Chen, L.-C., Papandreou, G., Schroff, F., Adam, H.: Rethinking atrous convolution for semantic image segmentation (2017)
26. Haralick, R.M., Shanmugam, K., Dinstein, I.: Textural features for image classification. IEEE Trans. Syst. Man. Cybern. **SMC-3**, 610–621 (1973)
27. Wang, J., Allebach, J.: Automatic assessment of online fashion shopping photo aesthetic quality. In: International Conference on Image Processing, pp. 2915–2919 (2015). https://github.com/szakrewsky/quality-feature-extraction
28. Hasler, D., Sabine, S.: Measuring colourfulness in natural images. In: Proceedings of SPIE - The International Society for Optical Engineering, pp. 87–95 (2003)
29. Li, F., Wu, J., Wang, Y., Zhao, Y., Zhang, X.: A color cast detection algorithm of robust performance. In: 2012 IEEE 5th International Conference on Advanced Computational Intelligence. ICACI 2012, pp. 662–664 (2012). https://github.com/hwp9527/color_cast
30. Robertson, A.R.: Computation of correlated color temperature and distribution temperature. J. Opt. Soc. Am. **58**, 1528 (1968). https://www.colour-science.org/
31. Hernández-Andrés, J., Lee, R.L., Romero, J.: Calculating correlated color temperatures across the entire gamut of daylight and skylight chromaticities. Appl. Opt. **38**, 5703 (1999). https://www.colour-science.org/
32. Talebi, H., Milanfar, P.: NIMA: neural image assessment (2017). https://github.com/idealo/image-quality-assessment

33. Mai, L., Le, H., Niu, Y., Liu, F.: Rule of thirds detection from photograph. In: Proceedings - 2011 IEEE International Symposium on Multimedia, ISM 2011, pp. 91–96 (2011). https://github.com/szakrewsky/quality-feature-extraction
34. Kumar, J., Chen, F., Doermann, D.: Sharpness estimation for document and scene images. In: Proceedings - International Conference on Pattern Recognition, pp. 3292–3295 (2012). https://github.com/umang-singhal/pydom
35. Narvekar, N.D., Karam, L.J.: A no-reference image blur metric based on the cumulative probability of blur detection (CPBD). IEEE Trans. Image Process. **20**, 2678–2683 (2011). https://github.com/0x64746b/python-cpbd
36. Su, B., Lu, S., Tan, C.L.: Blurred image region detection and classification. In: Proceedings of the 19th ACM International Conference on Multimedia - MM 2011, p. 1397. ACM Press, New York (2011)
37. Hanghang, T., Mingjing, L., Hongjiang, Z., Changshui, Z.: Blur detection for digital images using wavelet transform. In: 2004 IEEE International Conference on Multimedia & Expo. https://github.com/szakrewsky/quality-feature-extraction
38. Rakhshanfar, M., Amer, M.A.: Estimation of Gaussian, Poissonian-Gaussian, and processed visual noise and its level function. IEEE Trans. Image Process. 1–1 (2016). https://github.com/meisamrf/ivhc-estimator
39. Chen, G., Zhu, F., Heng, P.A.: An efficient statistical method for image noise level estimation. In: 2015 IEEE International Conference on Computer Vision (ICCV), pp. 477–485. IEEE (2015)
40. DeepLabv3 ResNet101. https://pytorch.org/hub/pytorch_vision_deeplabv3_resnet101/. Accessed 30 July 2020
41. Zhu, L., et al.: Bidirectional feature pyramid network with recurrent attention residual modules for shadow detection. In: Ferrari, V., Hebert, M., Sminchisescu, C., Weiss, Y. (eds.) ECCV 2018. LNCS, vol. 11210, pp. 122–137. Springer, Cham (2018). https://doi.org/10.1007/978-3-030-01231-1_8
42. CNN Weather Classification Models. https://github.com/666-zhf/weather-predicition, https://github.com/imaaditya-stack/Weather-image-classification, https://github.com/NgoJunHaoJason/weather-classification, https://github.com/berkgulay/WeatherPredictionFromImage. Accessed 30 July 2020
43. Jovic, A., Brkic, K., Bogunovic, N.: A review of feature selection methods with applications. In: 2015 38th International Convention on Information and Communication Technology, Electronics and Microelectronics (MIPRO), pp. 1200–1205. IEEE (2015)
44. Chandrashekar, G., Sahin, F.: A survey on feature selection methods. Comput. Electr. Eng. **40**, 16–28 (2014)
45. Lundberg, S., Lee, S.-I.: A unified approach to interpreting model predictions (2017). https://github.com/slundberg/sha
46. Chawla, N.V., Bowyer, K.W., Hall, L.O., Kegelmeyer, W.P.: SMOTE: synthetic minority over-sampling technique. J. Artif. Intell. Res. **16**, 321–357 (2002)
47. Hummel, G., Stütz, P.: Evaluation of synthetically generated airborne image datasets using feature detectors as performance metric. In: IPCV 2015, pp. 231–237 (2015)
48. Hummel, G., Stütz, P.: Using virtual simulation environments for development and qualification of UAV perceptive capabilities: comparison of real and rendered imagery with MPEG7 image descriptors. In: Hodicky, J. (eds.) Modelling and Simulation for Autonomous Systems. First International Workshop. MESAS 2014, vol. 8906, pp. 27–43 (2014). https://doi.org/10.1007/978-3-319-13823-7_4

Combining Epidemiological and Constructive Simulations for Robotics and Autonomous Systems Supporting Logistic Supply in Infectious Diseases Affected Areas

Walter David[1](✉), Federico Baldassi[2], Silvia Elena Piovan[3], Antony Hubervic[4], and Erwan Le Corre[4]

[1] Italian Army Training Specialization and Doctrine Command, 00143 Rome, Italy
walter.david@esercito.difesa.it

[2] Italian Joint NBC Defence School, 02100 Rieti, Italy
federico.baldassi@esercito.difesa.it

[3] University of Padua, 35122 Padua, Italy
silvia.piovan@unipd.it

[4] MASA Group, 75002 Paris, France
{antony.hubervic,erwan.lecorre}@masagroup.net

Abstract. It is very likely that the post-Covid-19 world will be significantly different from today. From the experience in fighting the pandemic we can identify lessons on the vulnerability of humans, logistics, and supply chain of vital strategic assets (e.g. medical equipment). This require to think about how to conduct operations in the future, investigate robotics and autonomous systems (RAS) to reduce the exposure while achieving operational improvement, and to assess if current doctrines need to undergo a review. Modelling and simulation play a significant role in analysis and training for scenarios that might include reacting and anticipating the unexpected, challenging our agility and resilience. Available constructive simulations have been designed primarily for training commanders and staff but often lack the ability to exploit the outcomes from predictive systems. The authors propose a novel approach considering the Spatiotemporal Epidemiological Modeler (STEM) for computing the epidemic trend. This tool has been linked with the MASA SWORD constructive simulation. STEM computed data enable the creation in SWORD of highly realistic scenarios in the context of infectious diseases, outbreaks, bioterrorism and biological defence where to model RAS, run the simulation, and analyse doctrine and courses of actions.

Keywords: Mathematical modelling of infectious diseases · Constructive simulation · COVID-19 · Robotics and Autonomous Systems (RAS) · Artificial Intelligence (AI) · Doctrine and courses of action analysis

J. Mazal et al. (Eds.): MESAS 2020, LNCS 12619, pp. 86–107, 2021.
https://doi.org/10.1007/978-3-030-70740-8_6

1 Introduction

1.1 COVID-19 Pandemic: The Rise of the Biological Threat

The World Health Organization declared [1] the coronavirus disease 2019 (COVID-19) outbreak, caused by severe acute respiratory syndrome coronavirus-2 (SARS-CoV-2), a pandemic on March 11, 2020 [2]. We are living one of the biggest crises in modern history and face unique challenges.

The outbreak's unusual severity has been determined by key drivers such as its exponential pace of transmission, fast global transportation, and crowded urban areas where more than one half of the world population live [3, 4]; these factors highlight the relevance of transportation networks and urban areas in the context of security.

Despite not physically damaging critical infrastructures [5] the pandemic rapidly disrupted the global air transportation network and caused immediate and severe economic damage [6], hitting hard entire sectors like services, tourism, education, culture, businesses conventions.

The adverse economic consequences of pandemics greatly exceed those of wars [7] and Covid-19 pandemic's effects are likely going to trigger the first contraction of global GDP since World War II [8].

A biologic threat is an infectious disease with the potential to spread and cause an outbreak. Infectious diseases are illnesses caused by germs (such as bacteria and viruses). Some infectious diseases require close contact between two people, like when people speak to each other. Other infectious diseases can only spread by germs carried in air, water, food, or soil, or by biting insects or by animals.

Biologic threats may spread naturally, as in a worldwide flu outbreak or be released intentionally in a bioterrorism attack. Anthrax was intentionally released in 2001 to cause harm.

Biological agents are usually viruses or bacteria which may take several days to make people sick once they are infected. If people are exposed to a biological agent, decontamination usually is not needed, but they may need medical evaluation and either antibiotics or vaccines.

The emergence of biological threats and the complexity of the countermeasures can be highlighted by the disruption to military training and the slow pace of operations. In fact, many militaries around the world have prioritized the protection of their most important assets, service women and men, with many of them sent to work from home, the implementation of e-learning and the support provided by military medical services, logistics and transportation units to national health and civil protection organisations.

From the experience in the fight against the pandemic we can identify lessons about the vulnerability of humans, traditional working procedures, lack of prepositioning and supply chain of vital strategic assets (e.g. medical equipment) and our reliance on systems that require large manning.

It is very likely that the post-Covid-19 world will be significantly different from today. These difficult times require to think about how to conduct operations in the future and to assess if current doctrines need to undergo a review, investigate if investing and deploying robotics and autonomous systems (RAS) could help reduce the exposure of humans while achieving operational improvement.

1.2 Aim and Purpose of the Paper

This paper aims to explore the use of epidemiological simulation in combination with artificial intelligence (AI) powered constructive simulation for supporting analysis of doctrine and courses of action in the deployment of robotics and autonomous systems (RAS) in the context of disease outbreak, bioterrorism and biological defence.

Technology can play a powerful role in the Covid-19 global challenge. Authors aim to investigate how modelling and simulation (M&S) and robotics and autonomous systems (RAS) could support contrasting biological threats.

In effect, M&S can play a significant role in analysis and training for scenarios [9, 10] that might include reacting and even anticipating the unexpected, challenging our agility and resilience.

The available legacy constructive simulations have been designed primarily for training commanders and staff but often lack the ability to exploit the outcomes from predictive systems, therefore authors want to explore if it is possible to combine epidemiological mathematical modelling with an innovative military simulation.

The goal is to make possible for the data of an epidemic scenario computed by a predictive tool to be used to develop training scenarios in a constructive simulation.

Authors have chosen the Covid-19 crisis as an example, due to its complexity and global disruptive effects. Literature, press, social media reports from China, Asian countries, European Union and United States have been carefully examined.

In this paper, the authors propose a novel approach taking in consideration the Spatiotemporal Epidemiological Modeler (STEM) for computing the epidemic trend. STEM has been chosen for its transparency, reproducibility, re-usability of models and for the availability of documentation and literature [11].

In the first part of the paper, a scenario of the Île-de-France region around Paris has been created on STEM, started a simulation of disease outbreak, run the simulation with and without countermeasures for 90 days.

In the second part of the paper, STEM computed data have been used as a reference to calibrate the artificial intelligence powered constructive simulation MASA SWORD to provide highly realistic scenarios in the context of disease outbreaks, bioterrorism and biological threats where to model RAS and human-robotic teams, to reduce human exposure, run the simulation, change the doctrine and courses of actions, run the same scenario again and check the differences.

2 Creating a Spatiotemporal Epidemiological Scenario

2.1 The Spatiotemporal Epidemiological Modeler (STEM)

The Spatiotemporal Epidemiological Modeler (STEM) is an open source, free, extensible, flexible, modular software framework, built by IBM and available on the platform of the Eclipse foundation. STEM has been designed for the creation of epidemiological scenarios and simulating epidemic emergencies, to study the phenomenon, possible preventive measures and countermeasures and develop effective management and control [12].

The tool has been used to track and, possibly, control outbreaks of infectious disease in populations, salmonella in Germany, influenza in Israel, H1N1 epidemic in Mexico, measles outbreaks in London, H7N9 avian influenza in China, dengue fever outbreaks, malaria, Ebola epidemic [11, 13].

STEM enables to create models for a country, a region, or the world, for performing complex spatial analysis with high-resolution data as well as model-based simulations with multiple populations (and even multiple diseases). Its models can contain detailed, complex subcomponents; they can be shared and validated [11].

Building a scenario with STEM requires the geographical, the population and the disease components. The available datasets describe the geography, transportation systems (including airports and roads), and population for 244 countries down to the administrative level 2, public denominator data from public and non-public sources, such as the CIA Fact Book, United Nations Environment Programme (UNEP), the World Health Organization (WHO), the US Census, the National Oceanic and Atmospheric Administration, and DIVA-GIS for use with any model [11].

The population is classified by the health status, computations are based on compartment models that assume an individual is in a particular state, either susceptible (S), exposed (E), infectious (I), or recovered (R), in classic SI(S), SIR(S), or SEIR(S) disease models, pre-coded with deterministic and stochastic variations. STEM simulates the models using ordinary differential equation solvers.

In an increasingly connected world, the vectors of infection can be quite complex, in STEM, the global air travel network (Fig. 1) is modelled like a fluid flow in an aqueduct network; passengers' flow among airports is calibrated based on actual data from passenger travels [11].

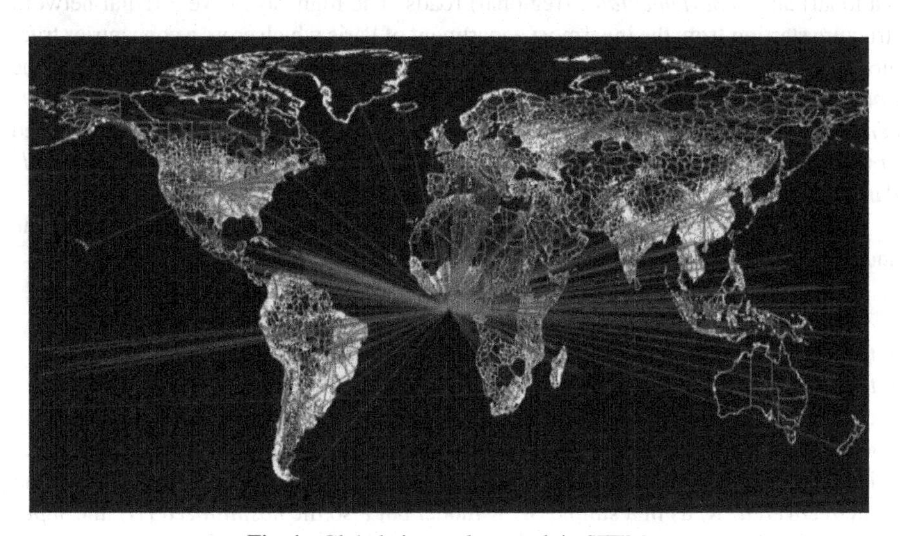

Fig. 1. Global air travel network in STEM.

STEM 4.0.0, released in 2019, has a graphic user interface and runs on Eclipse Equinox, industry standard framework supported by the Eclipse Foundation.

Two scenarios can be compared side-by-side across different dimensions, a collection of simulations can be run, based on a single scenario, modifying each simulation by varying one or more parameters to examine how this affects the model.

The output can be visualised on STEM maps, Google Earth, or graphic displays that plot data in time or in relation to other data values, and can be extracted in CSV format, during the time window established for the simulation.

2.2 The STEM Scenario

The paper then proceeds by considering the choices that have been taken by the authors.

For this study, the Île-de-France region has been chosen. The Île-de-France (12,012 km^2), is the region of northern France where Paris is located and is the most populated region in France with 12,278,210 inhabitants [14] and a population density of 1,000/km^2. This region is composed of 8 departments, 317 cantons and 1,281 municipalities. The central and the most populated department is Paris, around which there is a concentric ring of three highly urbanized departments (Hauts-de-Seine, Seine-Saint-Denis and Val-de-Marne), commonly known as the *petite couronne* ("small crown"). A second outer ring of four departments (Seine-et-Marne, Yvelines, Essonne and Val-d'Oise), known as the *grande couronne* ("large crown"), is largely rural.

Almost 80% of the region is occupied by agricultural land, forests and wetlands while the other 20% is urban. The altitude of the region varies between 10 and 200 m a.s.l. The major river draining the regions is the Seine which flows, with a meandering course, through the middle of the region. Minor rivers are also present and are often navigable.

The road network (Fig. 2) is composed of *autoroutes* (highways), *nationales* (national) and *departementales* (regional) roads. The highways have a radial network structure starting from the innermost department of Paris which cross a concentric structure centred on the capital and formed by four main rings. The innermost ring is the *Boulevard Périphérique*, around the city of Paris. The A86, also called *Paris super-périphérique*, forms a ring around the Paris urban area. The outer rings are the so called *Francilienne*, a partial ring, circa 50 km in diameter, and the *Grand contournement de Paris*, a wider loop bypassing Paris, still uncompleted.

The railway network is organized with a radial structure starting in Paris, where the underground network is denser.

Authors have made the following assumptions:

- the **initial cases** in the region have been set to 10 infected people;
- the **epidemic data** have been chosen from China, one of the worst in the world, using a modified SEIR epidemiological model which describes the time evolution of the number of infected individuals during a disease outbreak; such model takes into consideration not only the *susceptible (S),* the *exposed (E),* the *infected (I)* and the *recovered (R)* as in a simple SEIR model but also the *hospitalized (H),* the *super diffusers (P),* the *asymptomatic infected (A)* and the *deceased* [15];

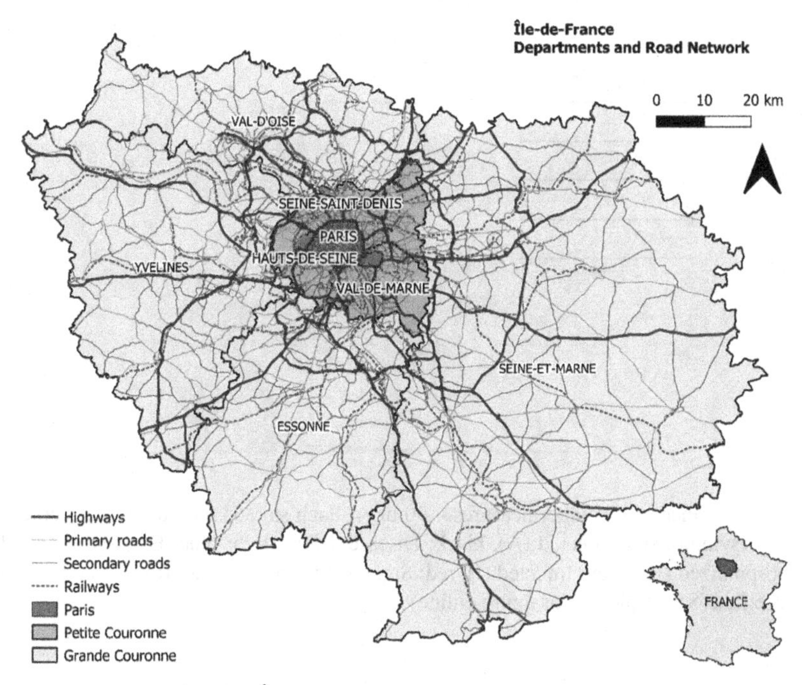

Fig. 2. 'Île-de-France highways, roads railways.

- the **countermeasures** have put in place, such as isolation, social distancing and the use of face masks. Different containment measures can be applied by the affected nation: shut airlines flights, shut inter-departmental routes, social-distancing [16], wearing face mask [16], isolation [16];
- the **activation times** of countermeasures have been selected as follows:
  - closure of regions' borders: 15 days after the beginning of the epidemic;
  - air transport block: 15 days after the outbreak starts;
  - isolation: 15 days after the outbreak starts;
  - individual protection systems: 20 days after the beginning of the epidemic;
  - social distancing: 15 days after the outbreak starts.

STEM simulations have been run on 90 days on the Île-de-France region (with real population data provided by the STEM database). The output has been visualized in the graphs for simulation performed without (Fig. 3) and with (Fig. 4) all countermeasures implemented to contain the spreading of the virus. The output of each simulation, in the appropriate folder, contains a series of files in comma separated values (CSV) format.

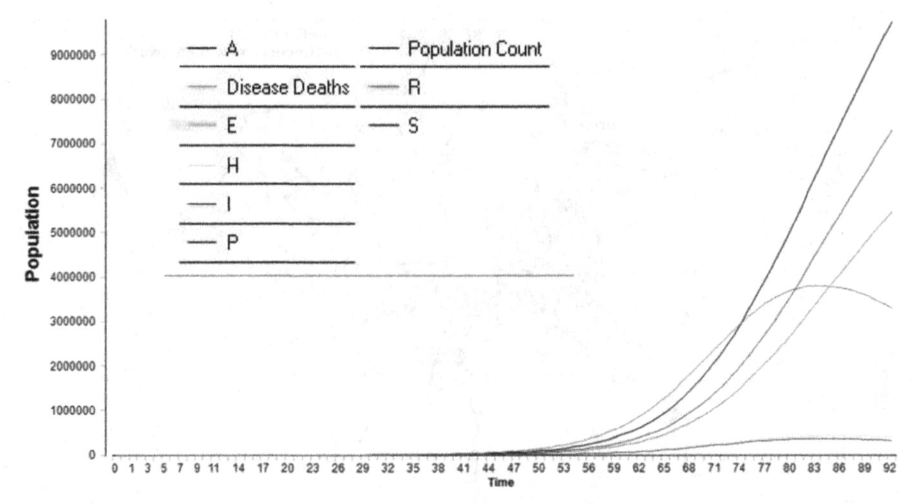

Fig. 3. Île-de-France - no countermeasures – 90 days. Each state is reported in a specific color and letter: Asymptomatic infected (A), black; Disease deaths, light blue; Exposed (E), yellow; People hospitalized (H), grey; Infected (I), red; Super diffusers (P), blue; Recovered (R), green; and Susceptible (S), purple (Color figure online)

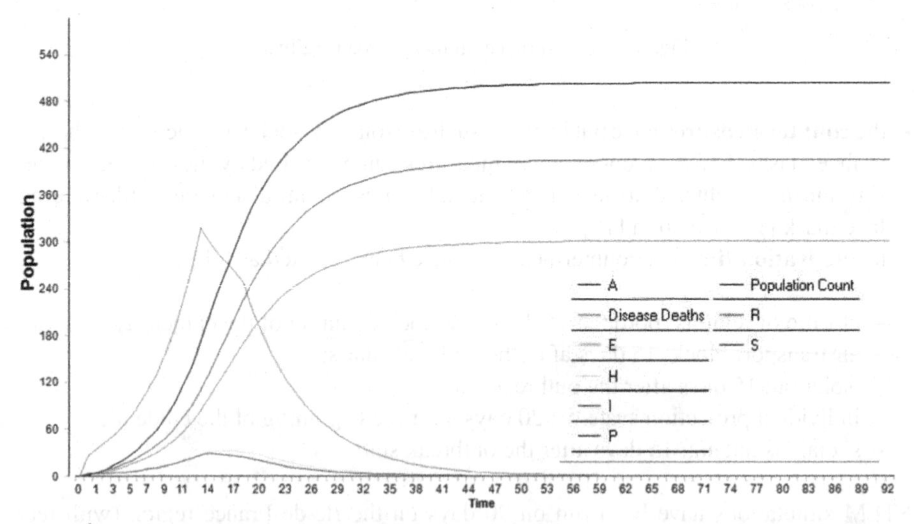

Fig. 4. Île-de-France region with countermeasures – 90 days. Each state is reported in a specific color and letter: Asymptomatic infected (A), black; Disease deaths, light blue; Exposed (E), yellow; People hospitalized (H), grey; Infected (I), red; Super diffusers (P), blue; Recovered (R), green; and Susceptible (S), purple. (Color figure online)

The outcomes of the spatial simulations are visualized in Fig. 5 without and with all countermeasures implemented.

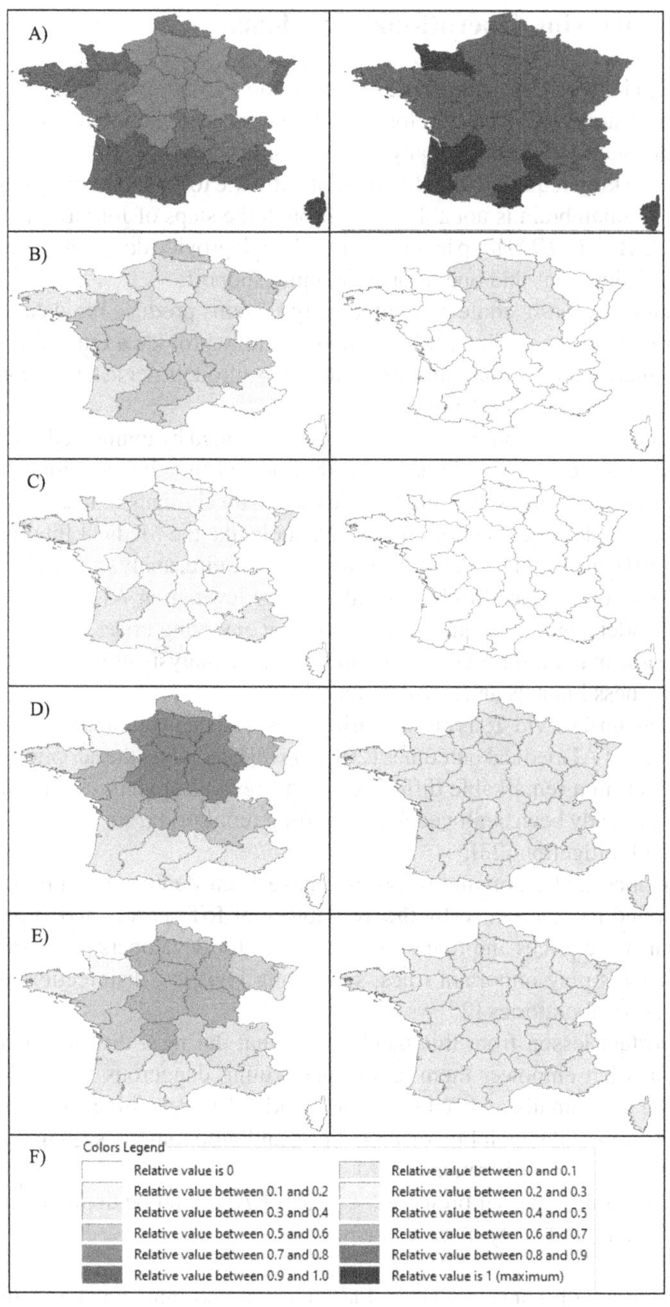

Fig. 5. STEM spatial simulation 90 days comparison without (left column) and with all countermeasures implemented (right column). A) Geographic distribution of the population of humans susceptible to COVID-19; B) Geographic distribution of the population of humans exposed to COVID-19; C) Geographic distribution of the population of humans infected by COVID-19; D) Geographic distribution of the population of humans recovered from COVID-19; E) Geographic distribution of the population of humans dead due to COVID-19. F) STEM colors legend and relative values assigned. (Color figure online)

3 RAS Addressing Operational Problems

The interaction between humans and machines is going to develop further. Robotics and Autonomous Systems (RAS) technology supported by artificial intelligence (AI) could be a game-changer in addressing many operational problems [9, 10].

Decision-making requires a combination of software tools and human judgment, but currently, the human brain is not able to go through the steps of John Boyle's *Observe-Orient-Decide-Act (OODA)* loop instantly [17]. AI will provide decision-making support in time-critical situations and operational planning support.

In a urban *warfighting* context, complex city systems produce big data that present a cognitive overload for soldiers; RAS technology may provide a crucial tool enabling militaries to increase the situational understanding through persistent reconnaissance and mapping, intelligence and preparation.

In a *disaster response* context, the information provided by unmanned aerial vehicles (UAV) and satellites can be supplemented with near/real-time information from internet of things (IoT) sensors and social media feeds to locate civilians in need [18].

In a biological defense scenario, innovative tools like the Medical Intelligence Platform (MIP) [19] can provide updated information to acquire, analyze, search and explore billions of data points related to biological risks by leveraging AI based on full natural language understanding to quickly identify new emerging crises, mitigate risks and support decision-making through case monitoring and analysis of medical discoveries, social and business impacts and citizens' emotions.

The fight against Covid-19 is already driving a speed up in automation, starting from the fast delivery of AI medical outcomes [20] and will accelerate the adoption of robotics [21, 22]. In fact, in a remarkable difference with previous epidemics, innovative technologies have already been deployed for supporting monitoring, surveillance, detection, prevention, and mitigation [23].

AI, unmanned and autonomous systems have been deployed to predict, identify and help respond [6], facilitated by the availability of IoT sensors and large data sets, affordable hardware prices and free online resources for developers [24]. Aerial drones and ground robots play important roles, supporting logistics, disinfection, delivery of blood samples to laboratories [25].

One important lesson from this pandemic is that during a disaster, robots do not replace humans but empower them, either performing dangerous tasks, or taking on tasks that free up humans and reduce their workload in repetitive jobs. Humans are still involved in the OODA loops of unmanned and autonomous systems; the crucial *meaningful human control* is retained [26].

In *warfighting* or in a civil protection context, RAS could be deployed in contaminated cities or in high density disease hotspots for:

- *logistic supply*, with convoys that include manned and unmanned and autonomous ground and aerial platforms;
- *sensing and detection* of chemical and radiological contamination, exploiting already available aerial and ground unmanned systems (with modular payloads);
- *reconnaissance and surveillance,* including approach movement, observation, threats detection, data collection, processing and analysis;

- *clearance of building*, including active observation, automated processing of information, occupation of buildings; processing of biometric data for threat recognition.

In urban warfare operations, tactical units can be equipped with small RAS; thus reducing the need for the traditional 6:1 attacker-to-defender ratio [27].

Platoons and squads can better avoid threats and clear objectives efficiently, while the sensors of small loitering unmanned aerial vehicle (UAV) collaborate with those of the unmanned ground vehicles (UGV) to provide enhanced situational awareness to (human) soldiers. Obviously, this would require the interoperability and the integration of all systems into command and control (C2) systems, to contribute to the common situational awareness.

The paper proceeds with the investigation about the possibility that the RAS technology could support the effective and efficient logistic supply in a scenario complicated by biological contamination with the aim to reduce the exposure of soldiers.

In our study; this required to build simulated robotic convoys to perform logistic supply operations in a large urban Covid-19 hotspot. Here we could perform operations as usual and get a forecast of the number of deployed soldiers ill from the disease (and need to accelerate the simulation) and perform the same operations with robots in order to compare the outcomes.

4 Creating a MASA SWORD Scenario

4.1 The MASA SWORD Constructive Simulation

To achieve our aims, we decided to consider the SWORD simulation as it is currently in use for the training of many armies.

SWORD is a constructive simulation that is natively used to immerse brigade and division command post staff in large-scale conflicts, stabilization operations, terrorist threats or natural disasters. SWORD simulates a diverse range of situations in realistic environments and, powered by Direct AI, lets trainees lead thousands of autonomous subordinate units (at platoon and company levels) on the virtual field.

Agents can receive operation orders and execute them without additional input from the players, while adapting their behavior accordingly as the situation evolves (Fig. 6).

Agents' behaviors are validated by doctrinal experts and end-users. Their model depends on algorithms that make agents perceive, move, communicate, shoot, as well as the capacities of underlying equipment, as described in a database.

As military officers at HQ or in the field do, each unit makes autonomous decisions based on the doctrine that describes the expected course of action and the results of each mission, its current mission, its operational state, the tactical situation, the terrain, the time elapsed.

4.2 Environment Model

The resolution and representation of the terrain used in the simulation must be related to the level of the scenario under study. For instance, a brigade-level exercise does not

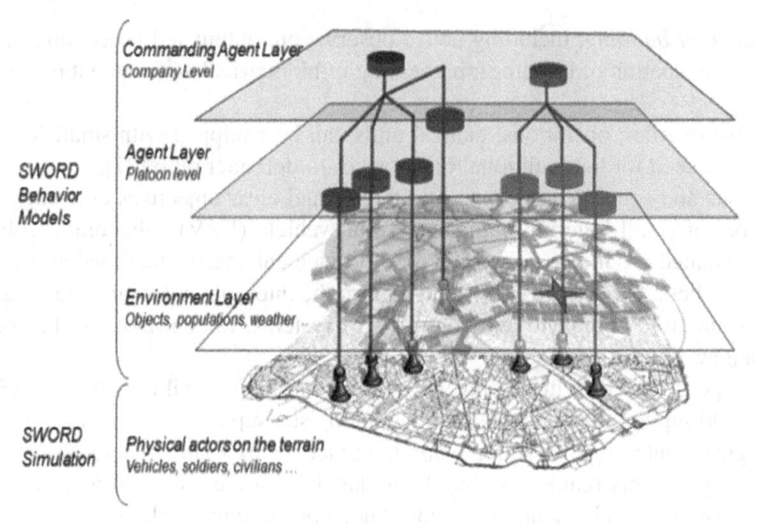

Fig. 6. SWORD layers.

require a highly detailed terrain representation for features like buildings. Cities are represented by homogeneous urban terrains, forests by wooded area, etc.

However, modern conflicts are increasingly associated with urban areas, peacekeeping operations and population management. This is even more true in civil security scenarios since nowadays the majority of people live in urban areas [3], increasing the need for simulation in such conditions.

In order to improve the realism of exercises in such contexts, the simulated urban terrain needs to be customized to include more detailed information. In SWORD, *urban blocks* (Fig. 7) are used to represent uniform sets of buildings with a unique *object* that simulated units can interact with. They are organized in districts and cities, and can be created manually but also automatically, depending on the terrain information (urban areas separated by roads, railroads or rivers).

These objects have multiple customizable attributes, such as: *Name, Colour, Height, Number of floors, Roof shape, Construction material, Occupation, Trafficability, Usages, Resource networks.*

All these attributes have a direct impact on the simulation:

- the detection capabilities of a unit are modified by the height, occupation and other characteristics of the urban area;
- the attrition and protection capabilities of a unit are impacted by the type of construction material surrounding it;
- the movement capacity of a unit inside an urban block depend on its trafficability and the size of the unit;
- usages define the hosting capacity of the urban block for inhabitants, depending on its usage (leisure, residence, office, etc.).

Fig. 7. Urban blocks are added to the terrain.

These constraints brought by operations in urban areas reflect on the doctrine implemented by tactical military units when operating in such an environment, establishing a specific *urban warfare* doctrine. This specific doctrine is introduced within SWORD to model how units adapt their speed and deployment in cities.

However, urban blocks are not sufficient to correctly model complex areas such as cities in regard of military and civil relief scenarios. Often these operations will be partly or totally dependent on resource management, like water or electricity.

In a military conflict, the control of the production and distribution points of essential resources is a major strategic stake; in a civil security scenario, the management of the distribution, production and maintenance of the resource networks have a direct influence on the satisfaction level of the affected population.

These problems can be rendered by the creation of resource networks which is represented by multiple nodes linked by connections. A node can be an urban block or an object, within addition of its normal characteristics the following properties:

- resource type;
- activation status;
- production capacity per tick (maximum and effective);
- consumption per tick (maximum and effective);
- stock capacity (initial, maximum and current);
- a flag "critical resource", that indicates if this resource has an influence on the functional state of the element, i.e. if this flag is activated, and the consumption for this resource is not satisfied, the element becomes ineffective.

A connection between two nodes is defined by a rate of flow (maximum and effective) for each resource. The simulation engine will check that the resource production of a source node is high enough to ensure the proper alimentation of the downstream consumer node.

In addition, the production, consumption and stock capacities of a node are proportional to its structural and functional state. The structural state of a node reflects its viability. The state can be decreased if the node is damaged by arson, sabotage, military action, etc., or increased if the node is repaired by an engineering unit. A node becomes ineffective if its structural state goes below a working threshold (30% per default).

The functional state depends on the structural state and the satisfaction of the consumption of critical resources. If the consumption of a critical resource is not satisfied, the node becomes ineffective. For example, if the electricity supply of a hospital is cut, the hospital can't be used anymore to manage injured people.

4.3 Unmanned Ground Vehicle (UGV) Behaviour Model

Although several levels of autonomy can be defined for autonomous systems (Annex A to CUAxS Disruptive Technology Assessment Game - DTAG Experiment Results and Conclusions), we decided to focus our work on autonomy level 6[1] systems.

The support for high level behaviors in MASA SWORD perfectly matches our goal and allowed us to implement the desired missions for the unmanned ground vehicles (UGV). We modeled the UGV as a unit programmed to execute operational mission instructions, in a similar way an officer would implement doctrine. Mission term, boundaries, phase lines, objectives, self-preservation are directly taken into by the model to decide the correct course of action.

The resulting UGV behavior model delivers a high level of autonomy and can be used to test several versions of our scenario (each one ran multiple times), without requiring inputs from an operator.

We also took advantage of the DirectIA behavior engine of SWORD to add self-protection behaviors to our UGV model. Indeed, while performing their main mission, the UGVs need to take into account their environment and react accordingly.

The "SelfProectionFromCrowd" behavior shown below (Fig. 8) programs the UGV to try to escape a potentially aggressive crowd (right part), while performing actions to repel it, e.g. sounding an alarm (left part). This protection behavior is run in parallel with the behavior of the main mission but with a higher priority. The behavior engine will ensure that the correct part of the overall behavior is executed depending on the conditions.

4.4 The SWORD Scenario

A logistic force is deployed in Paris city to support the civilian population by delivering food and water. To ensure that all the needs of population are covered, Paris has been divided in 10 areas.

A *combat train* (including a supply platoon and trucks, maintenance platoon and medical device) [28] is deployed in each area (Fig. 9). Combat trains have to deliver, in their responsible sectors, water and food from 4 distribution points.

[1] Based on its knowledge of a broader environment, the system can initiate automatically a mission the system gathers, filters, and prioritizes data. The system integrates, interprets data and makes predictions. The system performs final ranking. No information is ever displayed to the human. The system executes automatically and does not allow any human interaction.

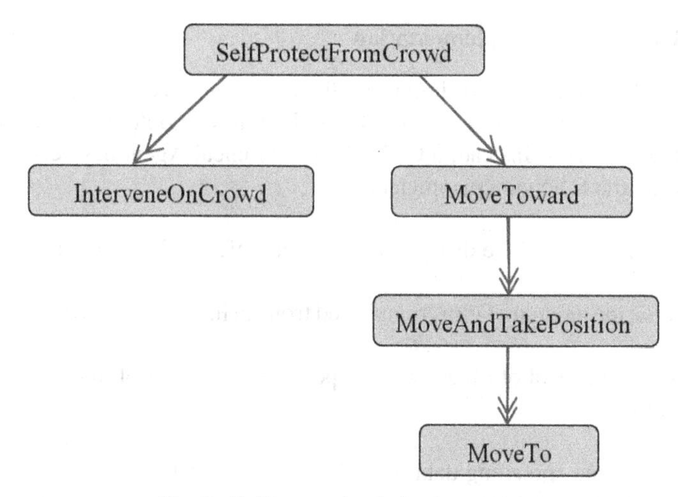

Fig. 8. Self-protection behavior model.

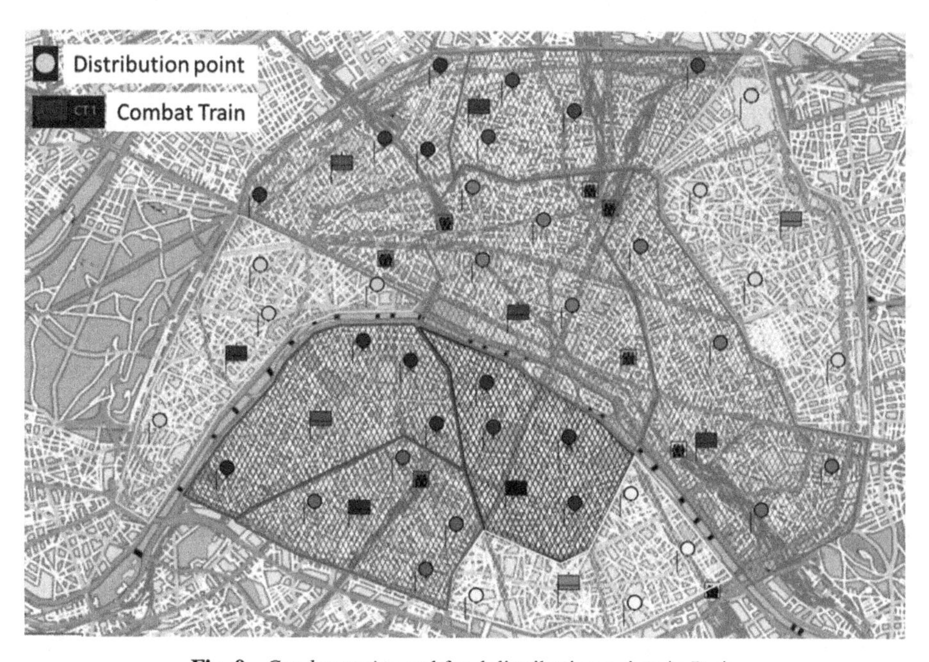

Fig. 9. *Combat trains* and food distribution points in Paris.

The needs of the civilian population have been determined so that each distribution point will be supplied once or twice a day by convoys issued from the combat trains. Combat trains will be refilled at the logistic base located in the south part of Paris.

4.5 SWORD Database Implementation

Although SWORD is capable of importing the data computed by STEM directly, this approach was not taken because it would not allow to compute the effects of the use of RAS in the scenarios. Instead, a COVID-19 Biological Agent has been created and described using the following parameters:

- *contamination distance*, the distance at which an infected human will infect another human;
- *transfer ratio*, the amount of virus transferred from an infected human to a non-infected one;
- *attrition level*: the level of illness with respect to the amount of ingested virus for an infected human.

The infection and spreading data from the STEM model were used to calibrate the SWORD COVID-19 agent. The same infection ratios from STEM have been achieved in SWORD using this approach:

- ratio (%) of asymptomatic people;
- ratio (%) of infected people without hospitalization;
- ratio (%) of infected people with hospitalization;
- ratio (%) of disease dead.

These four ratios were used during the analysis to compare the results of the execution of the options taken for our scenario.

Two types of robotics and autonomous systems (RAS) have been created in the SWORD database to meet the needs of this study. The first is an *autonomous load carrying vehicle*. It has the same characteristics (carrying capacity, fuel autonomy, etc.) as a load carrying vehicle manned by humans. The second is a *robot system* based on a human asset. Its objective is to replace humans in the distribution of food.

The *biological agent*, COVID-19, spreads from a distribution point, considered a *cluster*. The selected propagation mode is described as follows: in contact with the civilian population at a distribution point, the convoy is contaminated. Such convoy, when it comes back to the combat train, contaminates some of the personnel present in it. Which in turn, during a resupply at the logistic base, contaminates the members of this. Finally, other combat trains are also contaminated during their resupply at the logistics base.

5 Doctrine and COA Analysis with SWORD

For doctrine and equipment analysis SWORD is using the same methodology and technology. The equipment analysis methodology with SWORD goes as follows: first, a scenario is prepared, then it is run a set number of times (in Monte-Carlo batch mode) for each predefined criteria to analyze and, finally, the results are checked to find the differences between each course of action.

The doctrine analysis methodology is exactly the same. In our study case, we want to know the impact of implementing autonomous vehicles to replace manned vehicles, which is a huge doctrinal shift.

Two courses of action (COA) have been selected. For each of them, the following assumptions are made:

- COA 1 in which all distribution operations are handled by human soldiers; this course of action is the reference scenario against which the results of a second course of action will be compared to check its effectiveness;
- COA 2 in which autonomous vehicles are used to replace humans when distributing food and water.

For each of them, the following assumptions are made:

- no protective measures against the virus are taken;
- each individual is tested every day;
- for each case, asymptomatic (A), infected without hospitalization (I), the staff are quarantined, are not replaced and do not return to their unit after quarantine. Hospitalized staff do not return to their unit after receiving treatment and completing their convalescence.

5.1 Analysis of Courses of Action 1 and 2

Course of Action (COA) 1. The logistic units ensuring the distribution of resources to the civilian population are made up of conventional equipment manned by (human) soldiers.

The analysis of this COA will make it possible to determine from which point the mission can no longer be carried out due to lack of personnel able to transport the needs of the population. It is considered that the mission cannot be maintained when the operational state (i.e. material and human capacities) reaches a level of 70% of the initial state. The capacity indicators will be set on 3 combat trains, each of them having a different contamination date.

Course of Action (COA) 2. This COA includes the same initial elements as the COA 1, with the difference that the Command, in view of the decrease in the capacity to fulfill the mission, decides to set up autonomous vehicles.

Those vehicles replace the load carrying vehicles which are supplying the civilian population and those allowing the replenishment of the combat train at the logistics base. Robots replace the soldiers ensuring the distribution of food to the population.

In this COA, it is also considered that no protective measures are taken against the virus, robots and autonomous vehicles are therefore not disinfected at each rotation performed. There may therefore remain some traces of virus on the RAS and thus infect staff (in 5% of cases).

For both courses of action, three executions of the simulation were performed, covering an extent of approximately one month. The simulation use of random number generation introduces variations in the outcome of each execution. The average values of the monitored variables were then computed for each COA. Figure 10 represents the random selection performed by SWORD to apply the outcomes from STEM.

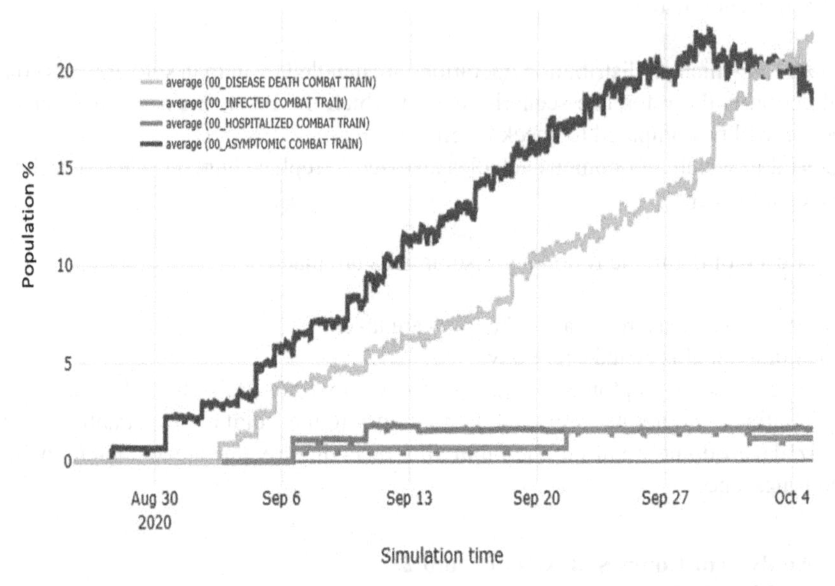

Fig. 10. Human infection in combat trains with SWORD using STEM model.

SWORD outcomes are slightly weaker than those of STEM. The analysis of STEM data shows that the first significant impacts on the population appear after 56 days. In fact, SWORD simulation runs from the 25th of August to the 3rd of October (representing 40 days).

In SWORD, the operational state of a unit takes into account its equipment and staff, therefore, to obtain the most realistic outcomes possible, a random breakdown factor on the equipment is also taken into account.

This random breakdown requires the establishment of a maintenance chain to repair the equipment and make it available again for its original unit. These breakdowns therefore influence the operational state of the logistic unit and therefore the fluctuations of the curve.

Outcomes from COA 1. We observe that the operational state gradually decreases to the value of 70% of the initial capacity of the combat trains around October 1st (Fig. 11).

We can therefore deduce that **without protective measures** for the soldiers, **the mission cannot be maintained efficiently after this date.**

Outcomes from COA 2. Unlike COA 1 (where no action is taken), in COA 2 the Command decides to deploy **robotics and autonomous system (RAS)** as soon as the first tests reveal soldiers affected by any form of the disease. This in order to minimize the contact between the logistics support force and the civilian population.

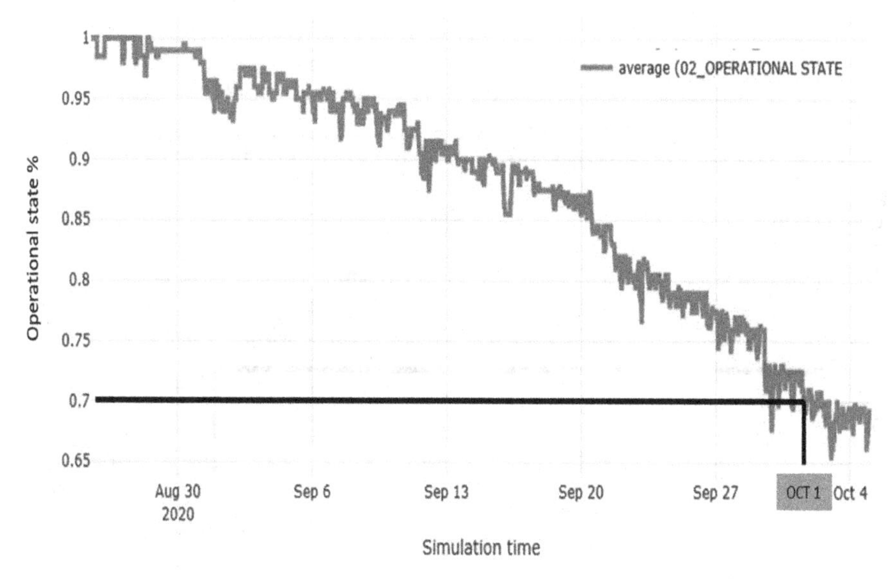

Fig. 11. Logistic force operational state.

The decision taken is to implement *autonomous vehicles* instead of load carrying vehicles and *robots to replace the staff* ensuring the distribution of food and water. Support is provided by RAS units from September 3 for the combat train n. 4.

To keep elements of comparison in the study, only one combat train is equipped with RAS. The effects are convincing. The different curves are immediately inflected for the combat train n. 4 (Fig. 12).

The number of soldiers in contact with the virus is reduced by the deployment of robots and autonomous systems to ensure upstream supply and also ensure downstream replenishment (Fig. 13).

5.2 Doctrine Analysis

The deployment of robotics and autonomous systems (RAS) in the context of a pandemic will minimize the human contact and therefore minimize the risk of contamination.

In the context of the scenario, the logistical distances are small and therefore allow humans to maintain maximum control over the RAS.

Moreover, the delivery and distribution of water and food takes place under very favorable conditions. There is no reaction from the civilian population, whether favorable or not. But it is not unthinkable to have reactions from the population against devices carrying out the support.

We must consider this possibility. To prevent this possible type of reaction, one of the possibilities of using this type of equipment would be the combination of humans/robots as a squad. The human acting as a team leader can thus modify and/or control the orders programmed in the preparation of a mission or the behaviors induced by artificial intelligence.

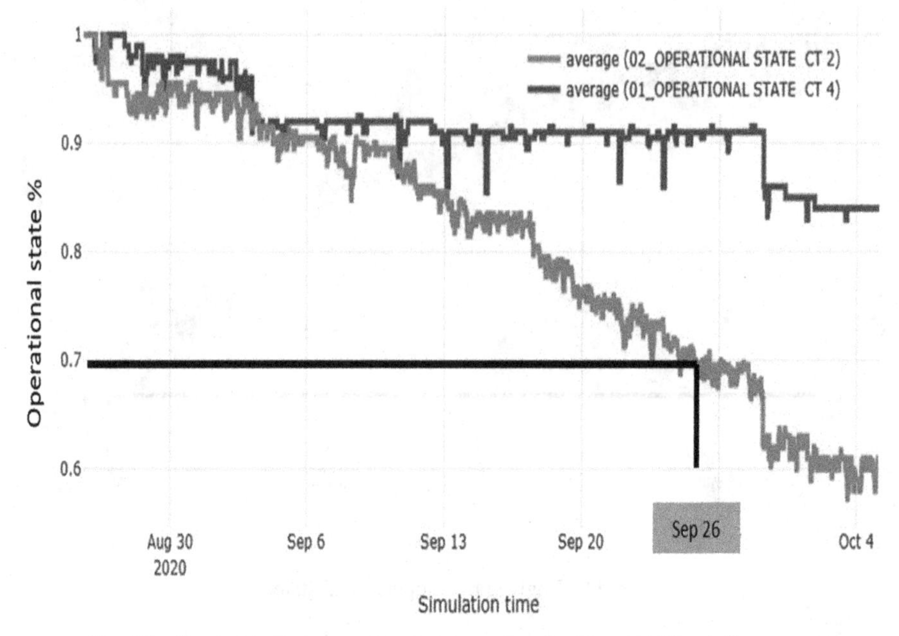

Fig. 12. Combat trains' operational state with and without RAS reinforcement.

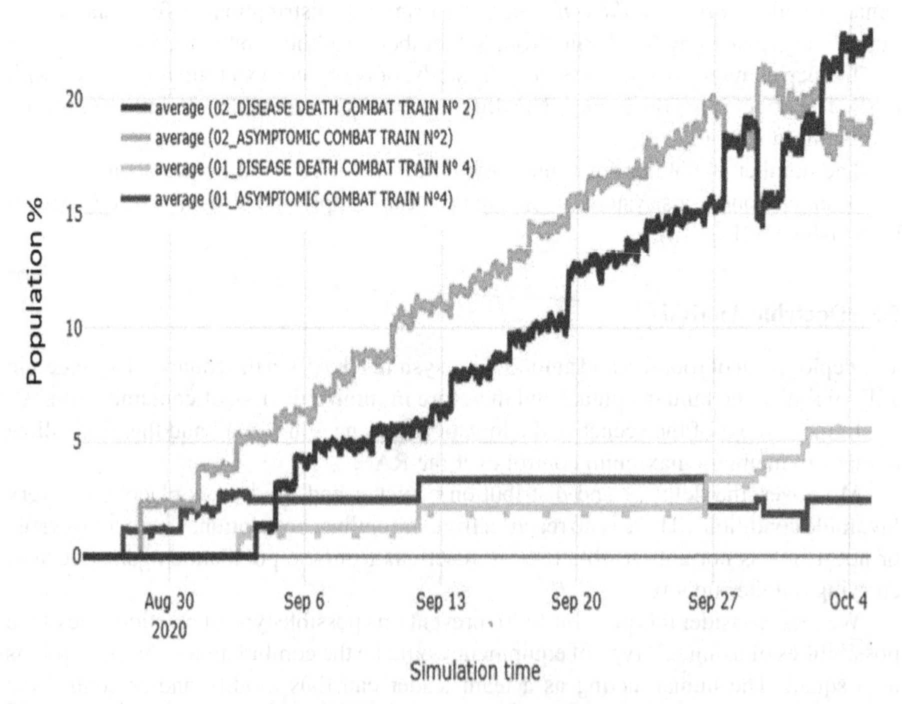

Fig. 13. Human infection with SWORD in combat trains with and without RAS reinforcement.

6 Conclusions

This challenging year 2020 provides opportunities to learn [25]. Very likely, the post-Covid-19 world will be significantly different from the pre-Wuhan period with the pandemic crisis acting as a driver making it easier to accept automation as a part of our lives [21, 22, 29].

In a novel approach to contrast biological threats by deploying robotics and autonomous systems (RAS), authors have linked the Spatiotemporal Epidemiological Modeler (STEM) with the AI-powered constructive simulation system MASA SWORD.

The difficulties regarding STEM setting up have been several. Particularly, each disease and its related agent needs a specific and proper model in order to describe as better as possible its behavior, consequently the authors had to conduct a deep scientific literature research to select the most precise and detailed mathematical model and related epidemiological parameters for Covid-19.

Moreover, STEM is an open source software, designed to help scientists and public health officials, that needs a skilled end user that, beyond to research and select the best disease model and related parameters, is able to build up the scenario and to obtain and evaluate the simulation outcomes.

Last but not least, the difficulty to overcome when integrating STEM data into SWORD for this type of study is the scale of the population treated by STEM (several million people) while for conventional military units, it works on a scale of a few thousand at most. It is therefore more useful to work on the ratio than on the numbers in raw data. The ratio is therefore sometimes quite difficult to transpose and requires several scenario executions to obtain an acceptable correlation between the two tools.

Those data must, of course, be adapted to what this type of software can assimilate and to the aim and type of use (e.g.: analysis, staff training).

In conclusion, despite the technical challenges, our study shows that the **epidemiological data** supplied by STEM can be used **as a reference to calibrate** a constructive simulation tool like MASA SWORD to provide highly realistic scenarios.

This calibration could be achieved because the scenario was limited in space, time and scope. In particular, the effect of hospitalization and treatment of the soldiers were not taken into account.

Furthermore, we focused on **highly autonomous RAS** which limit the interaction between the soldiers and the population but do not represent the current state of the art of RAS in particular when dealing with crowds.

Finally, as a matter of the fact, the simulations and the outcomes obtained demonstrate this and, in particular, taking as an example, the **number of soldiers in contact with the virus can be reduced** by the choice to deploy RAS convoys to ensure upstream logistics supply to the civilian population in the affected areas and also to ensure downstream replenishment.

References

1. World Health Organization: Department of Communicable Disease Surveillance and Response. Consensus document on the epidemiology of severe acute respiratory syndrome (SARS). https://www.who.int/csr/sars/WHOconsensus.pdf?ua=1. Accessed 12 Aug 2020

2. Petersen, E., et al.: Comparing SARS-CoV-2 with SARS-CoV and influenza pandemics (2020). https://doi.org/10.1016/S1473-3099(20)30484-9. https://www.thelancet.com/journals/laninf/article/PIIS1473-3099(20)30484-9/fulltext#back-bib. Accessed 20 Aug 2020
3. United Nations Department of Economic and Social Affairs: World Urbanization Prospects Revision (2018). https://population.un.org/wup/Publications/Files/WUP2018-Highlights.pdf
4. Paradox Engineeering: When smart technologies combat Covid-19 and contribute to urban. https://www.pdxeng.ch/2020/03/31/smart-technologies-Covid-19-urban-resilience/. Accessed 20 Aug 2020
5. Barnett, D.J., Rosenblum, A.J., Strauss-Riggs, K., Kirsch, T.D.: Readying for a post–COVID-19 world. The case for concurrent pandemic disaster response and recovery efforts in public health. J. Public Health Manag. Pract. **26**(4), 310–313 (2020). https://doi.org/10.1097/PHH.0000000000001199
6. Collins, A., Florin, M.-V., Renn, O.: COVID-19 risk governance: drivers, responses and lessons to be learned. J. Risk Res. (2020). https://doi.org/10.1080/13669877.2020.1760332
7. Jordà, Ò., Singh, S.R., Taylor, A.M.: Longer-run economic consequences of pandemics. Federal Reserve Bank of San Francisco Working Paper 2020-09. https://doi.org/10.24148/wp2020-09
8. The Economist: Economist Intelligence Unit 2020: The long recovery to 2019 GDP levels, 18 June 2020. https://www.eiu.com/n/the-long-recovery-to-2019-gdp-levels/. Accessed 21 July 2020
9. Hodicky, J., Prochazka, D.: Modelling and simulation paradigms to support autonomous system operationalization. In: Mazal, J., Fagiolini, A., Vasik, P. (eds.) MESAS 2019. LNCS, vol. 11995, pp. 361–371. Springer, Cham (2020). https://doi.org/10.1007/978-3-030-43890-6_29. ISBN 978-3-030-14984-0
10. Hodicky, J., Prochazka, D., Prochazka, J.: Automation in experimentation with constructive simulation. In: Mazal, J. (ed.) MESAS 2018. LNCS, vol. 11472, pp. 566–576. Springer, Cham (2019). https://doi.org/10.1007/978-3-030-14984-0_42. ISBN 978-303014983-3.2
11. Douglas, J.V., et al.: STEM: an open source tool for disease (2019). https://doi.org/10.1089/hs.2019.0018. Accessed 19 Aug 2019
12. IBM: Public Health Research - The Spatiotemporal Epidemiological Modeler (STEM) Overview. https://researcher.watson.ibm.com. Accessed 15 July 2020
13. Baldassi, F., et al.: Testing the accuracy ratio of the Spatio-Temporal Epidemiological Modeler (STEM) through Ebola haemorrhagic fever outbreaks. Epidemiol. Infect. **144**(7), 1463–1472 (2015). https://doi.org/10.1017/S0950268815002939
14. Ined: Institut national d'études démographiques. https://www.ined.fr. Accessed 04 July 2020
15. Ndairou, F., Area, I., Nieto, J.J., Torres, D.F.M.: Mathematical modeling of COVID-19 transmission dynamics with a case study of Wuhan. Chaos Solitons Fractals **135** (2020). https://doi.org/10.1016/j.chaos.2020.109846
16. Chu, D.K., Akl, E.A., Duda, S., Solo, K., Yaacoub, S., Schunemann, H.J.: Physical distancing, face masks, and eye protection to prevent person-to-person transmission of SARS-CoV-2 and COVID-19: a systematic review and meta-analysis **395** (2020). https://doi.org/10.1016/S0140-6736(20)31142-9. www.thelancet.com
17. Anderson, W.R., Husain, A., Rosner, M.: The OODA loop: why timing is everything. Cognitive Times, December 2017. https://www.europarl.europa.eu/cmsdata/155280/WendyRAnderson_CognitiveTimes_OODA%20LoopArticle.pdf. Accessed 15 June 2020
18. David, W., et al.: Giving life to the map can save more lives. Wildfire scenario with interoperable simulations. Adv. Cartogr. GIScience Int. Cartogr. Assoc. **1**, 4 (2019). https://doi.org/10.5194/ica-adv-1-4-2019
19. Expert System: https://expertsystem.com/products/medical-intelligence-platform/. Accessed 04 Aug 2020

20. Grumelard, S., Bisca, P.M.: Can humanitarians, peacekeepers, and development agencies work together to fight epidemics? 24 April 2020. https://blogs.worldbank.org/dev4peace/can-humanitarians-peacekeepers-and-development-agencies-work-together-fight-epidemics. Accessed 15 June 2020
21. Hooda, D.-S., in General's Jottings: Lessons from pandemic: robotics and readiness for info warfare in Times of India, 19 April 2020. https://timesofindia.indiatimes.com/blogs/generals-jottings/lessons-from-pandemic-robotics-and-readiness-for-info-warfare/. Accessed 18 June 2020
22. Howard, A., Borenstein, J.: AI, Robots, and Ethics in the Age of COVID-19, 12 May 2020. https://sloanreview.mit.edu/article/ai-robots-and-ethics-in-the-age-of-Covid-19/. Accessed 13 July 2020
23. Ting, D.S.W., Carin, L., Dzau, V., Wong, T.Y.: Digital technology and COVID-19. Nat. Med. **26**, 459–461 (2020). https://doi.org/10.1038/s41591-020-0824-5
24. David, W., Pappalepore, P., Stefanova, A., Sarbu, B.A.: AI-powered lethal autonomous weapon systems in defence transformation. Impact and challenges. In: Mazal, J., Fagiolini, A., Vasik, P. (eds.) MESAS 2019. LNCS, vol. 11995, pp. 337–350. Springer, Cham (2020). https://doi.org/10.1007/978-3-030-43890-6_27
25. Murphy, R.R., Adams, J., Gandudi, V.B.M.: Robots are playing many roles in the coronavirus crisis – and offering lessons for future disasters, 22 April 2020. https://theconversation.com/robots-are-playing-many-roles-in-the-coronavirus-crisis-and-offering-lessons-for-future-disasters-135527. Accessed 13 June 2020
26. International Committee of the Red Cross (ICRC): Autonomy, artificial intelligence and robotics: technical aspects of human control, Geneva, August 2019 (2019). https://www.icrc.org/en/document/autonomy-artificial-intelligence-and-robotics-technical-aspects-human-control. Accessed 13 June 2020
27. US Army: The U.S. Army robotic and autonomous systems strategy. https://www.tradoc.army.mil/Portals/14/Documents/RAS_Strategy.pdf. Accessed 14 Aug 2020
28. US Army: Army Techniques Publication (ATP) 3-90.5 (2016). https://armypubs.army.mil/ProductMaps/PubForm/Details.aspx?PUB_ID=106018. Accessed 17 Aug 2020
29. FT Editorial Board: The role of robots in a post-pandemic world, 21 May 2020. https://www.ft.com/content/291f3066-9b53-11ea-adb1-529f96d8a00b. Accessed 15 June 2020

M&S of Intelligent Systems - R&D and Application

Numerical Experiment on Optimal Control for the Dubins Car Model

Stanislav Frolík[✉], Matej Rajchl, and Marek Stodola

Faculty of Mechanical Engineering, Brno University of Technology, Technická 2896/2, 616 69 Brno, Czech Republic
{stanislav.frolik,matej.rajchl}@vutbr.cz,
Marek.Stodola@vut.cz

Abstract. This paper deals with optimal control of the Dubins car model, whose configuration space is three dimensional smooth manifold. There are computed vector fields that generate solvable Lie algebra isomorphically equivalent to tangent space in each point of the manifold. Then an optimal control problem for finding the shortest curve between two points on the manifold is formulated. The problem is analytically solved for the nilpotent approximation of the fields and fixed points of the symmetries are found. To these fixed points multiple optimal trajectories can lead. The problem is also solved by using numerical simulations and these are compared with the analytical solutions.

Keywords: Dubins car model · Lie algebra · Optimal control problem · Pontryagin maximum principle · Nilpotent algebra · Numerical simulations

1 Introduction

We study optimal control of planar mechanisms based on Lie groups, see classical books [1–4] or papers [7,10–12]. The key mathematical apparatus in optimal control is Pontryagin maximum principle (PMP) [1,2]. Using PMP we get system of ordinary differential equations of the first order (SODE1), whose solution is the solution of the optimal control problem. However, at some points the optimal solutions lose their optimality. One of the cases is such point to which two optimal trajectories lead. These points can be found as fixed points of symmetries [2]. The next problem is that the system usually cannot be solved analytically or finding its solution is complicated, so it is necessary to use a numerical procedure. In this article we have chosen a simple mechanism, so we can find the analytical solution of the system and compare it with the procedure based on numerical

The first author was supported by grant FSI-S-20-6187 of the FME 2020 at Brno University of Technology. The second author was supported by solution grant FSI-S-20-6407 of the FME 2020 at Brno University of Technology. The third author was supported by solution grant FV20-25 science Fund of the FME 2020 at Brno University of Technology.

J. Mazal et al. (Eds.): MESAS 2020, LNCS 12619, pp. 111–122, 2021.
https://doi.org/10.1007/978-3-030-70740-8_7

simulations. The aim was to find out if by numerical procedures we can find more optimal solutions leading to the fixed points of symmetries and only one optimal solution leading to a point that is not a fixed point of symmetries (Fig. 1).

One of the simpliest models in control theory is the Dubins car [6,15,16,18]. This model can be viewed as a rigid body with four wheels, which can not slip sideways, but can move backwards and forwards and change its orientation on a plane. Thus its configuration space is three-dimensional smooth manifold

$$\mathcal{C} = \{q = (X, \theta) | (x, y) \in \mathbb{R}^2, \theta \in \mathbb{S}^1\} = \mathbb{R}^2 \times \mathbb{S}^1.$$

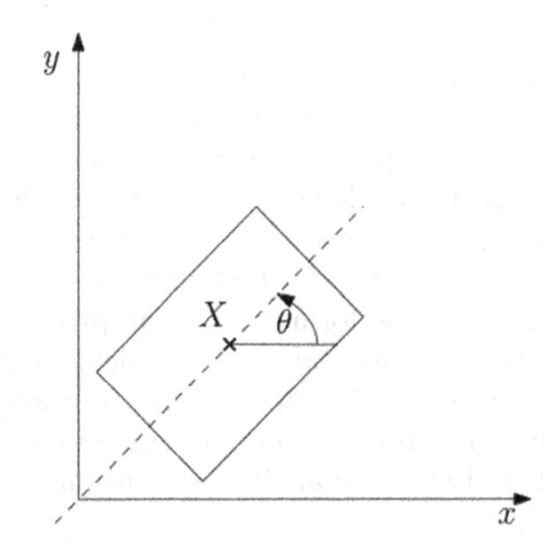

Fig. 1. Representation of a simple car in $\mathcal{C}$.

2 Differential Kinematics of the Mechanism

In this section we want to describe differential kinematics of the mechanism. Similar issues can be found for example in [9].

The non-slip condition gives us following equation, referred as Pfaffian constraint:

$$(\dot{x}, \dot{y}) \cdot \vec{n} = -\dot{x} \sin\theta + \dot{y} \cos\theta = 0.$$

This equation holds iff

$$\begin{aligned}\dot{x} &= u_1 \cos\theta, \\ \dot{y} &= u_1 \sin\theta, \\ \dot{\theta} &= u_2.\end{aligned}$$

for some $u_1, u_2 \in \mathbb{R}$, to which we usually refer as control parameters. Notice that any movement back and forth can be expressed by

$$\begin{aligned}\dot{x} &= u_1 \cos\theta, \\ \dot{y} &= u_1 \sin\theta, \\ \dot{\theta} &= 0\end{aligned}$$

and any change of the orientation can be expressed as

$$\dot{x} = 0,$$
$$\dot{y} = 0,$$
$$\dot{\theta} = u_2.$$

It is obvious that we obtain two vector fields

$$V_1 = \cos\theta\partial_x + \sin\theta\partial_y,$$
$$V_2 = \partial_\theta.$$

yielding a first-order control system

$$\dot{q} = u_1 V_1 + u_2 V_2.$$

3 Lie Algebra of the System

Math theory of Lie algebras can be found in [1,2].

Definition 1. *A Lie algebra is a vector space* $\mathfrak{g}$ *over field* $\mathbb{R}$ *with binary operation* $[\cdot,\cdot] : \mathfrak{g} \times \mathfrak{g} \to \mathfrak{g}$ *called Lie bracket satisfying the following axioms:*

$$[ax + by, z] = a[x, z] + b[y, z],$$
$$[z, ax + by] = a[z, x] + b[z, y],$$
$$[x, x] = 0,$$
$$[x, [y, z]] + [z, [x, y]] + [y, [z, x]] = 0,$$

for all $a, b \in \mathbb{R}$ *and for all* $x, y, z \in \mathfrak{g}$.

Remark 1. Bilinearity and alternativity imply anticomutativity [1]:

$$[x, y] = -[y, x].$$

There can be defined the Lie bracket of vector fields that satisfies the axioms from Definition 1 [1]:

Definition 2. *Lie bracket of vector fields* X, Y *on manifold* M*:*

$$[X, Y] = X(Y(f)) - Y(X(f)), \quad \textit{for all } f \in C^\infty(M),$$

where $X(f)$ *is a function whose value at a point* p *is the directional derivative of* f *in the direction* $X(p)$.

In each point $p \in C$, the tangent space T_pC is generated by vectors $V_1(p), V_2(p)$ and $V_{12}(p)$. On the other hand the vectors fields V_1, V_2 and V_{12} generate solvable Lie algebra with multiplication Table 1, see [5].

$$\begin{aligned} V_{12} &:= [V_1, V_2] = \sin\theta\partial_x - \cos\theta\partial_y, \\ V_{121} &:= [V_{12}, V_1] = 0, \\ V_{122} &:= [V_{12}, V_2] = V_1. \end{aligned} \tag{1}$$

Table 1. Multiplication table of systems Lie algebra

$[\cdot,\cdot]$	V_1	V_2	V_{12}
V_1	0	V_{12}	0
V_2	$-V_{12}$	0	V_1
V_{12}	0	$-V_1$	0

4 Optimal Control of the System

In optimal control problems, we want to find the control $u(t)$ and a corresponding trajectory $q(t)$ for which:

$$\begin{aligned}
&\dot{q}(t) = f(q(t), u(t)), \quad q(t) \in \mathcal{M}, \quad u \in U \subset \mathbb{R}^m, \\
&q(0) = q_0, \quad q(t_1) = q_1, \quad t_1 \text{ fixed or free}, \\
&J(u) = \int_0^{t_1} f_0(q(t), u(t))dt \to \min,
\end{aligned} \tag{2}$$

where $\mathcal{M}$ is smooth manifold and $f : \mathcal{M} \times U \to \mathbb{R}^n$ and $f_0 : \mathcal{M} \times U \to \mathbb{R}$ are known functions [1,2].

Definition 3. *Let $h \in T^*\mathcal{M}$ be a covector and $\nu \in \mathbb{R}$ a parameter, then function*

$$H_u^\nu(h) = \langle h, f(q,u) \rangle + \nu f_0(q,u)$$

is called Hamiltonian of the optimal problem (2).

Theorem 1 *(Pontryagin maximum principle on smooth manifolds). Let $\tilde{u}(t), t \in [0, t_1]$, be an optimal control in the problem (7.1) – (7.3) with fixed time t_1. Then there exists a Lipschitzian curve $h_t \in T^*_{\tilde{q}(t)}\mathcal{M}, t \in [0, t_1]$, and a number $\nu \in \mathbb{R}$ such that*

$$\begin{aligned}
\dot{h}_t &= \overrightarrow{H_{\tilde{u}}^\nu}(h_t), \\
H_{\tilde{u}(t)}^\nu(h_t) &= \max_{u \in U} H_u^\nu(h_t), \\
(h_t, \nu) &\neq (0,0), \quad t \in [0, t_1]. \\
\nu &\leq 0.
\end{aligned}$$

Proof can be found in [1].

Remark 2. If $\nu < 0$, then we can normalize the Hamiltonian by the choice $\nu = -1$ [1].

In our cases we want to minimize length of curve $q \in \mathcal{M}$ between two fixed points k_0 and k_1. Minimizing of the energy of a curve implies minimizing of its length, so our optimal control problem has the following form:

$$\dot{q} = u_1 V_1 + u_2 V_2 = u_1 \begin{pmatrix} \cos\theta \\ \sin\theta \\ 0 \end{pmatrix} + u_2 \begin{pmatrix} 0 \\ 0 \\ 1 \end{pmatrix}, \quad q \in \mathcal{C}, \quad (u_1, u_2) \in \mathbb{R}^2,$$

$$q(0) = k_0, \quad q(t_1) = k_1,$$

$$J(u) = \int_0^{t_1} (u_1^2 + u_2^2) dt \to \min. \tag{3}$$

Hamiltonian of the problem has the form

$$H_u^{\nu}(h) = u_1 h_1 + u_2 h_2 + \frac{\nu}{2}(u_1^2 + u_2^2).$$

If $\nu = -1$, optimality conditions $\frac{\partial H}{\partial u_1} = 0$, $\frac{\partial H}{\partial u_2} = 0$ give

$$H = \frac{1}{2}\left(h_1^2 + h_2^2\right).$$

For covectors h_i holds

$$\dot{h}_i = \{H, h_i\},$$

where $\{\cdot, \cdot\}$ is Poisson bracket [2]. It gives following SODE1 [8]:

$$\begin{aligned} \dot{h}_1 &= h_3 h_2, \\ \dot{h}_2 &= -h_3 h_1, \\ \dot{h}_3 &= -h_2 h_1, \\ \dot{x} &= h_1 \cos\theta, \\ \dot{y} &= h_1 \sin\theta, \\ \dot{\theta} &= h_2. \end{aligned} \tag{4}$$

The system (4) can be analytically solved [19], but the solution is formulated in the language of Jacobi eliptic functions that are hard to use in implementations. The next step is compute nilpotent aproximation of the vector fields V_1 and V_2 [14] and formulate the optimal control problem of them.

5 Optimal Control on Nilpotent Lie Algebra

As already mentioned, the next step is compute nilpotent aproximation of the vector fields V_1 and V_2 [14] and formulate the optimal control problem of them. Similar procedure is used for example in [13,17].

It can be easily shown that the Taylor expansion of the commutator of the flows of vector fields V_1 and V_2 is an infinite sum. Now we can show an algorithm, in which we approximate these vector fields in such a way that mentioned Taylor expansion will be finite. Let $q_0 = (0, 0, 0)$. We would like to have such vector fields that

$$V_i(q_0) = \begin{pmatrix} \delta_1^i \\ \delta_2^i \\ \delta_3^i \end{pmatrix}$$

where δ_j^i is Kronecker delta. Suggest following $(x, y, \theta) \rightarrow (y_1, y_2, y_3)$ transformation.

$$\begin{aligned} y_1 &= x \\ y_2 &= \theta \\ y_3 &= -y \end{aligned} \tag{5}$$

Thus vector fields $\tilde{V}_1, \tilde{V}_2$ in the new coordinates are

$$\tilde{V}_1 = \begin{pmatrix} \cos y_2 \\ 0 \\ -\sin y_2 \end{pmatrix}, \quad \tilde{V}_2 = \begin{pmatrix} 0 \\ 1 \\ 0 \end{pmatrix}$$

and with well known trigonometric approximation in q_0 neighborhood $\sin x \approx x, \cos x \approx 1$ yields

$$N_1 = \begin{pmatrix} 1 \\ 0 \\ -y_2 \end{pmatrix}, \quad N_2 = \begin{pmatrix} 0 \\ 1 \\ 0 \end{pmatrix}.$$

This yields nilpotent Lie algebra with the multiplication Table 2 which is called Heisenberg.

Table 2. Multiplication table of nilpotent Lie algebra

$[\cdot, \cdot]$	N_1	N_2	N_{12}
N_1	0	N_{12}	0
N_2	$-N_{12}$	0	0
N_{12}	0	0	0

Now we can reformulate the optimal control problem:

$$\begin{aligned} &\dot{q} = u_1 N_1 + u_2 N_2, \quad q \in \mathcal{C}, \quad (u_1, u_2) \in \mathbb{R}^2, \\ &q(0) = k_0, \quad q(t_1) = k_1, \\ &J(u) = \int_0^{t_1} (u_1^2 + u_2^2) dt \rightarrow \min. \end{aligned} \tag{6}$$

We can follow the same steps as in the chapter 4 and we get the system whose solution is solution of the optimal control problem (6). The system has the following form [8]:

$$\begin{aligned}
\dot{h}_1 &= h_3 h_2, \\
\dot{h}_2 &= -h_3 h_1, \\
\dot{h}_3 &= 0, \\
\dot{y}_1 &= h_1, \\
\dot{y}_2 &= h_2, \\
\dot{y}_3 &= -y_2 h_1
\end{aligned} \tag{7}$$

The system (7) can be analytically solved and the procedure can be found in [8]. The analytical solution of the covectors has the form

$$\begin{aligned}
h_1 &= C_2 \sin(C_1 t) + C_3 \cos(C_1 t), \\
h_2 &= C_3 \sin(C_1 t) - C_2 \cos(C_1 t), \\
h_3 &= C_1.
\end{aligned}$$

Let's choose $k_0 = (0, 0, 0)$, then $y_1(0) = y_2(0) = y_3(0) = 0$ and for $C_1 \neq 0$ the analytical solution of y_1, y_2 and y_3 is

$$\begin{aligned}
y_1 &= \frac{1}{C_1}(C_3 - C_2 \sin(C_1 t) - C_3 \cos(C_1 t)), \\
y_2 &= \frac{1}{C_1}(C_2 - C_2 \cos(C_1 t) + C_3 \sin(C_1 t)), \\
y_3 &= -\frac{1}{4C_1^2}(2C_1(C_2^2 + C_3^2)t - 4C_2 C_3 \cos(C_1 t) + 2C_2 C_3 \cos(2C_1 t) \\
&\quad - 4C_2^2 \sin(C_1 t) + (C_2^2 - C_3^2)\sin(2C_1 t) + 2C_2 C_3),
\end{aligned}$$

and for $C_1 = 0$ the solution of y_1, y_2 and y_3 is

$$\begin{aligned}
y_1 &= C_3 t, \\
y_2 &= -C_2 t, \\
y_3 &= \frac{1}{2} C_2 C_3 t^2.
\end{aligned}$$

With respect to (5) for $C_1 \neq 0$ the analytical solution of x, y, θ is

$$x = \frac{1}{C_1}(C_3 - C_2 \sin(C_1 t) - C_3 \cos(C_1 t)), \tag{8}$$

$$\begin{aligned}
y = \frac{1}{4C_1^2}&(2C_1(C_2^2 + C_3^2)t - 4C_2 C_3 \cos(C_1 t) + 2C_2 C_3 \cos(2C_1 t) \\
&- 4C_2^2 \sin(C_1 t) + (C_2^2 - C_3^2)\sin(2C_1 t) + 2C_2 C_3)
\end{aligned} \tag{9}$$

$$\theta = \frac{1}{C_1}(C_2 - C_2\cos(C_1 t) + C_3\sin(C_1 t)), \tag{10}$$

and for $C_1 = 0$ the solution of x, y, θ is

$$\begin{aligned} x &= C_3 t, \\ y &= -\frac{1}{2}C_2 C_3 t^2, \\ \theta &= -C_2 t. \end{aligned}$$

6 Symmetries

For each point k_0 we want to find the points in which the optimal curves from this point lose their optimality. One of the possibilities is such points to which two optimal trajectories lead. Because it's control on groups, we can choose the origin as the starting point. Then we can find these points as fixed points of symmetries that preserve the origin [2].

We can find explicitly all infinitesimal symmetries of the sub-Riemmanian structure $(N, \mathcal{N}, r)$. Indeed, we are interested in vector fields v such that $\mathcal{L}_v(\mathcal{N}) \subset \mathcal{N}$ and $\mathcal{L}_v(r) = 0$. This gives us a system of pde's that can be solved explicitly in the case of the left-invariant nilpotent structure.

Theorem 2. *Infinitesimal symmetries of the left-invariant sub-Riemmanian structure $(N, \mathcal{N}, r)$ form a Lie algebra generated (over $\mathbb{R}$) by vector fields*

$$\begin{aligned} t_0 &:= \theta\partial_x + \frac{\theta^2 - x^2}{2}\partial_y - x\partial_\theta \\ t_1 &:= \partial_x \\ t_2 &:= x\partial_y + \partial_\theta \\ t_3 &:= \partial_y \end{aligned}$$

In particular, t_0 generates the isotropic subalgebra at $(0, 0, 0)$. Fields $t_i, i = 1, 2, 3$ are translations that reflect the Heisenberg structure and coincide with the right-invariant fields.

Proof can be found in [8].

Corollary 1. *If the origin $(0, 0, 0)$ is the fixed point of symmetries then the points $(0, s, 0), s \in \mathbb{R}$ are the only fixed points of symmetries.*

Remark 3. That's because only the vector field t_0 preserves the origin and the vector field also preserves points $(0, s, 0), s \in \mathbb{R}$ [8].

7 Numerical Simulations

The system (7) can be solved numerically, by choosing an initial condition $[x_0, y_0, \theta_0, h_{10}, h_{20}, h_{30}]$ and using a standard numerical solver for ordinary differential equations (e.g. Runge-Kutta). This produces a discrete approximation of the real solution $x(t), y(t), \theta(t), h_1(t), h_2(t), h_3(t)$. To further demonstrate this point a few specific examples will be provided in the following sections.

As a first example we would like to show that we can obtain more than one solution of the system which represents a curve starting in the point $k_0 = (0, 0, 0)$ and ends in fixed point of symmetries $k_1 = (0, y_{t_1}, 0)$ at time $t = t_1$. We can choose any arbitrary point k_1 and time t_1, but for simplicity of our first example we chose $k_1 = (0, 10, 0)$ and $t_1 = 1$. The problem becomes finding the suitable initial conditions for the system (7), which end up in the final specified point. This means we can reformulate the problem as numerical minimization problem. Assume we define $s_2 = (x_{t_1}, y_{t_1}, \theta_{t_1})$ as numerical solution of the system (7) at time $t = t_1$, then the minimization problem can be expressed as

$$\min_{h_{10}, h_{20}, h_{30}} ||s_2 - k_1||_2^2$$

We start with random combination of h_{10}, h_{20}, h_{30} and use Nelder-Mead algorithm to iteratively update our estimate and find those h_{10}, h_{20}, h_{30} for which $s_2 \approx (0, 10, 0)$. In the Fig. 2 we can see solution with initial values $h_{10} = -9.1679$, $h_{20} = 6.4533$, $h_{30} = 6.2833$.

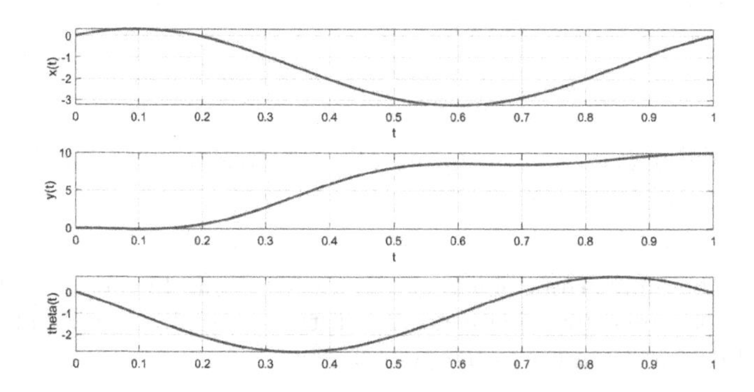

Fig. 2. Solution for $h_{10} = -9.1679$, $h_{20} = 6.4533$, $h_{30} = 6.2833$

In the Fig. 3 we can see solution with initial values $h_{10} = -13.0881$, $h_{20} = 14.3621$, $h_{30} = 18.8482$.

In fact we should be able to find infinitely many solutions to this problem. We repeat the process with different random combinations of h_{10}, h_{20}, h_{30} using a pseudo-random number generator with normal distribution. The result of this process can be seen in Fig. 4. The reason, why the points form circles, is that if the end point is $(0, y_{t_1}, 0)$, the Eqs. (8) and (9) imply $C_1 = \frac{2K\pi}{t_1}, K \in \mathbb{Z}$. Then

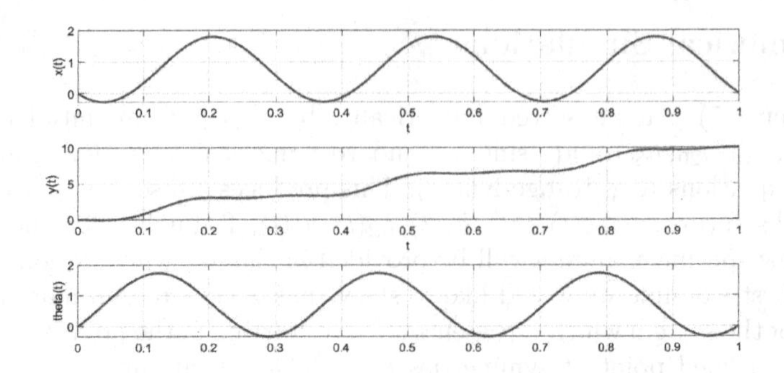

Fig. 3. Solution for $h_{10} = -13.0881$, $h_{20} = 14.3621$, $h_{30} = 18.8482$

the Eq. (10) implies that $2y_{t_1}C_1 = C_2^2 + C_3^2$. We can see these facts in the figure as well as on the concrete choices of $h_{10}.h_{20}, h_{30}$.

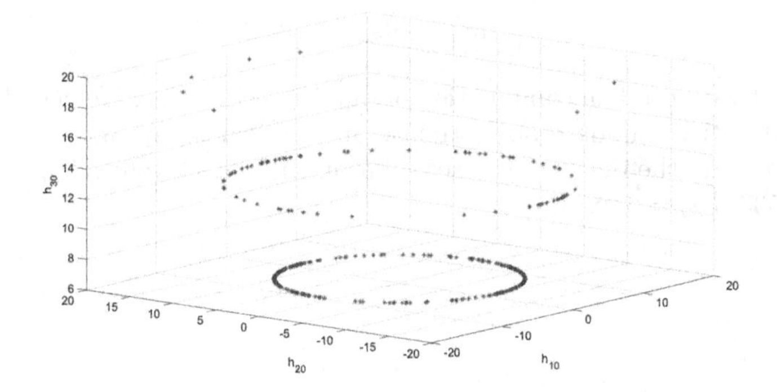

Fig. 4. Set of $h_{10}.h_{20}, h_{30}$ for which the solution is suitable

Remark 4. The number of points decreases when going further away from the origin. This is caused by using a normal distribution with the pseudo-random generator for generating the initial points.

As another example we set $k_0 = (0, 0, 0)$ and $k_1 = (0, 0, \theta_{t_1})$, where k_1 is not a fixed point of symmetries. After repeating the process of the minimization described above we get only one solution for each value of θ_{t_1}. No matter which initial estimate was chosen the solution always converged to a single point. The solution for $\theta_{t_1} = 3$, $h_{10} = 3$, $h_{20} = 0$, $h_{30} = 0$ we can see in the Fig. 5.

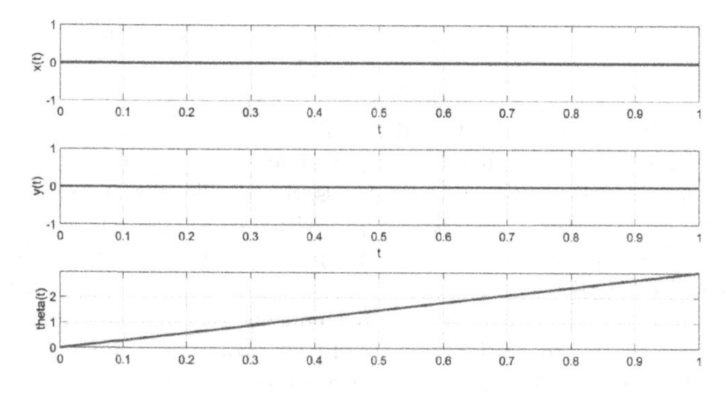

Fig. 5. Solution for $k_1 = (0, 0, \theta_{t_1})$

8 Conclusion

In this paper we described forward kinematics of the Dubins car model on three-dimensional smooth manifold. Then we computed vector fields that generate tangent space and Lie algebra in each point of the manifold. Afterwards an optimal control problem for finding the shortest curve between two points in the manifold was formulated. Using the Pontryagin's maximum principle we obtained the system of ordinary differential equations whose solution is the solution of the problem. However, the solution had complicated form, so we reformulated the problem on nilpotent approximation of the fields and found the solution of the approximated system. The next step was to find fixed points of symmetries, because in these points the optimal curves can lose their optimality. Finally we simulated the system numerically and we compared the analytical solution with the numerical. As a first example we chose the end point as a fixed point of symmetries and as a second example we did not. The numerical solution was found using Runge-Kutta algorithm, followed by numerical optimization.

In the fixed point of symmetries the numerical solution gave more optimal curves. If the point wasn't a fixed point of symmetries, we would find only one optimal curve, so the results of our numerical experiments correspond with the analytical solution and the numerical procedure can be used for more complex mechanisms where the analytical solution of the system cannot be found. However, numerical solutions cannot prove mathematical theorems, on the other hand they can give useful information in the proving.

References

1. Agrachev, A.A., Sachkov, Y.L.: Control Theory from the Geometric Viewpoint. Encyclopaedia of Mathematical Sciences, Control Theory and Optimization. Springer, Heidelberg (2006). https://doi.org/10.1007/978-3-662-06404-7
2. Agrachev, A.A., Barilari, D., Boscain, U.: A Comprehensive Introduction to Sub-Riemannian Geometry. Cambridge Studies in Advanced Mathematics. Cambridge University Press, New York (2019)

3. Bloch, E.D.: A First Course in Geometric Topology and Differential Geometry. Birkhauser, Boston (1997). ISBN 3764338407
4. Bloch, A.M.: Nonholonomic Mechanics and Control. IAM, vol. 24. Springer, New York (2003). https://doi.org/10.1007/b97376
5. Calin, O., Chang, D.E., Greiner, P.C.: Geometric Analysis on the Heisenberg Group and Its Generalizations. International Press, Somerville (2007). ISBN 0821843192
6. Dubins, L.E.: On curves of minimal Length with a constraint on average curvature, and with prescribed initial and terminal positions and tangents. Am. J. Math. **79**(3), 497–516 (1957)
7. Frolík, S.: Note on signature of trident mechanisms with distribution growth vector (4,7). In: Mazal, J. (ed.) MESAS 2018. LNCS, vol. 11472, pp. 82–89. Springer, Cham (2019). https://doi.org/10.1007/978-3-030-14984-0_7
8. Hrdina, J., Návrat A., Zalabová, L.: Geometric control theory of vertical rolling disc using symmetries - arXiv preprint arXiv:1908.03352 (2019). arxiv.org
9. Hrdina, J., Vašík, P.: Notes on differential kinematics in conformal geometric algebra approach. In: Matoušek, R. (ed.) Mendel 2015. AISC, vol. 378, pp. 363–374. Springer, Cham (2015). https://doi.org/10.1007/978-3-319-19824-8_30
10. Hrdina, J., Zalabová, L.: Local geometric control of a certain mechanism with the growth vector (4,7). J. Dyn. Control Syst. **26**, 199–216 (2020). https://doi.org/10.1007/s10883-019-09460-7
11. Hrdina, J., Vašík, P., Návrat, A., Matoušek, R.: Nilpotent approximation of a trident snake robot controlling distribution. Kybernetika **53**(6), 1118–1130 (2017). https://doi.org/10.14736/kyb-2017-6-1118
12. Hrdina, J., Vašík, P., Návrat, A., Matoušek, R.: CGA-based robotic snake control. Adv. Appl. Clifford Algebras **27**(1), 621–632 (2016). https://doi.org/10.1007/s00006-016-0695-5
13. Hůlka, T., Matoušek, R., Dobrovský, L., Dosoudilová, M.: Optimization of snake-like robot locomotion using GA: serpenoid design. Mendel **26**(1), 1–6 (2020)
14. Jean, F.: Control of Nonholonomic Systems: From Sub-Riemannian Geometry to Motion Planning. SM. Springer, Cham (2014). https://doi.org/10.1007/978-3-319-08690-3
15. Michor, P.W.: Topics in Differential Geometry. Graduate Studies in Mathematics, vol. 93. American Mathematical Society, Providence (2008)
16. Murray, R.M., Zexiang, L., Sastry, S.S.: A Mathematical Introduction to Robotic Manipulation. CRC Press, Boca Raton (1994)
17. Návrat, A., Vašík, P.: On geometric control models of a robotic snake. Note di Matematica **37**, 119–129 (2017). https://doi.org/10.1285/i15900932v37suppl1p119
18. Reeds, J.A., Shepp, L.A.: Optimal paths for a car that goes both forwards and backward. Pac. J. Math. **145**, 367–393 (1990)
19. Sachkov, Y.L., Moiseev, I.: Maxwell strata in sub-Riemannian problem on the group of motions of a plane. ESAIM: COCV **16**, 380–399 (2010)
20. Selig, J.M.: Geometric fundamentals of robotics. Monographs in Computer Science, 2nd edn. Springer, New York (2005). https://doi.org/10.1007/b138859

M&S Based Testbed to Support V&V of Autonomous Resources Task Coordinator

Giovanni Luca Maglione[1(✉)], Luca Berretta[1], Sasha Blue Godfrey[1], Savvas Apostolidis[2], Athanasios Kapoutsis[2], Elias Kosmatopoulos[2], and Alberto Tremori[1]

[1] NATO STO Centre for Maritime Research and Experimentation, Viale San Bartolomeo, 400, La Spezia, Italy
{giovanni.maglione,luca.berretta,sasha.godfrey, alberto.tremori}@cmre.nato.int

[2] Centre for Research and Technology, Hellas, Thermi Thessaloniki – Central Directorate, 6th km Charilaou-Thermi Rd, P.O. Box 60361, 57001 Thermi, Thessaloniki, Greece
{sapostol,athakapo,kosmatop}@iti.gr

Abstract. Political instability around the world continues to place a significant emphasis on border control. Monitoring these borders requires persistent surveillance in a variety of remote, hazardous and hostile environments. While recent developments in autonomous and unmanned systems promise to provide a new generation of tools to assist in border control missions, the complexity of designing, testing and operating large-scale systems limits their adoption.

A seam of research is developing around the use of Modelling and Simulation (M&S) methodologies to support the development, testing and operation of complex, multi-domain autonomous systems deployments. This paper builds upon recent progress in the use of M&S to conduct Verification and Validation (V&V) of complex software functionalities.

Specifically, the authors have designed and developed an HLA (High Level Architecture) interoperable M&S testbed capability applied in support of the European Union's ROBORDER H2020 project. V&V has been completed on the Autonomous Resources Task Coordinator (ARTC) software, a module that will be employed in live demonstrations to automatically design missions for heterogeneous autonomous assets.

The development and the employment of the interoperable simulation capability is discussed in a scenario designed to test the ARTC. The scenario involves aerial (fixed wing and rotary wing) and underwater assets mounting Electro-Optical/Infra-Red (EO/IR) cameras and pollution detection sensors. Asset and sensor performance is affected by realistic environmental conditions.

The M&S-based test-bed capability has shown the correct operation of the ARTC, efficiently communicating the key findings to a range of stakeholder groups. The work has resulted in the creation and testing of an interoperable, modular, reusable testbed capability that will be reused to further support the wide-spread adoption of autonomous and unmanned systems in a range of operations.

Keywords: Modelling and simulation · Autonomous systems · Verification and validation · Testing · Algorithms development

J. Mazal et al. (Eds.): MESAS 2020, LNCS 12619, pp. 123–138, 2021.
https://doi.org/10.1007/978-3-030-70740-8_8

1 Introduction

This paper describes the work performed for the evaluation work package of H2020 ROBORDER Project and reports it in accordance to the standardised practice described in the DSEEP (Distributed Simulation Engineering and Execution Process) process [1], steps 3 to 7 (Fig. 1).

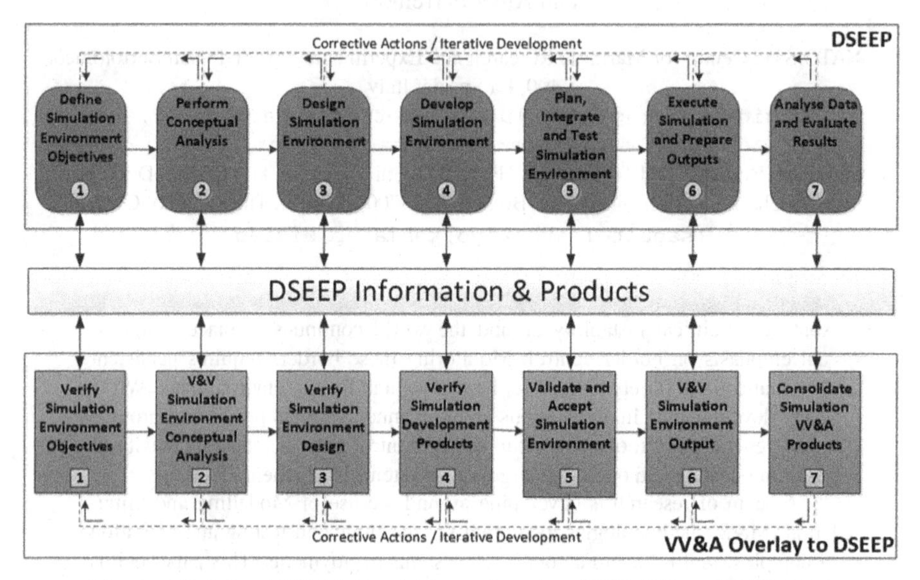

Fig. 1. DSEEP and VV&A Overlaid steps based on IEEE 1730 and IEEE 1516

European external border control activities nowadays strongly rely on human personnel and traditional manned assets such as aircraft, vessels or cars. The operational environment challenges the boots on the ground, which often need to operate in harsh conditions such as snow, high wind, rough sea state and deep mud. The diversity of the threats and the increased availability of low cost technological solutions pose a serious challenge for Law Enforcement and Border Agencies. All these factors make current operations both expensive and resource-intensive.

Significant technological advances have been made by European countries and companies in recent years, especially in the field of autonomous vehicles, to address the challenges mentioned above. Nonetheless, nations still lack solutions to address the complexity of the overall picture. The European Union and NATO are looking at integrating capabilities, creating complex architectures of multi-domain manned and unmanned systems and enhancing Command and Control by adopting Augmented and Virtual Reality. Within the framework of the ROBORDER H2020 project, the authors are involved in developing and demonstrating an autonomous border surveillance platform with unmanned multi-domain vehicles.

The resulting capability should satisfy tomorrow's need for reducing the workload on border authorities and provide the ability to remotely sense, evaluate and track incidents at the borders. Early incident assessment based on enhanced data acquisition means a safer operational environment for the deployed human personnel.

The complexity of the solution may require additions to traditional design processes. There is also a need of assessing how the design of such a heterogeneous solution meets the expectations and operational requirements of the customers and users. Modelling and Simulation (M&S) is the methodology used to support all phases of the life cycle of systems, system of systems and concepts.

Until few years ago M&S has been used for ROBORDER like case studies to support the evaluation of the dynamics of specific vehicles or specific sensors performances or algorithms [2, 3], while only recent trends showed the application simulations of complex multi-domain heterogeneous surveillance systems [4] with the inclusion of realistic weather models [5–7] for supporting the evaluation of systems of systems performances.

In this project, simulation has been adopted to demonstrate its potential as a tool to evaluate system performance in large-scale scenarios with realistic operational conditions. In addition, an M&S-based testbed capability, developed since the early stages of the architecture lifecycle allows de-risking the design, while the integration of interoperable simulation with a platform permits the analysis of the algorithms developed to support future real operations. In this first iteration, simulation has fulfilled all these goals by supporting the verification and validation of algorithms and software components developed by technical partners. To demonstrate the platform adaptability, a set of Pilot Use Cases (PUCs) provided by the end-users were identified to test the performance against a range of conditions. Each PUC represents sea and/or land scenarios where the ROBORDER platform will be demonstrated.

The M&S environment complies with High Level Architecture (HLA) interoperability standards [8] and is designed according to IEEE and NATO best practices. It consists of a federation of simulators including virtual and constructive components and is able to run both in real and fast time.

2 Simulation Environment

The areas of development required for the ROBORDER M&S capability models are elicited from the analysis of the project objectives and requirements, the ROBORDER components (platforms, sensors and algorithms) and the characteristics of the maritime scenarios addressed in this paper. The modules identified (Fig. 2) are described in the subsections that follow.

2.1 Platforms

A set of platform models has been created with the aim of providing the simulation entities (Fig. 3) with kinematic behaviours, guidance and effect of the environment. Kinematic data are shared over High Level Architecture (HLA) interface and consumed by the others federates.

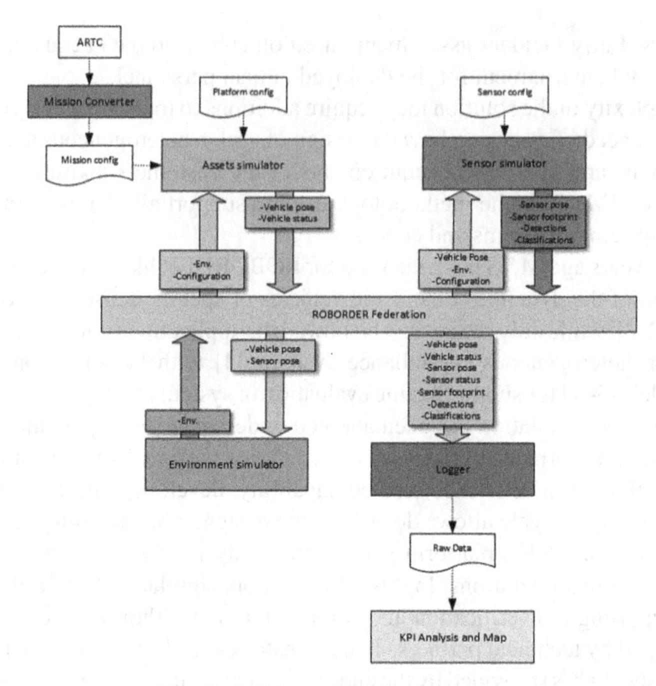

Fig. 2. M&S capability architecture

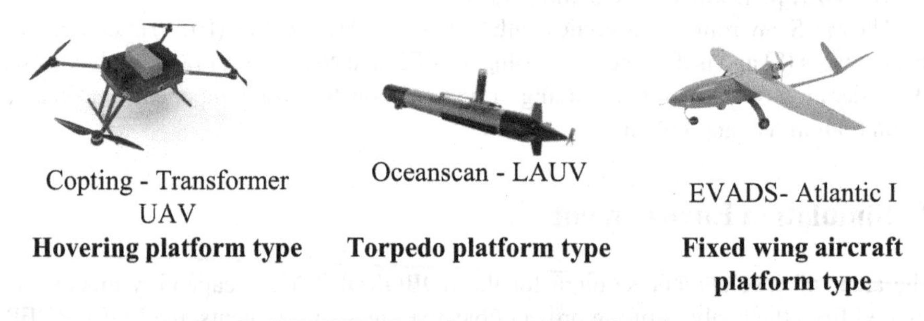

Fig. 3. ROBORDER platform considered for this paper

Kinematic Model

Platforms have been grouped into 3 types, based on their kinematic model properties:

- **Hovering platform (Copting UAV):** In the kinematic model, all axes (Fig. 4) can be actuated independetly of each other degree of freedom. There are no cross couplings between axes. Pitch and roll (axes 5 & 6) motions are assumed to be minimal.
- **Torpedo platfom (Oceanscan-UUV):** Yaw and pitch (axes 4 & 5) motions are a function of forward speed (axis 3). Vertical speed is a function of forward speed and pitch angle. Heave and sway (axes 1 & 2), as well as roll motions (axis 6), are assumed to be minimal.

- **Fixed wing aircraft (Evads-UAV):** Vertical position is a function of forward speed and pitch angle. Yaw and pitch (axis 4 & 5) motions are a function of forward speed (axis 3). Yaw rate (axis 4) is a function of forward speed and roll (axis 3 & 6). Motion along vertical and lateral axes (axis 1 & 2) are assumed to be minimal.

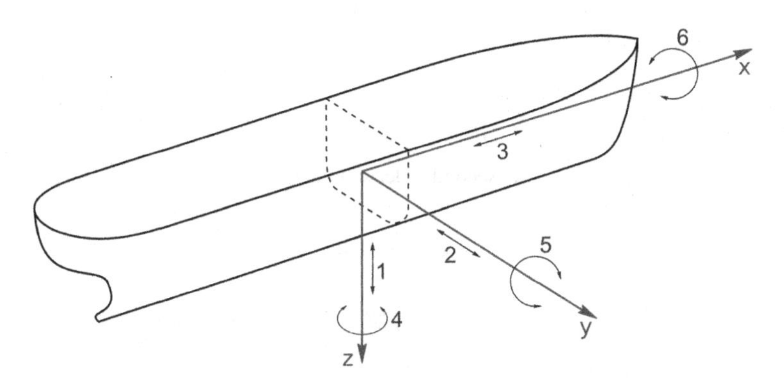

Fig. 4. Platform kinematic model (left handed) axis set

The kinematic model includes the physics of the platform and its low-level control system, merged into a single mathematical model. The control system allows the platform to follow target commands of speed, heading and altitude/depth.

For all three-platform types, the longitudinal speed and pitch degrees of freedom have been implemented as independent, second order functions. An additional independent, second order function has been implemented for the heading (hovering and torpedo platforms) and roll (fixed wing platform). Both Torpedo and Fixed wing asset types assume forward speed is always greater than zero.

The equation and the block diagram detailing these transfer functions are depicted in Fig. 5 with the following characteristics:

- Dumping: Critical dump to avoid overshooting while following the target command;
- Settling time: validated with actual telemetry data;
- Saturation: In order to avoid exceeding the maximum rate of change of each variable, saturation values have been added.

This approach allows modelling the systems in a straightforward way, relying only on interactions with system providers and the limited data that can be inferred by the telemetry.

An additional step has been done for the altitude/depth command. Since the vehicles are able to change depth only by adjusting pitch angle, a PID closed-loop control system (see Fig. 6) has been developed around the pitch degree of freedom.

Platform Guidance

The guidance module is in charge of providing the kinematic module with the proper steering commands to perform the assigned mission. The steering commands are:

$$s^2Y(s) + asY(s) + bY(s) = X(s)$$

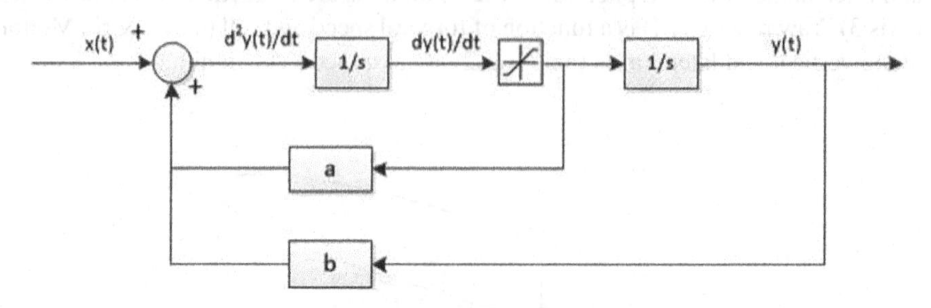

Fig. 5. Second order saturated system

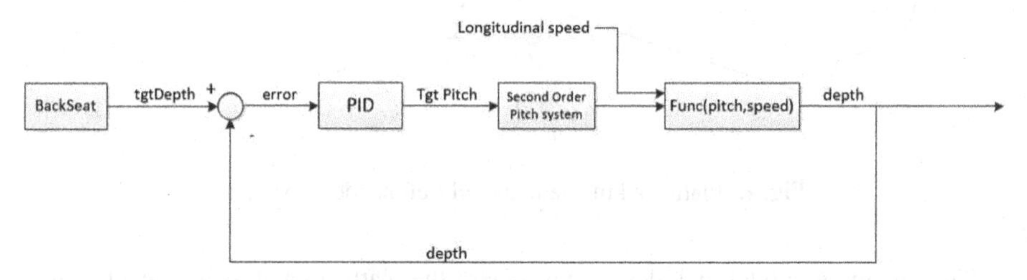

Fig. 6. Altitude/depth feedback control loop

- Target forward speed
- Target heading
- Target altitude/depth

Once the kinematic module receives the reference steering commands from the guidance module, it is able to follow them, since it implements both the physics and the required low-level controller.

The guidance module is able to handle waypoint missions: the mission is a sequence of waypoints each one having a set of common parameters like: geographical position, altitude to be reached, speed to be kept, next waypoint, etc. This mission is supposed to be accomplished moving in straight line from one waypoint to the next one. However for each waypoint it is possible to specify how to link two consecutive straight lines: passing over the specified waypoint and thus overshooting the transition to the following (fly-over) one or anticipating the transition before reaching the waypoint and thus shortcutting the path and linking smoothly to the next straight leg (fly-by).

Each kinematic model is implemented to take into account sea current (for marine vehicles) and wind. The model is able to compute the speed and course relative to the air/water, Speed over Ground and Course over Ground.

The model has two working principles to compensate the cross track error, based on the navigation sensors on board the vehicle:

- The vehicle estimates its position (e.g. INS): the guidance sets the heading to the direction of the next waypoint, this behavior is implemented for underwater platforms (Fig. 7a).
- The vehicle has feedback on its position (e.g. GPS): guidance compensates cross track error instant by instant, this behavior is implemented for fixed wing and rotary wing platforms (Fig. 7b).

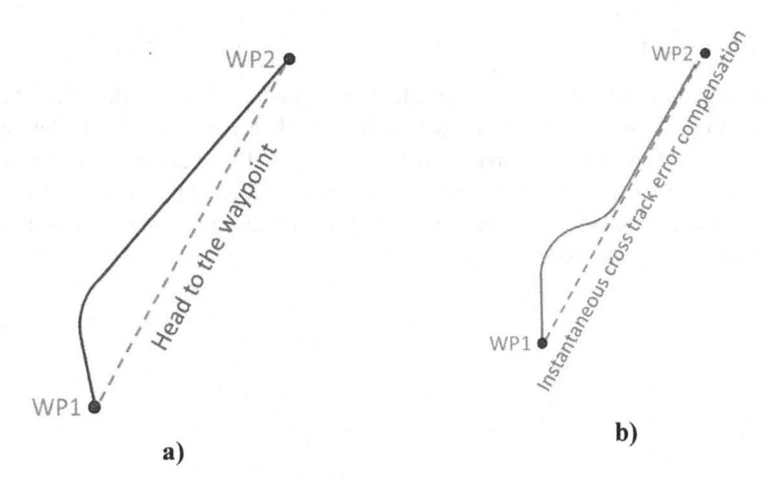

Fig. 7. Guidance models implemented to compensate for environmental effects

Further environmental effects relate to the limited conditions for Launch and Recovery of the platforms. Operational limit conditions are inferred from surveys with vehicle providers. For instance, the UAV provider states that the platform is unable to operate if the wind speed is higher than 20 m/s. This will trigger a mission abort in case the limit is exceeded.

2.2 Sensors

A set of sensor models has been created with the aim of providing the simulation sensors (Fig. 8) with field of view, detection and classification performance and effect of the environment. Detection and classification data are shared over High Level Architecture (HLA) interface. These data are used as an input to models from across the federation.

Two models have been developed to compute detections and classifications. The first model reproduces directional (frustum shaped) or omnidirectional (sphere shaped) field of view, using colliders in Unity 3D. This module is also in charge of providing the information for the computation of the sensor coverage of the mission area. The second module is in charge of simulating the performance of the detections and classifications algorithms. The process is triggered when targets are within a sensor's field of view. The detection and classification module is out of scope for the experiment described in this paper and hence its description is not included.

OMST – pollution detection sensor
Omnidirectional sensor type

EVADS/Copting – EO/IR sensor
Directional sensor type

Fig. 8. ROBORDER sensors considered for this paper

2.3 Environment

The environment is divided into four gridded zones, the Air Column, the Water surface, the Water Column and the seabed. Each zone is broken into a series of 'data cubes' that contain all of the relevant environmental attributes. These cubes are referred to by row, column and, in case of the air and water columns, layer values (Fig. 9). The model contains publicly-available data sets that allow historical and forecast environmental conditions to be represented.

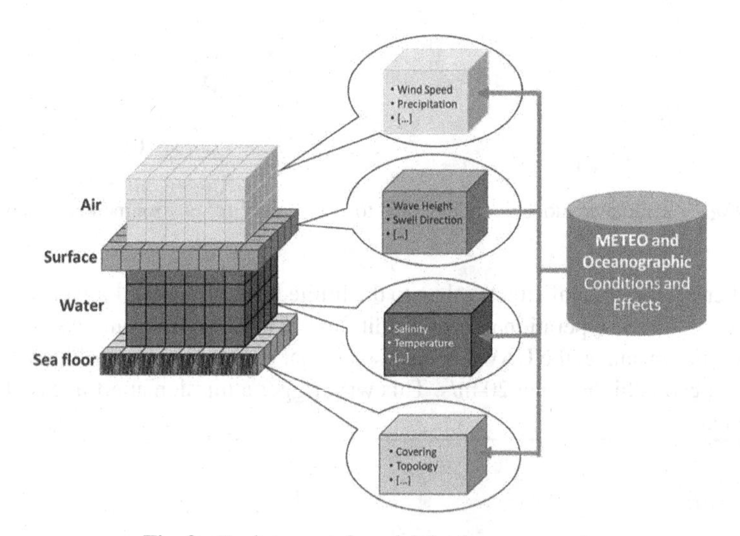

Fig. 9. Environmental model database approach

2.4 Autonomous Resource Task Coordinator Integration

The Autonomous Resource Task Coordinator (ARTC) module is responsible for designing the trajectories for all the operational vehicles, having as objective to cooperatively cover (map + monitor) a user-defined region of interest (ROI). Before the trajectory calculation, the module incorporates the following user-defined parameters:

1. The ROI to be covered in a form of polygon in WGS84 coordinates
2. Areas (inside the previously defined ROI) that should be neglected from the mission. These areas could correspond to subparts of the ROI where i) autonomous operation is not allowed (e.g. actively covered by personnel), ii) the morphology of this area is not suitable for autonomous operation (e.g. presence of obstacles), or iii) the underlying information is not important/already known (e.g. large empty spaces). Therefore, any operation inside these zones would be either unsafe or a waste of resources.
3. The number of vehicles that are going to be deployed along with their operational capabilities (e.g. battery level, maximum flight time, etc.) and their initial positions.
4. The required level of details in the acquired sensors' readings that is expressed with the scanning density parameter. In simple words, scanning density denotes the distance between two sequential sweeps.

As first step, ARTC discretizes the ROI into a number of grid-cells based on the user-defined scanning density. Special attention has been paid to properly place (position + orientation) these cells inside the ROI, having as objective to maximize the part of ROI that will be covered, if one just visits these cells. Based on this idea, an optimization process utilizing the Simulated Annealing [9] algorithm is proceeded, to maximize an optimization index that is directly connected with the coverage of the ROI (Fig. 10).

Fig. 10. ARTC functionality example

Having an optimal representation of the ROI on grid, DARP algorithm [10] takes over to carefully design the trajectories for all the autonomous vehicles so as i) all the grid-cells are covered, ii) all autonomous vehicles are efficiently utilized, iii) without any unwanted overlap, and, iv) with the minimum possible number of turns that slow down the mission and increase the energy demands. It should be highlighted that, the service does not provide "blindly" equal paths for all the autonomous vehicles, but incorporates each vehicle's characteristics (3rd bullet from the user-defined parameters) to produce paths tailored to their operational capabilities. Finally, the paths are designed in such a way to be ready for execution not only once, in a case where just a "snapshot" of the operational area is needed, but also in a continuous manner, in a form of patrolling the whole area, capable of monitoring the evolution of dynamically-evolving phenomena.

Such an energy-aware design of paths, but most importantly the efficient utilization of a team of autonomous vehicles, can be of paramount importance in the time-critical applications of ROBORDER, where complete mission awareness is needed as soon as possible.

3 Scenario

In ROBORDER, the scenario was developed in close collaboration with project partners. The development of the scenario was carried out in an iterative process, using NAF L2-L3 (scenario overview in Fig. 11) and L6 (logical sequence of events and interactions in Fig. 12) views to aid communication between communities.

The scenario, located in the bay of Thessaloniki in the Aegean Sea, has been created together with CERTH to provide quantitative means for assessing the performance of the mission planner, supporting the V&V of the ARTC. The scenario includes Fixed wing UAVs and UUVs deployed in a coverage mission.

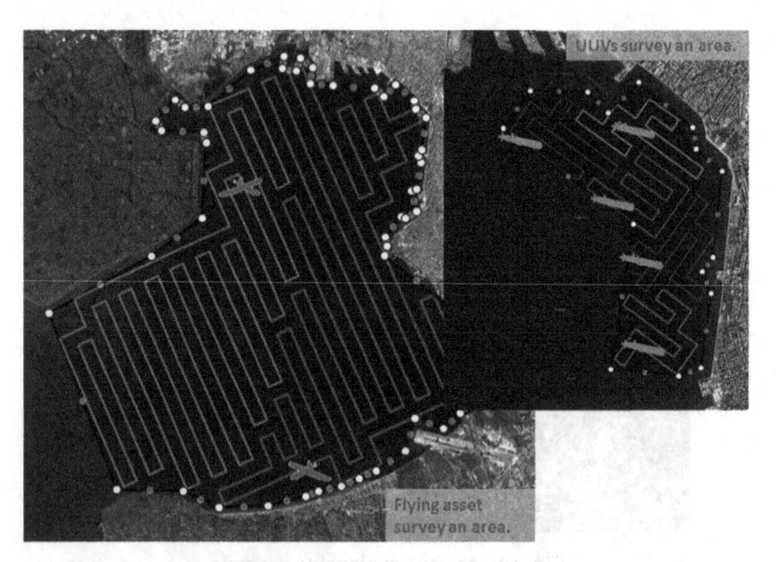

Fig. 11. NATO architectural framework L2-L3 view of the scenario

4 Results

4.1 Simulators Verification and Validation

The Verification and Validation of the models developed for the federation was conducted via a series of test runs specific for each model.

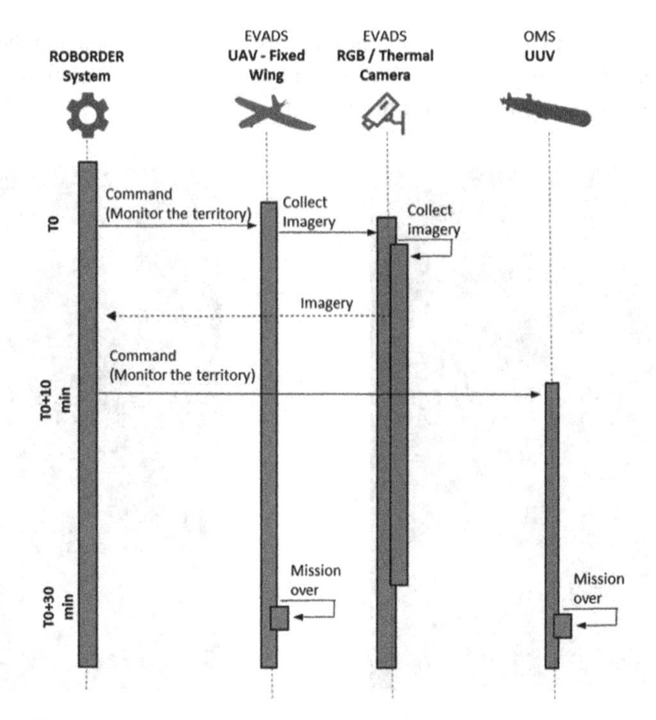

Fig. 12. NATO architectural framework L6 view of the scenario

Platform

Assets federate validation was performed by comparing the real telemetry data with simulated trajectories. The platform providers were asked to provide a mission executed by the vehicle and the relative telemetry as ground truth. The mission was executed by the simulated asset.

For the fixed wing asset type, the validation of the kinematic model has been done with the Atlantic I vehicle, provided by EVADS. The choice of the asset was made based on the completeness of the dataset provided by the assets provider; the one for Atlantic I was the most suitable for validating the kinematic model. Fig. 13 shows the ground truth (green) and the simulated trajectory (red) for the Atlantic I. Table 1 shows the error in the simulated trajectories; the performances of the models are evaluated computing:

- Maximum error: projecting the trajectories on the horizontal plane, the maximum length of the segment orthogonal to the ground truth intersecting the simulated trajectory.
- Average error in the curve: average distance (computed as above) measured where the model has the lowest accuracy (the curve between two straight paths).
- Average error over minimum mission legs distance: the ratio between the error approximating the curvature radius and the minimum curvature radius in a mission.
- Mission duration error: the error between the duration of the real and the simulated mission.

EVADS assessed that the asset simulator is very accurate reproducing the behaviour of the platforms, validating the asset simulator. Figure 13 shows the ground truth (green) and the simulated trajectory (red) for the Atlantic I.

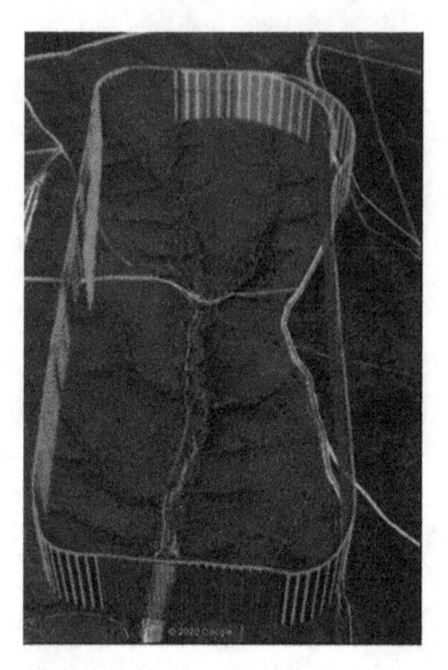
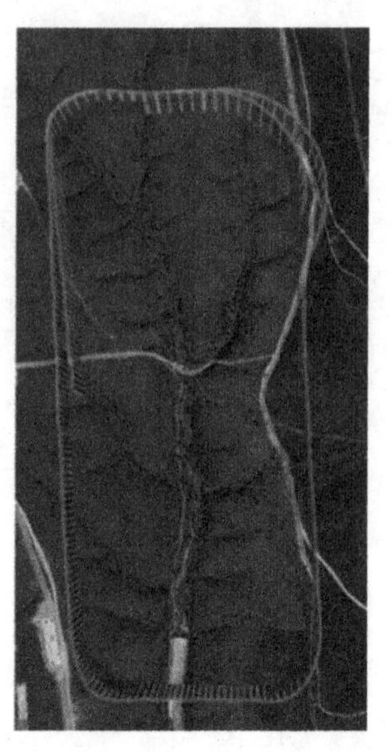

Fig. 13. Asset validation for the Atlantic I vehicle, in perspective (left) and from above (right) (Color figure online)

Table 1. Errors in kinematic model

Maximum error [m]	Average error in the curve [m]	Average error over minimum mission legs distance	Mission duration error
34.58	9.46	1.89%	2.16%

Torpedo asset type model is based on CMRE accredited models. The model was previously validated using vehicles similar to the LAUV, specifically the Bluefin Muscle; it is thus assumed that no further validation is needed for the torpedo asset type.

Sensors
The sensor models are based on the reproduction of behaviours based on configuration files customizable by simulation users.

The key activities in the test of the model have been verification tests to ensure that the dimensions representing the fields of view have a 100% match with the values provided by the technical partners.

Environment
The meteorological and oceanographic model was based on the representation of previously validated, standardised datasets.

With this in mind, the key activities in the test of the model were the verification tests to ensure that the imported data values matched those provided to the distributed simulation environments.

4.2 Experimental Set up for the ARTC

The experimental set up included two sets of missions for testing UUV and UAV missions. The main objective was to evaluate the feasibility of the missions, measuring the intersections that may occur while following the simulated trajectories, because of the way UxVs pass through turning points.

For the first set, five different missions were generated (Table 2). All of them are utilizing five UUVs, however with different scanning densities, from 50 to 300 m and different turning modes (fly-by/fly-over).

The second set of missions (Table 3) involve fixed-wing UAVs. These simulations were helpful in order to decide the best platform guidance to employ, the number of UAVs that could, or should, be utilized to cover an area and the appropriate distance between the designed trajectories (scanning density).

The evaluation of the performance for both sets is obtained through both the visualization of the trajectories of the simulated assets (example in Fig. 14) and from the analysis of the values of a set of performance indicators, down-selected from project's KPI set. The KPIs used are:

- KPI_1: Intersection. The number of times two platforms are violating the safety distance between them.
- KPI_2: Intersection duration. For each intersection, the time in seconds the platforms are violating the safety distance between them.
- KPI_3: Mission length. The amount of time in hours it takes each vehicle from the start to the end of its mission.
- KPI_4: Mission coverage. The percentage of the mission area covered by the sensor of each vehicle.

Table 2. UUV simulations

Missions:	1	2	3	4	5
Scanning Density [m]	50	100	150	200	300
Number of UUVs	5	5	5	5	5
Turn-point kind	Fly-by, Fly-over	Fly-by, Fly-over	Fly-by, Fly-over	Fly-by, Fly-over	Fly-by, Fly-over

Table 3. UAV simulations

Missions	1	2
Scanning density [m]	500	1000
Number of UAVs	1, 2	1, 2
Turn-point kind	Fly-by, Fly-over	Fly-by, Fly-over

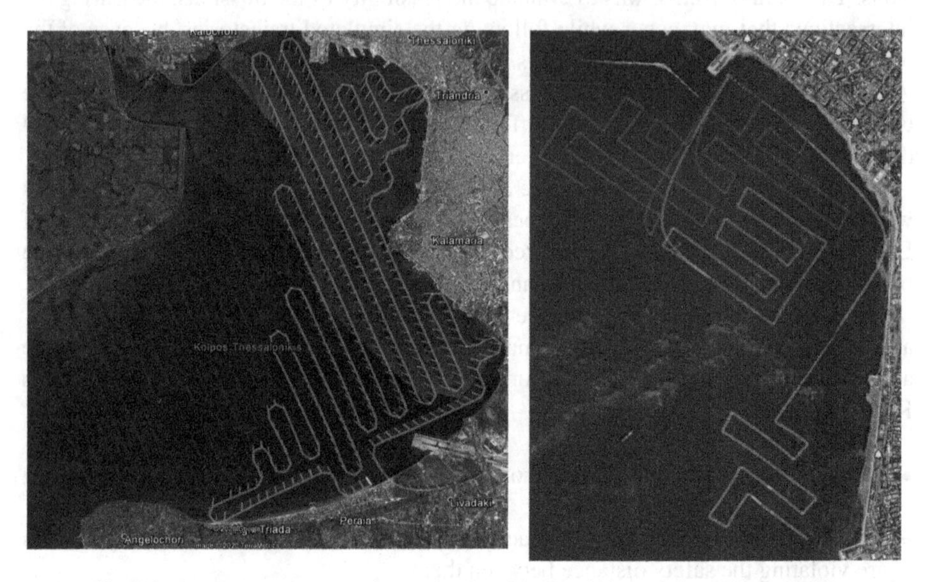

Fig. 14. Example trajectory of two simulated UAVs (left) and five UUVs (right)

4.3 Discussion of the Results

The analysis of the results of the simulations has been performed checking the values of the KPIs and investigating the trajectories of the simulated assets; the analysis showed the following:

- Trajectories intersections: Intersections may happen during transit from the launching location to the first mission waypoint. It is suggested to carefully identify the launching position based on the portion of the mission area covered by each vehicle or in case this is not possible to shift the launch time in order to avoid them. The simulations also identified that intersections might happen during the execution of the missions. This could happen for missions that were designed with no respect for the kinematic limitations of the involved vehicles (e.g. minimum turn capability). Simulation ca be used to identify missions with collision risk and reject them, as they are not safe to be executed in real world scenarios.
- Mission duration: none of the results suggested that nominal vehicle autonomy was exceeded.

- Mission coverage: simulation results identified that the mission area is completely covered, regardless of the distance between mission legs. This suggests that portions of the mission area are redundantly covered. Further investigation on this aspect showed that redundant coverage seems to be caused by unrealistic assumptions on both sensors field of view and vehicles trajectory computation within the ARTC. In order to shine a light on this phenomenon, it was suggested to include an additional KPI computing the percentage of the mission area that is covered more than once, as well as a visual representation of this phenomenon. This should allow CERTH to identify the optimal distance between mission legs, minimizing the redundant coverage.

5 Conclusions and Way Ahead

The paper describes the design and use of the M&S-based test bed capability developed for the H2020 ROBORDER Project. The work done demonstrates the potential of M&S as a tool to evaluate system performance in large-scale scenarios in realistic operational conditions. The M&S environment complies with High Level Architecture (HLA) interoperability standards and is designed according to NATO best practices.

The developments described were driven by the maritime scenario developed to support the Verification and Validation of the Autonomous Resources Task Coordinator, a mission planning and management software module developed by CERTH, for safe and efficient autonomous missions. The outcome of the research showed that simulations can significantly help to adjust the missions generated by the ARTC in order to achieve the desired coverage, without wasting operational resources, in realistic conditions that will be faced in a real-life experiment.

The results also identified areas of further development, i.e. the inclusion of KPIs for computing the coverage redundancy and the visualization of area coverage on a map for each vehicle. CMRE team is currently working on those aspects and new results are expected to be included in follow-up simulations.

CMRE team is currently working on enabling project partners to configure and run their own specific scenarios and support dedicated experiments for their specific goals.

Next steps involve the implementation of the feedback provided by CERTH to further study the performance of the ARTC module for both test scenarios and on specific scenarios that will be demonstrated in the project live demonstrations in Portugal, Hungary and Greece.

References

1. IEEE: IEEE Recommended Practice for Distributed Simulation Engineering and Execution Process (DSEEP) (2011)
2. Slater, D.M., Jacyna, G.M.: Evaluating detection and estimation capabilities of magnetometer-based vehicle sensors. In: Signal Processing, Sensor Fusion, and Target Recognition XXII, Baltimore, USA (2013)
3. Aseeri, M., Ahmed, M., Shakib, M., Ghorbel O., Hussein, S.: Detection of attacker and location in wireless sensor network as an application for border surveillance. Int. J. Distrib. Sensor Netw. **13** (2017)

4. Lee, S., et al.: Design and development of a DDDAMS-based border surveillance system via UVs and hybrid simulations. Expert Syst. Appl. **128**, 109–123 (2019)
5. Erba, S., Causone, F., Armani, R.: The effect of weather datasets on building energy simulation outputs. Energy Procedia **134**, 545–554 (2017)
6. Herrera, M., et al.: A review of current and future weather data for building simulation. Build. Serv. Eng. Res. Technol. **38**(5), 602–627 (2017)
7. Young, J.E., Crowell, A., Fabian, A.: Modelling weather in simulation and analysis. In: Integrated Communications, Navigation and Surveillance Conference (ICNS) (2013)
8. IEEE: 1516–2010 Standard for Modeling and Simulation (M&S) High Level Architecture (HLA)--Framework and Rules. IEEE (2010)
9. Kirkpatrick, S., Gelatt, C.D., Vecchi, M.P.: Optimization by simulated annealing. Science **220**(4598), 671–680 (1983)
10. Kapoutsis, A.C., Chatzichristofis, S.A., Kosmatopoulos, E.B.: DARP: divide areas algorithm for optimal multi-robot coverage path planning. J. Intell. Rob. Syst. **86**(3–4), 663–680 (2017). https://doi.org/10.1007/s10846-016-0461-x

Replacement Possibilities of the Medium-Size Truck Transport Capability by UAVs in the Disturbed Logistics Infrastructure

Petr Tulach[1,2], Pavel Foltin[2(✉)], Martin Gesvret[2], and Daniel Zlatník[3]

[1] Logio, s.r.o., Prague, Czech Republic
petr.tulach@logio.cz
[2] Department of Logistics, Faculty of Military Leadership, University of Defence in Brno, Brno, Czech Republic
{petr.tulach,pavel.foltin,martin.gesvret}@unob.cz
[3] Multinational Logistics Coordination Centre, Prague, Czech Republic
zlatind@army.cz

Abstract. The article deals with the issue of material distribution under the conditions of uncertainty in disturbed or damaged logistics infrastructure due to natural disaster or extensive industrial accident. Possibilities of material distribution using a medium-size freight truck were analysed under the condition of the limited functionality and availability of logistics infrastructure. Requirements for the load capacity, flying range and number of Unmanned Aerial Vehicles (UAVs) that could replace distribution of material by trucks within a damaged logistics infrastructure were identified based on the theoretical model through discrete simulation.

Keywords: Supply chain · Delivery capabilities · Logistics infrastructure · Unmanned Aerial Vehicles (UAV) · Simulation and modelling

1 Introductory Characteristics

Current environment is characterized by high dynamics of changes and interdependence of affecting factors. At the same time, overall societal dynamics is significantly affected by weather, natural anomalies and social, economic, cultural and security challenges (Foltin et al. 2015). Natural disasters and adverse weather conditions represent one of significantly important sources of possible disruptions (Vlkovsky et al. 2018) and negatively affected availability of logistics infrastructure for possible relief and humanitarian distribution (Dvorak et al. 2017). From the point of view of global supply chains, the most frequent natural disasters and extraordinary events not inflicted by humans are caused by weather conditions in 41%, disruptions in transport infrastructure in 27%, wildfires in 15% and environmental disasters in 9% of cases (BCI 2018). Adverse weather conditions and consequences of natural disasters have direct influence on:

- Affected inhabitants, especially from the perspective of potential loss of lives and possible impacts on human lives;

J. Mazal et al. (Eds.): MESAS 2020, LNCS 12619, pp. 139–153, 2021.
https://doi.org/10.1007/978-3-030-70740-8_9

- Property, especially economical loss in relation to the necessity to provide for the recovery of damaged infrastructure;
- Limited accessibility of the infrastructure and distribution system;
- Side-effects of the supply chain disruption, especially leverage effect and domino effect on the available logistics infrastructure.

It is possible to identify partial short-time negative impacts on the functionality and availability of individual infrastructure elements from the point of view of negative consequences of natural impacts on the availability of logistics infrastructure (Bernatik et al. 2013, Rehak and Novotny 2016). On the other hand, extreme consequences could course wide, deep, sometimes also really huge and permanent disruption of the distribution capability availability (Sedlacik et al. 2014), together with disrupting consequences to decision making process (Hodicky et al. 2020). The most adverse weather consequences in the EU countries within the 1980–2016 time period were the 2002 floods in Central Europe (over EUR 20 billion), the 2003 drought and heat wave (almost EUR 15 billion), the 1999 winter storm and October 2000 flood in Italy and France (EUR 13 billion), all expressed in 2016 values (European Environment Agency 2018).

Disasters affecting at least one million people are not frequent in Europe. Solely three such large disasters occurred between 2006 and 2015: wildfire in Macedonia in 2007, flood in the Czech Republic in 2013, and another flood in Bosnia-Hercegovina in 2014 (Guha-Sapir et al. 2017). Disasters affecting more than 100,000 people in EU were more frequent: 6 disasters during the 2006–2015 period. Future negative evolution trend and consequences of adverse weather effects have also been predicted. From approximately 2,700 deaths a year between 1981–2010, mean value of 151,500 deaths a year in 2071–2100 period have been estimated (Forzieri et al. 2017).

Seriousness of consequences of these disasters requires the following two main types of measures (Urban et al. 2017):

- Preventive measures from the point of view of stockpiles: to stock up and prepare stocks to cover all needs of different crisis scenarios, with disadvantage of tied-up personnel, material and resources enabling storing and maintaining capabilities to respond to scenarios with low frequency but large impact;
- Post-crisis measures, i.g. those after a disaster: to help affected population as quickly as possible through urgent assistance to reduce the crisis impact and suffering of the affected population.

The system of disaster relief and humanitarian aid distribution is focused on the quick distribution mainly within the first 72 h (Kovács and Spens 2007). The reason for quick distribution of material aid is to decrease the deprivation of the affected population and specially to save lives and material values. Similar supply chains goals are expected within humanitarian and military operations also (EU Military Committee 2012).

The range of possible reactions in order to mitigate the effects of crisis situation is getting much wider and the speed of reaction can be significantly increased due to the development of new technologies and bringing them to life (Foltin et al. 2018). This could be mainly the case of the use of UAVs (Unmanned Aerial Vehicles) and UGVs (Unmanned Ground Vehicles) which allow for goods transport to required positions without personnel on-board. Some UAVs have already been used in agriculture. On the market, it is possible to find wide supply of UAVs with different cargo weight capacities (even with 30 kg and more), speed, flying range or built-in system for optimization of the flight route (Zhang 2019). Furthermore, some UAVs with particular characteristics are under development, e.g. for military purposes, where the request for carrying capacity, speed and range are really high and the requirements for the operational use in wide range of weather condition are broad as well.

In the military field of UAVs research, they already proved themselves both in crisis situation for reconnaissance and observing and in combat as an air-support (Hodicky and Prochazka 2020). Logistics solutions are already available in prototype versions to test the application approaches, e.g. Israeli Defence Forces (IDF) has been testing a concept of UAV used for material and ammunition replenishment as a Vertical Take-Off and Landing (VTOL) type (Grohman 2016). On the other hand, there is a wide range of real-life applications, especially for agricultural or industrial purposes. Agriculture drones load capacity ranges from 10–30 kg for liquid and loose fertilizers with spraying cycle time of 15–18 min. These drones have higher level of autonomy, altitude microwave radars and obstacle avoidance sensors.

Based on previously mentioned findings, an appropriate next step in the research carried out was identified as focusing on some former disaster analyses based on real situations, creating a case study and analysing potential scenarios of possible improvement in increasing the distribution. Created scenario for further analysis is described in the following Sect. 2, the model and its background are presented in Sects. 3 and 4.

2 Possible Scenario of Humanitarian Distribution Logistics

Current trends in technology development, especially in the area of unmanned air and ground vehicles significantly affect their practical utilization potential for logistics purposes and humanitarian aid.

The idea to carry out research into the possibility of replacing freight trucks as the means of transport by the UAVs with higher level of autonomy (or using the combination of them) for the humanitarian aid distribution was driven by several reasons: mainly in order to shorten reactivity time and reach the required points with material as soon as possible when infrastructure is affected (damaged) by ongoing crisis. The supposed application of UAVs with higher level of autonomy offers the possibility to keep the number of required operators minimal.

For the purpose of testing the idea and comparing performance of medium-size freight trucks and UAVs, a real location in the Czech Republic was chosen. This particular location of Jizerske mountains (Jizerské hory) was chosen because it was affected by flash floods in 2010 and Czech Army garrison in Liberec was involved within the crisis system in providing humanitarian aid and transportation of necessary material and equipment due to the fact that bridges and communications were severely damaged.

Selected location (municipality of Hejnice, Libverda and Bílý Potok) hit by floods are separated from the city of Liberec by a mountain range and it is possible to reach it from two directions. This location enabled to consider different scenarios of constrains on communications. Gradual worsening of situation and gradual restoration of communications were considered in creating the scenarios.

The model consists of 3 basic transportation scenarios:

- Variant T: transport using solely medium-size freight trucks;
- Variant U: transport using solely UAVs;
- Variant C: combination of medium-size freight trucks and UAVs usage.

3 Research Goal, Methodology and Limitations

Research goal was to gain information about the humanitarian aid distribution during ongoing humanitarian crisis from the *Base* to required places within defined number of affected personnel, routes and different kind of means of transport in time for further decision-making process about the best or acceptable settings of the crisis planning.

Conceptual Modelling for Discrete-Event Simulation was the main concept for preparing the model of the situation and further experiment simulation run according to the pre-defined variants above (Robinson 2004). The overall methodological approach is based on the Distributed Simulation Engineering and Execution Process (DSEEP) (Topcu et al. 2016), in generally recommended following steps:

1. defining federation objectives;
2. performing conceptual analysis;
3. designing federation;
4. developing federation;
5. planning, integrating and testing of the federation;
6. executing federation and preparing outputs;
7. analysing data and evaluating the results.

The first step of DSEEP was preparing the model focused on designing it to be consistent with the real state (towns, villages and river banks) and later the roads between them, which were identified as the possible transport routes. Other roads were not implemented in the model in order to simplify it according to its aim and to decrease the conceptual mistake possibility due to high density of roads and transport nodes.

The main storage of the humanitarian aid, for the purpose of the research marked as the *Base,* and the requested points of delivery, marked as *Drop Points* together with their characteristics were defined simultaneously in order to make the design of the model concept complete. Moreover, means of transport were defined and descripted.

When the concept of the model was prepared, the second, third and fourth steps of DSEEP methodology were applied, when research questions and simulation objectives were formulated in order to make the behaviour of all elements more accurate and to prepare further experiments:

- How many vehicles is needed to meet all requirements and how does the change in the vehicle number used for transportation affect the time?
- How will the time change if UAVs are used instead of medium-size freight trucks or both means of transport are used? What is the best possible combination of those transport means?

Following fifth step of DSEEP in the process was the model validation. It means running the simulation within a pre-defined period of time in order to receive the simulation results and compare them to the results of computing by different methods with the same input parameters as the model. This step allowed the team to uncover some system failures and mistake, especially in some value setting, counting methods or missing/wrong connection between particular elements. When that happened and mistakes were found, the model was updated and adjusted and another validation run was carried out. This process was repeated as required until the moment when the whole model was considered to be valid.

After the validation of the model, the sixth step of DSEEP was applied, when simulation and experiments runs followed. Every run or experiment was carried out several times in order to obtain statistically reliable results due to the implementation of probability values within the behaviour of the model elements.

Finally the seventh step of DSEEP was applied, after successful finishing the runs, all the results were analysed and visualized via diagrams and graphs for their presentation and explication. Information obtained can be used for further decision-making process about the best available scenario as well as the objectives and requirements set at the beginning of modelling.

It was crucial to pay attention to several very important factors and facts about modelling and simulation, which were: the level of the model details and several limitation or assumption under which the model was built (Sherman 2003). These factors had to be carefully discussed and agreed on, mainly during the problem-solving analysis or during the conceptual model preparation phase, e.g. before creating its computer design or programming it. Each further change can be challenging or time consuming to be implemented into the already existing model.

The level of model details allows to design the model as simple as possible according to pre-defined objections, required outputs without losing important elements and links between them, however avoiding increasing the complexity of the model without any valuable output.

As mentioned above, several research questions were formulated and objective of the model with several limitations or assumptions was set:

- Scheduling for the workers involved in the transportation (drivers or UAV operators) was not limited by any working time or breaks. It was assumed that there is sufficient manpower deployed at the *Base* to be able to work 24 h per day on several shifts;
- Objectives of the model were to find answers for the number of the means of transport in order to provide transportation of the humanitarian aid to the *Drop Points* regardless of its costs. It was considered that the state is able to provide crisis budget to cover all necessary expenses;
- Number of the vehicles or drones was not limited;
- Contents of the Humanitarian aid kit was not important – it was considered that the *Base* is ready to dispatch requested amount of the "packs" as ordered and that unlimited stock is available and that it was already provided.
- There is not included further evolving rainy and windy weather conditions immediately after studied consequences of flash floods in model. These bad weather conditions do not limited possible UAVs' capabilities and applications in considered model.

– Due to the testing of the new use of the concept of transport UAVs, for which serial production has not yet been realized, the economic aspects of individual scenarios are not considered.

4 Computing Model Creation and Its Modifications

Simulation program SIMIO, as a tool for discrete event simulation, was chosen for computer modelling of the situation when the conceptual model was created. The need of creation and description of the elements and objects during preparation of the model concept, as listed in the Table 1, was identified.

Table 1. Identification of elements and objects for model purposes

Object	Role	Parameters
Base	- Supply base, which is a source of transported material	- Capacity
Drop Point	- Place where the material is requested and received for further redistribution - This object simulates a place where affected people are picking-up the humanitarian packages	- Rate of request generation, according to population of the place
Route	- Line between two nodes in the simulation model, creating the network of communications - It can be a physical route (for trucks) or an air route for UAVs	- Networking
T-810	- Medium-size freight truck, capable of driving both on concrete and unpaved roads	- Speed: 50 km/h - Payload: 5.5 t - Time to load/unload: 20 min
UAV	- Unmanned Aerial Vehicle, capable of flying on air routes	- Speed: 80 km/h - Payload: 100 kg - Time to load/unload: 5 min
Nodes	- Representing brides on the roads which can be affected by ongoing flood	- Logical value: passable/im passable

Network of Communications

Routes possibly available to be used during the simulated floods were created in the simulation model as a network of nodes and roads. To be accurate, the ESRI maps provided by the simulation software were used and the network was created in them with nodes representing the junctions and routes representing the roads and bridges (Esri 2018). For the purpose of creating different scenarios of bridges passability, different networks were created (a list of routes in the simulation software) which represent the following changes to the road passability in time, presented in Table 2.

Table 2. Road networks identification

Time	Network condition	Network name
0–24	All bridges passable	All roads
24–48	4 bridges impassable	Offbridge4
48–72	2 bridges impassable	Offbridge2
72–96	All bridges passable	All roads

The first 24 h represent the state of third stage of flood emergency, when the preparation for upcoming events takes place and *Drop Points* are supplied with initial supply. The following 24 h represent the time when the bridges collapse or are declared impassable and transport must find alternative routes. In another 24 h 2 bridges were declared passable or repaired. In the created scenario of events it took another 24 h to repair the remaining 2 bridges.

Process of Events in the Simulation

Events in the simulation start with creation of entity representing request to transport one humanitarian package to the *Drop Point*. A time stamp was applied to the entity's *Time Created* state at its creation. The entity was then transferred to the *Base* object, where it waited until a transporter was available to load it. Each loading of an entity took 1 min. Transporter waited for 60 min or until it was loaded up to 50% and left to deliver entities to their respective destinations. Unloading time was 1 min per entity. The whole system of process events is presented in Fig. 1.

Fig. 1. Process of events in the simulation

Changes in the network during the scenario were made by 4 timer elements (object in simulation software which triggers an event or process) and 4 processes which changed the current network for the freight truck object to respective network shown in Table 3 when the timer triggers.

During the simulation run, the following responses were observed:

- Time to address the request - calculated from the difference in the time of creation and the time of arrival at the *Drop Point*;
- Utilization of transporters (either trucks or UAVs) calculated as percentage of time spent dealing with transportation (loading, transporting, unloading, refuelling/changing battery) and the total time available.

Model Inputs

The requests for humanitarian packages were simulated using random exponential distribution with varying parameter of time between arrivals. Parameters are expressed in the Table 3 below. Hejnice and Bílý Potok lie on both sides of the river, so they were split into R (right) and L (left) (Fig. 2).

Table 3. Parameters of exponential distribution in time

Time	Libverda	Hejnice R	Hejnice L	Ferdinandov	Bílý Potok R	Bílý Potok L
0–24	8.5	2.7	2.7	13.4	10.4	10.4
24–48	67.9	21.3	21.3	107.1	83.4	83.4
48–72	34.0	10.7	10.7	53.5	41.7	41.7
72–96	67.9	21.3	21.3	107.1	83.4	83.4
96–120	6.8	2.1	2.1	10.7	8.3	8.3

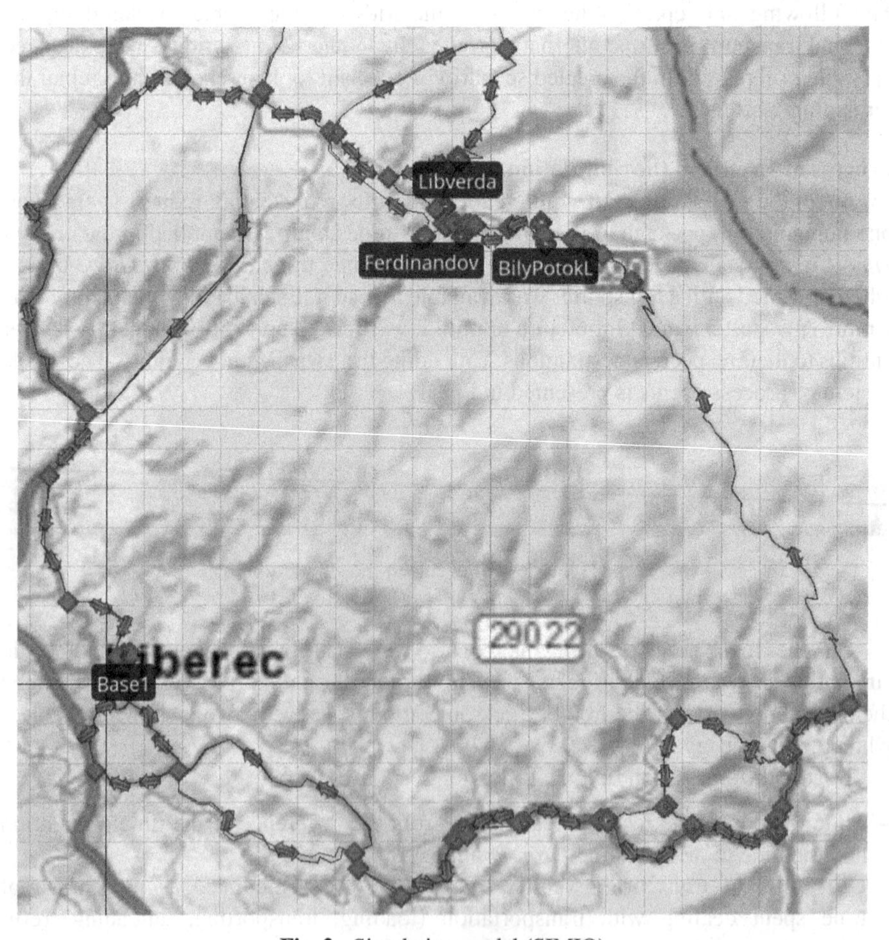

Fig. 2. Simulation model (SIMIO)

A set of scenarios was created to study the dependence of the time to serve the request on the number of UAVs and freight trucks and their availability. Each scenario was run 10 times to reach confidence levels for the results.

Results and Discussion

Average time to serve and utilization of freight trucks and UAVs were compared as model outputs in terms of the model created. The results from simulation tool SIMIO come in the form of SMORE plots. SIMIO Measure of Risk & Error (SMORE) plot displays both the estimated expected value of a scenario and multiple levels of variability behind the expected value. A SMORE plot consists of a Mean, Confidence Interval for the Mean, Upper Percentile Value, Confidence Interval for the Upper Percentile Value, Lower Percentile Value, Confidence Interval for the Lower Percentile Value, Median, Maximum Value, and Minimum Value, see Fig. 3.

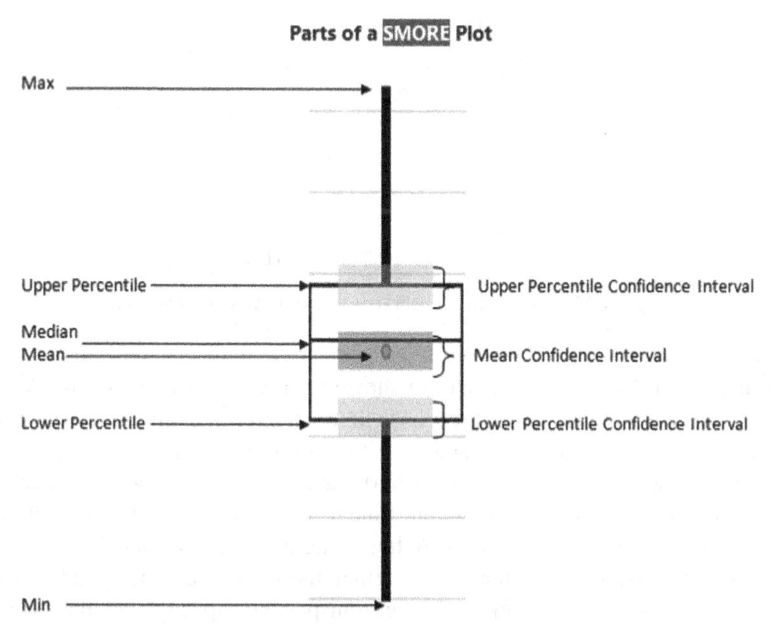

Fig. 3. Sample of the presentation of the results. Source: (Sturrock 2013)

Altogether, 50 replications of simulation model were used to evaluate the scenarios with regard to different random events.

In the chart below, comparison of scenarios with UAVs (Dx) and freight trucks (Tx) can be seen.

Usage of 6 UAVs was compared to use of six medium-size trucks. The observed response of the model (maximum time to serve) was consistent after adding more than 6 vehicles with decreasing utilization which drops to range between 20 and 25%. Using more vehicles did not provide additional benefit in decreasing the reaction time to transport requests.

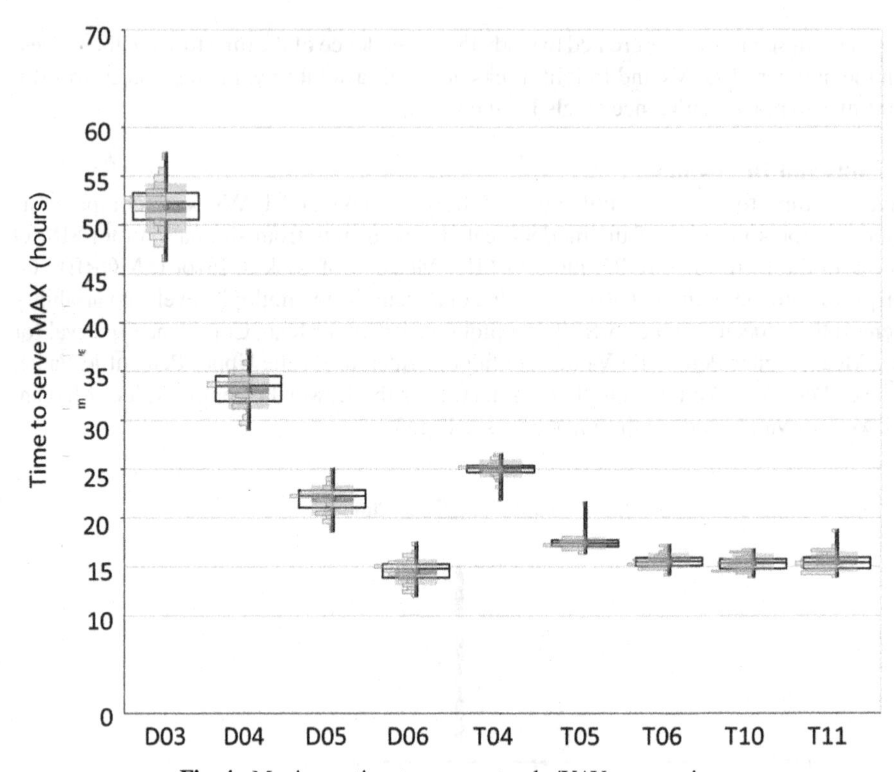

Fig. 4. Maximum time to serve – trucks/UAVs comparison

The maximum time to serve response showed that with 6 UAVs or 6 trucks it was possible to achieve less than 20 h of response time. The average time to serve was lower with UAVs - 4.7 h on average compared to 8.7 h for 6 trucks (Fig. 4).

The results show that by using UAVs shorter average time to serve can be achieved, however, when considering the maximum time to serve, use of both UAVs and trucks are on the same level. This is caused by higher agility of UAVs and less time spend waiting for enough material to load when utilize the UAV. It can be stated that UAVs are more agile, however, they lack the higher transporting capacity, which is needed on days, when there are many requests for transport.

The idea of utilizing strengths of both modes of transport was considered. Combined scenarios to measure if addition of limited number of UAVs has positive impact on overall performance, was created.

From Fig. 8 it can be found out that combined fleet of 4 medium-size freight trucks and 3 UAVs can achieve better results than 6 medium-size freight trucks in average time to serve, while only slightly increasing the maximum time to serve. Overall results are highlighted in Table 4.

It is important to mention the utilization of both types of vehicles. As can be seen on Fig. 6 and Fig. 7, UAV's utilization decreases with the number of vehicles in fleet and decreases very strongly when used in combined fleets with the trucks. The reason for such behaviour comes from the fact that UAVs are used as a flexible part of the fleet and if not necessary, are left waiting if it is possible to meet the demand using standard truck (Figs. 4, 5 and 9).

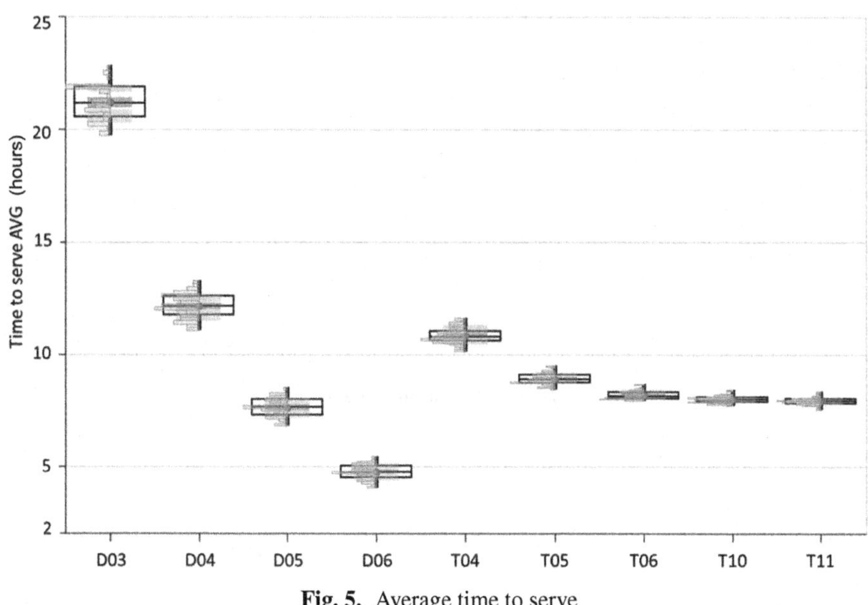

Fig. 5. Average time to serve

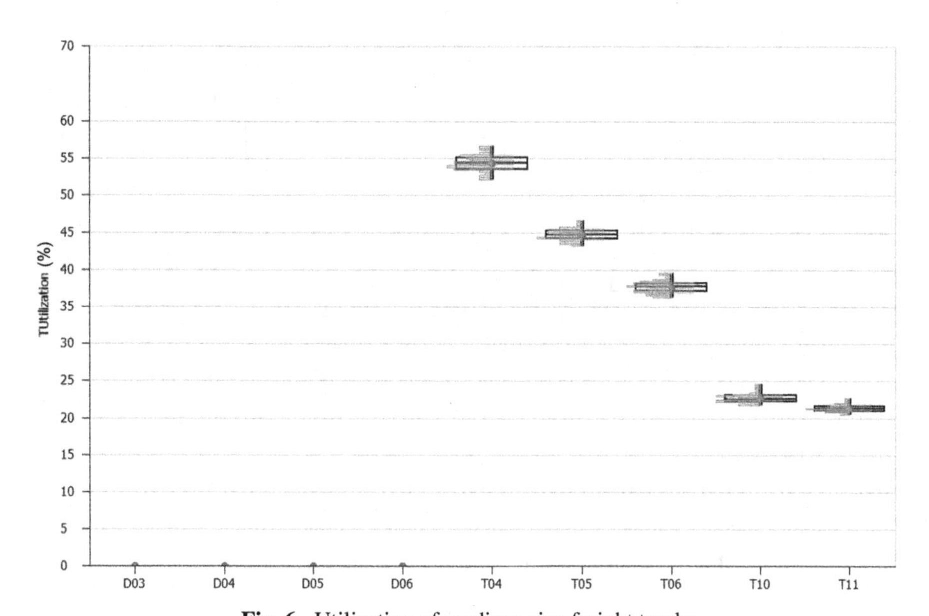

Fig. 6. Utilization of medium-size freight trucks

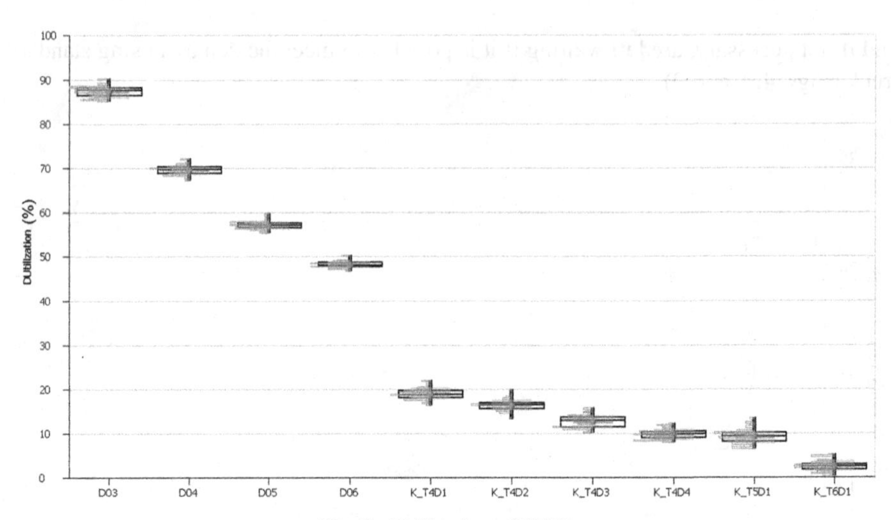

Fig. 7. Utilization of UAVs

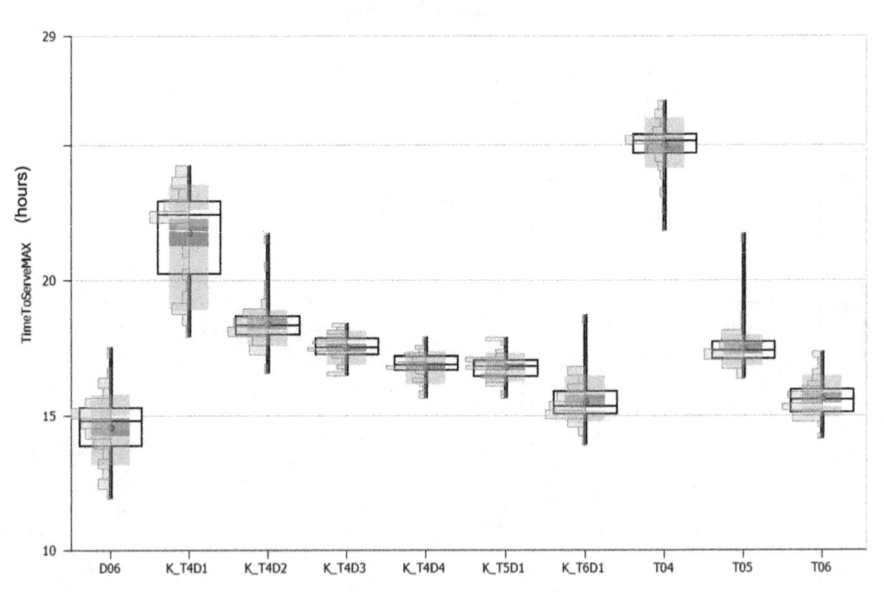

Fig. 8. Time to serve - Maximum Combined approach

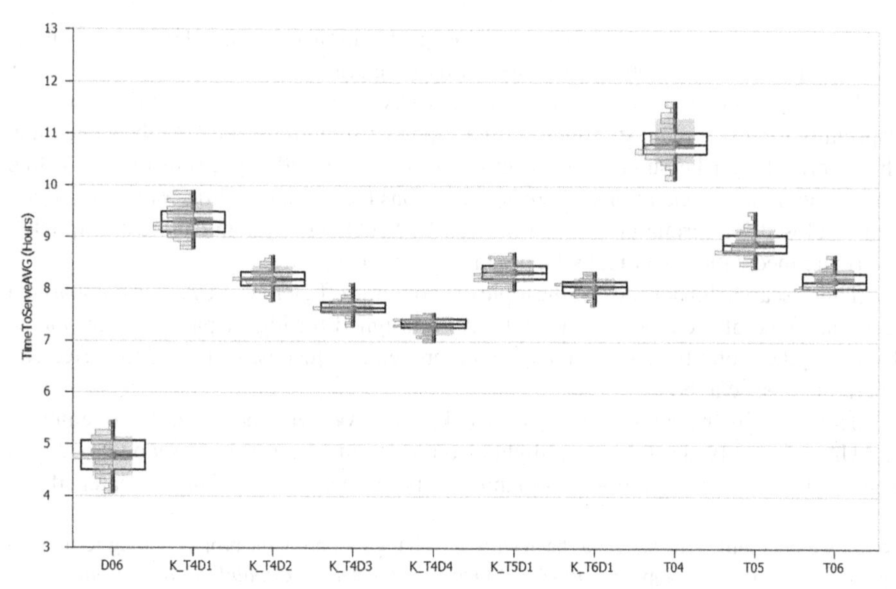

Fig. 9. Time to serve - Average Combined approach

Table 4. Overall model results comparison

Scenario	Time to Serve MAX	Time to Serve AVG	UAV utilization [%]	Truck utilization [%]
D03	51.9	21.2	87.4	–
D04	33.2	12.1	69.7	–
T04	25.0	10.8	–	54.4
D05	22.0	7.6	57.1	–
K_T4D1	21.8	9.3	19.0	51.8
K_T4D2	18.4	8.2	16.4	49.9
K_T4D3	17.5	7.6	12.7	49.0
T05	17.4	8.9	–	44.8
K_T4D4	16.8	7.3	9.8	48.6
K_T5D1	16.8	8.3	9.2	43.7
T06	15.6	8.2	–	37.7
K_T6D1	15.5	8.0	2.6	38.5
T11	15.5	7.9	–	21.4
T10	15.4	8.0	–	22.8
D06	14.6	4.8	48.2	–

5 Conclusions

The model used in the research proved that there are possibilities to improve humanitarian logistics using the UAVs. The limited payload of UAV represents both its strength and weakness. It was observed, that adding the UAVs to the fleet of medium-size trucks can significantly reduce the average time to serve to the transportation requests. The trucks seem more efficient when the volume of transport is higher than that UAVs can handle, which in the scenario applies to the first 24 h of the crisis.

Using combinations of freight trucks and UAVs can help to reduce time to serve especially for lesser requests and make the supply system more agile. At the same time, this approach can be suitable to be utilized under the wide spectrum of crises, like anti-epidemiologic operations where some regions (e.g. municipalities or city districts) can be closed for certain period of time and UAV delivery system can help to quickly distribute medication, medical kits, or similar urgent supplies.

This research did not address the topic of economic efficiency of UAVs in comparison to trucks. One of the reasons was that UAVs designed for higher payload transport in harsh weather conditions have not yet been proven and just estimates for the price and operating costs can be used.

The idea of using combined fleets of trucks and UAVs with the possibility of employing UAVs for only short distance flights from previously established secondary *Bases* deployed from the freight trucks is planned to be further developed and researched.

Acknowledgement. The paper has been written with the support of the project of long-term strategy of organization development: ROZVOLOG: *Development of Capabilities and Sustainability of Logistics Support* (DZRO ROZVOLOG 2016–2020) funded by the Ministry of Defence of the Czech Republic.

References

BCI: The BCI. BCI Supply Chain Resilience Report 2018. The BCI, Caversham, The Business Continuity Institute, 1 September 2018. https://www.thebci.org/. Accessed 18 Nov 2018

Bernatik, A., Senovsky, P., Senovsky, M., Rehak, D.: Territorial risk analysis and mapping. In: 14th Symposium on Loss Prevention and Safety Promotion in the Process Industries, LP 2013, vols. I and II, AIDIC SERVIZI SRL, Florence, pp. 79–84 (2013). https://doi.org/10.3303/CET1331014

Dvorak, Z., Sventekova, E., Rehak, D., Cekerevac, Z.: Assessment of critical infrastructure elements in transport. In: TRANSBALTICA 2017: Transportation Science and Technology, pp. 548–555. Elsevier Science, Vilnius (2017). https://doi.org/10.1016/j.proeng.2017.04.413

ESRI: World street map, 2 October 2018

EU Military Committee: EU Concept for Reception, Staging, Onward Movement and Integration (RSOI) for EU-led Military Operations. EU Military Commettee, Brussels (2012)

European Environment Agency: Economic losses from climate-related extremes (European Union), 27 February 2018. European Environment Agency. https://goo.gl/uxzsj7. Accessed 19 Nov 2018

Foltin, P., Gontarczyk, M., Swiderski, A., Zelkowski, J.: Evaluation model of the companies operating within logistic network. Arch. Transp. **36**(4), 21–33 (2015). https://doi.org/10.5604/08669546.1185196

Foltin, P., Vlkovský, M., Mazal, J., Husák, J., Brunclík, M.: Discrete event simulation in future military logistics applications and aspects. In: Mazal, J. (ed.) MESAS 2017. LNCS, vol. 10756, pp. 410–421. Springer, Cham (2018). https://doi.org/10.1007/978-3-319-76072-8_30

Forzieri, G., Cescatti, A., Batiste de Silva, F., Feyen, L.: Increasing risk over time of weather-related hazards to the European population: a data-driven prognostic study. LANCET Planet. Health **2017**(5), e200–e208 (2017). https://doi.org/10.1016/S2542-5196(17)30082-7

Grohman, J.: Cormorant: Robotický létající náklaďák pro vojáky i civilisty. Hybrid.cz, 21 Nov 2016. https://www.hybrid.cz/cormorant-roboticky-letajici-nakladak-pro-vojaky-i-civilisty. Accessed 27 Nov 2018

Guha-Sapir, D., Hoyois, P., Wallemacq, P., Below, R.: Annual Disaster Statistical Review 2016: The Numbers and Trends. The International Disaster Database (2017). https://www.emdat.be/sites/default/files/adsr_2016.pdf. Accessed 10 Sep 2018

Hodicky, J., Prochazka, D.: Modelling and simulation paradigms to support autonomous system operationalization. In: Mazal, J., Fagiolini, A., Vasik, P. (eds.) MESAS 2019. LNCS, vol. 11995, pp. 361–371. Springer, Cham (2020). https://doi.org/10.1007/978-3-030-43890-6_29

Hodicky, J., Özkan, G., Özdemir, H., Stodola, P., Drozd, J., Buck, W.: Dynamic modeling for resilience measurement: NATO resilience decision support model. Appl. Sci. **10**(8), 1–10 (2020). https://doi.org/10.3390/app10082639

Kovács, G., Spens, K.: Humanitarian logistics in disaster relief operations. Int. J. Phys. Distrib. Logist. Manage. **37**(2), 99–114 (2007). https://doi.org/10.1108/09600030710734820

Rehak, D., Novotny, P.: Bases for modelling the impacts of the critical infrastructure failure. Chem. Eng. Trans. **2016**(1) (2016). https://doi.org/10.3303/CET1653016

Robinson, S.: Simulation: The Practice of Model Development and Use. Wiley, Chichester (2004)

Sedlacik, M., Odehnal, J., Foltin, P.: Classification of terrorism risk by multidimensional statistical methods. In: International Conference on Numerical Analysis and Applied Mathematics (ICNAAM). American Institute of Physics Inc. (2014). https://doi.org/10.1063/1.4912948

Sherman, N.: A Stochastic Model for Joint Reception, Staging, Onward Movement, and Integration (JRSOI). Air Force Institute of Technology, Wright-Patterson Air Force Base (2003)

Sturrock, W.K.-J.-D.: Simio and Simulation: Modeling, Analysis, Applications: Economy. CreateSpace Independent Publishing Platform, Scotts Valley (2013)

Topcu, O., Durak, U., Oguztuzun, H., Yilmaz, L.: Distributed Simulation. Springer, New York (2016). https://doi.org/10.1007/978-3-319-03050-0

Urban, R., Oulehlová, A., Malachová, H.: Computer simulation - efficient tool of crisis management. In: International Conference Knowledge-Based Organization, pp. 135–141. "Nicolae Balcescu" Land Forces Academy, Sibiu (2017)

Vlkovsky, M., Koziol, P., Grzesica, D.: Wavelet based analysis of truck vibrations during off-road transportation. In: The 14th International Conference on Vibration Engineering and Technology of Machinery (VETOMAC XIV), MATEC Web of Conferences, Lisbon (2018).

Zhang, F.: Electronic consultations to usability of TT Aviation Technology Co., Ltd products, 18 September 2019. (P. Foltin, Interviewer)

Experimental Leg Inverse Dynamics Learning of Multi-legged Walking Robot

Jiří Kubík, Petr Čížek(✉), Rudolf Szadkowski, and Jan Faigl

Department of Computer Science, Faculty of Electrical Engineering, Czech Technical University in Prague, Technická 2, 166 27 Prague 6, Czech Republic
{kubikji2,petr.cizek,szadkrud,faiglj}@fel.cvut.cz
https://comrob.fel.cvut.cz/

Abstract. Rough terrain locomotion is a domain where multi-legged robots benefit from their relatively complex morphology compared to the wheeled or tracked robots. Efficient rough terrain locomotion requires the legged robot sense contacts with the terrain to adapt its behavior and cope with the terrain irregularities. Usage of inverse dynamics to estimate the leg state and detect the leg contacts with the terrain suffers from computational complexity. Furthermore, it requires a precise analytical model identification that does not cope with adverse changes of the leg parameters such as friction changes due to the joint wear, the increased weight of the leg due to the mud deposits, and possible leg morphology change due to damage. In this paper, we report the experimental study on the locomotion performance with machine learning-based inverse dynamics model learning. Experimental examining three different learning models show that a simplified model is sufficient for leg collision detection learning. Moreover, the learned model is faster for calculation and generalizes better than more complex models when the leg parameters change.

1 Introduction

Enhanced terrain traversability of multi-legged robots [15] stems from their relatively complex morphology but comes at the cost of complex locomotion control [19]. A critical part of the multi-legged locomotion control is the robot state estimation, including timely and reliable tactile sensing to detect the leg contact with the ground or obstacles [2,4,10]. The leg contact detection and leg-state estimation, i.e., assessing whether the leg is supporting the body or not, are essential in maintaining the attitude of the robot in complex terrains [2,8], and for the accuracy of the legged-odometry [3,4,10,12]. Further, the foot-contact detection is utilized to synchronize oscillations in controllers based on neural oscillators [1,5] or to trigger reflexive behaviors [5,7].

Model-based locomotion control methods [17] use inverse dynamics model in a contact detection [9]. Their applicability in real-world mobile robotic applications may be cumbersome due to the difficulty of accurately determining the various kinematic and dynamic parameters of such analytical models. It

J. Mazal et al. (Eds.): MESAS 2020, LNCS 12619, pp. 154–168, 2021.
https://doi.org/10.1007/978-3-030-70740-8_10

can be especially expected in deployments with increasingly complex scenarios [11,16], where robots might struggle in challenging environments, and their characteristics might change significantly. Automated parameter identification and online adaptation of models are beneficial strategies because they can capture non-stationarities in the mechanical properties of the robot [18]. Such non-stationarities include the adverse changes of the leg parameters, e.g., friction changes due to the joint wear, increased weight of the leg due to the mud deposits, or leg morphology change because of damage.

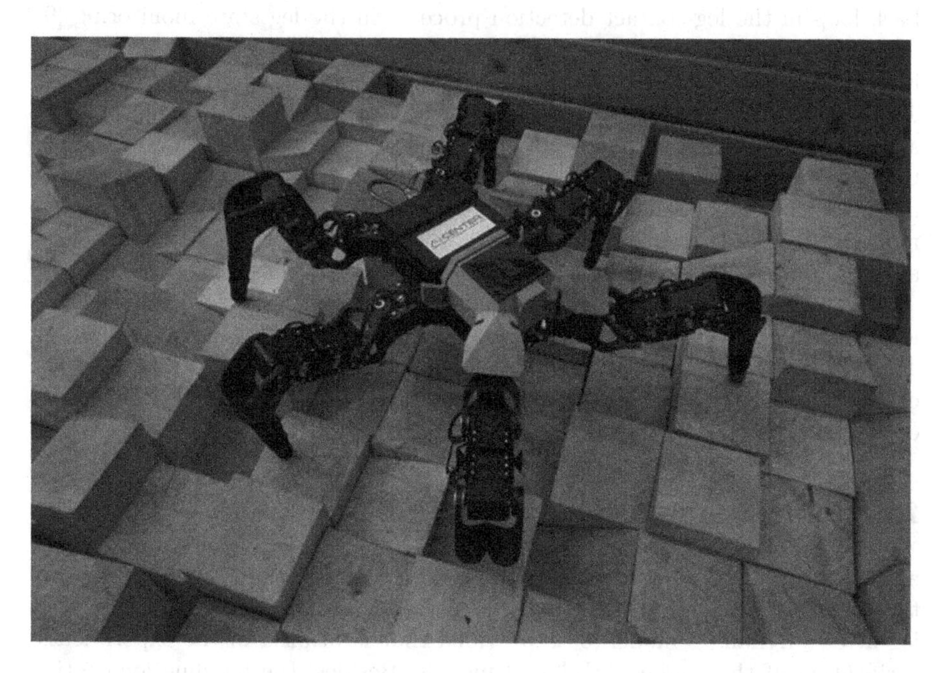

Fig. 1. Hexapod walking robot SCARAB (Slow-Crawling Autonomous Reconnaissance All-terrain Bot) used for the experimental study of machine-learned inverse dynamics.

This work reports on the experimental study of the machine learning (ML) based inverse dynamics model learning to the locomotion performance of a small affordable hexapod walking robot SCARAB shown in Fig. 1. The employed ML approaches include linear regression, second-order polynomial regression, and a three-layered neural network, each learned by the real motion data collected using the experimental hexapod walking platform. The linear and polynomial regression can be considered as statistical methods; however, in the experimental evaluation, we prefer lightweight techniques suitable for online model learning instead of methods that require extensive training datasets [14]. The performance of the learned models has been examined with a focus on the following aspects.

1. Comparison of ML-based inverse dynamics models with the baseline analytical model [6].

2. Model performance w.r.t. the size of the training dataset.
3. Computational complexity of model learning and prediction.
4. Robustness of the learned model to a non-stationary environment.
5. Performance of the collision detection integrated with the ML-based model compared to the baseline approach [6].

The main challenge of the addressed problem is to learn the leg inverse dynamics model to predict the future state of the leg using the current motion command and the current leg state. The prediction is used to close the feedback loop in the leg-contact detection process via the leg state monitoring [9]. ML approaches are well applicable to the addressed problem, as the leg state is influenced by numerous factors, including the previous trajectory of the leg. The performed experiments indicate that the learned dynamic model provides a similar performance of inverse dynamics regression and collision detection as the baseline dynamic model [6], but it is computationally less demanding. The results show a more reliable prediction of the learned model than the baseline model when the leg parameters change, which supports the idea of ML-based adaptive online incrementally learned locomotion controller.

The remainder of the paper is organized as follows. Sect. 2 details the studied problem and briefly presents the baseline model [6] used in the evaluation. The examined ML regressors are briefly described in Sect. 3. Results on the experimental deployment are reported in Sect. 4. Finally, Sect. 5 is dedicated to concluding remarks.

2 Problem Statement

addressed problem is to learn the robot leg inverse dynamics model to predict the collision-free motion of the leg. In this section, a background of the robot leg inverse dynamics model [6] is provided that is utilized in the experimental verification of the studied ML-based models. Besides, the baseline locomotion controller is briefly described in Sect. 2.2.

2.1 Leg Inverse Dynamics Model

The inverse dynamics can be modeled analytically using Euler-Lagrange formulation [17] for the vector of the generalized n-dimensional coordinates $\mathbf{q} = \{\theta_1, \theta_2, \cdots, \theta_n\}$, corresponding to the leg joint angles

$$\boldsymbol{D}(\boldsymbol{q})\ddot{q} + \boldsymbol{C}(\boldsymbol{q}, \dot{\boldsymbol{q}})\dot{q} + \boldsymbol{G}(\boldsymbol{q}) = \boldsymbol{\tau}, \tag{1}$$

where $\boldsymbol{D}(\boldsymbol{q})$ is the *inertia matrix* of the chain of the rigid bodies, $\boldsymbol{C}(\boldsymbol{q}, \dot{\boldsymbol{q}})$ is a tensor representing the *centrifugal* and *Coriolis* effects induced on the joints, $\boldsymbol{G}(\boldsymbol{q})$ is the vector of moments generated at the joints by the *gravitational acceleration*, and $\boldsymbol{\tau}$ is the vector of actuation torques at the respective joints. All the terms $\boldsymbol{D}(\boldsymbol{q}), \boldsymbol{C}(\boldsymbol{q}, \dot{\boldsymbol{q}})$, and $\boldsymbol{G}(\boldsymbol{q})$ depend on a set of parameters that has to be identified. The most influencing parameters, w.r.t. the precision of the inverse dynamic model, are the leg links inertia matrices and the estimated frictions in the leg joints. Due to the complexity of the calculation and measurement of the inertia matrices, simplified models such as point mass and rigid-rod models are used for the model calculation, which introduces error into the prediction of the inverse dynamics. Moreover, the inertia matrices are most influenced by the non-stationarities that may occur during the robot deployment.

In our particular case of SCARAB, the servo motors provide only the position feedback. Furthermore, the torque nor the electric current is measured, which can be utilized for joint torque estimation. Therefore, an additional step in the inverse dynamics modeling is necessary. The real behavior of the actuator composed of the motor and reduction gear is modeled together with the underlying servo motor controller. The dynamic model is given by

$$J\ddot{q}^M + B\dot{q}^M + F(\dot{q}^M) + R\tau = K\,V, \tag{2}$$

where q^M is the rotor position angle before the reduction, J is the rotor inertia, B is the rotor damping, F is the sum of the static, dynamic, and viscous frictions that depend on the current rotor speed, R is the gearbox ratio, τ is the servo motor torque, K is the back electromotive force, and V is the motor voltage. The appropriate values of J, B, F, R, and K have to be experimentally identified using the real servo motor and the values specified in the manufacturer datasheet.

The servo motor controller is modeled as the P-type position controller, which sets the voltage as $V = k_P \cdot err$, where k_P is the controller gain, and err is the difference between the set position and the current position of the actuator. The controller operates with 1 kHz frequency. The complete model of the leg inverse dynamics in the joint angles can be derived by substituting (2) into (1).

The major issue of the analytical inverse dynamic model is the numerous joint-related and link-related parameters that have to be identified before using the inverse dynamics model. The identification process and parametrization of the baseline model are detailed in Sect. 4.

2.2 Hexapod Robot Locomotion Controller

The inverse dynamics model is utilized in the position tracking controller [6] that executes the leg trajectory step-by-step. At each step, the controller reads the current joint angles and compares them to the predicted values provided by the inverse dynamic model. The actuator is iteratively commanded with a new desired position θ_{des}, and the tracking continues until the difference between the real measured position θ_{real} and the position estimated by the model θ_{est} is above the threshold ϵ_{thld} that indicate a tactile event is recognized. This simple principle allows for terrain negotiation and rough terrain locomotion even with affordable multi-legged platforms with the position feedback only. However, the performance of the locomotion controller tightly depends on the precision of the inverse dynamics model and identification of its parameters. Therefore, we aim to employed ML-based techniques for estimating the leg inverse dynamics model to avoid the cumbersome identification of the parameters needed in the analytical model. The ML-based methods considered in our experimental study are described in the following section.

3 Learning-Based Inverse Dynamics Models

The main motivation behind using the ML-based model of the leg inverse dynamics is to overcome the cumbersome identification of the analytical model. For the considered SCARAB, 18 sets of joint and 18 sets of link parameters have to be found. Additional parameter changes are introduced by the non-stationary electrical and mechanical characteristics of the servo motor that change due to the heating up, gearbox wear-out, and variations in the link shape and mass caused by imperfect 3D printing and also environmental effects. A robust robotic system should overcome parameter variations, but the analytical inverse dynamic model lacks such ability as it requires an online parameter identification step of the adaptive control [17]. Therefore we prospect ML techniques to learn the model.

In the experimental evaluation, we focus on lightweight ML techniques that do not require extensive training datasets like deep-learning-based techniques [14]. We consider three ML approaches: (i) Ordinary Least Squares regression (OLS) further referred to as the *linear regressor*; (ii) Ordinary Least Squares regression with second-order polynomial features denoted the *polynomial regressor*, and (iii) three-layer feedforward neural network with Rectified Linear Unit (ReLU) activation function further referred to as *ReLU regressor*. The used learning input is formed from the n most recent triplets of the discrete position measurements accompanied by the triplets of the desired positions set to the servo motors further considered with the known baud rate. The regressors are trained to predict the leg dynamics for m steps to the future. The second-order differential equations (2) are used for the robot leg dynamics. The value of the dynamic variables can be estimated from at least three recent position samples, but we use $n = 4$ the most recent measurements. The expected leg position is predicted two-steps-ahead $m = 2$, as possible delay can occur in the data flow pipeline, and predictions into a more distant future are losing accuracy.

The regressors have been implemented in Python with the Scikit-learn library [13] for the linear and polynomial regressors, whereas the ReLU regressor uses Chainer framework [20] with 100 neurons in the hidden layer and Leaky ReLU activation function. The ReLU regressor hidden layer size has been selected randomly as the hyper-parameter search would require extensive testing. The main aim of this work is to experimentally validate the concept of inverse dynamics learning for the small legged robot. The performance of regressors compared to the baseline analytical model [6] is reported in the following section.

4 Experimental Evaluation

The performance of the three regressors of the leg inverse dynamics has been validated in the experimental deployment scenarios with the hexapod walking platform SCARAB shown in Fig. 1. SCARAB is an affordable six-legged robot with 18 controllable degrees of freedom, actuated by 18 Dynamixel AX-12A servo motors. Three servo motors per each leg are named according to the entomology nomenclature (from the body to foot-tip): coxa, femur, and tibia. Each Dynamixel AX-12A actuator enables position control with the internal P-type controller and provides reading its current position at the limited rate of 1000 Hz. All the experiments have been performed using the laptop computer with the dual-core Intel Core i5-3320M CPU @ 2.60 GHz, 16 GB RAM without GPU acceleration, running Ubuntu 18.04 Bionic Beaver operating system with the ROS melodic.

Table 1. Mechanical properties of SCARAB

Product	Variable	Measurement	Unit	Description
Coxa	a_{c}	52	mm	Coxa link length
CoM 1	a_{cc}	25	mm	Coxa link center of mass position
Mass 1	m_{c}	20	gm	Coxa link mass
Femur	a_{f}	66	mm	Femur link length
CoM 2	a_{cf}	20	mm	Femur link center of mass position
Mass 2	m_{f}	115	gm	Femur link mass
Tibia	a_{t}	132	mm	Tibia link length
CoM 3	a_{ct}	50	mm	Tibia link center of mass position
Mass 3	m_{t}	62	gm	Tibia link mass

The baseline analytical model [6] is parameterized using mechanical properties as in Table 1 utilized to calculate $\boldsymbol{D}, \boldsymbol{C}$, and $\boldsymbol{G}$ of (1). The rigid rod simplified model has been used to calculate the inertia matrices. The dynamic model defined by (2) has been parameterized by values from experimental identification based on measured two reference positions for the actuator moving forth and back without load for different control voltage. The identified minimum voltage is $v_{\min} = 0.5\,\mathrm{V}$ that defines the maximal static friction as $F \simeq (k/R_{\mathrm{a}})v_{\min}$, where $k = 3.07 \cdot 10^{-3}\mathrm{N\,m\,A^{-1}}$ is the back EMF constant, and $R_{\mathrm{a}} = 6.5\,\Omega$ is the motor resistance, which can be found together with the gearbox ratio $R = 1/254$ in the actuator datasheet. The values of the parameters have been estimated using the minimum square root method with Euler's method employed in the solution of (2). The identified parameters of the Dynamixel AX-12A are listed in Table 2.

The experimental examination of the regressors is based on the off-line processed datasets collected using SCARAB. A single leg data has been utilized as all the legs share the same morphology apart from minor differences in the servo motor orientation and offset angles. Nine datasets have been collected, capturing different leg movements with various induced non-stationarities that alter the leg parameters. The individual datasets listed in Table 3 and the made leg modifications are depicted in Fig. 2.

Table 2. Dynamic model parameters of the Dynamixel AX-12A

Parameter	Value	Unit
J	$1.032 \cdot 10^{-7}$	$\mathrm{Kg\,m^2}$
B	$3.121 \cdot 10^{-6}$	N m s
F	$2.369 \cdot 10^{-4}$	N m
R	$3.937 \cdot 10^{-3}$	-
K	$3.912 \cdot 10^{-3}$	$\mathrm{N\,m\,A^{-1}}$

Table 3. List of collected datasets

#	Dataset name	Dataset size [n]	Induced modifications
1	Vanilla	55 333	No modifications
2	halved(t)	27 859	Weight of tibia link reduced by 12 g (see Fig. 2a)
3	weight(t)	27 033	Weight of tibia link increased by 31 g (see Fig. 2b)
4	weight(f)	27 603	Weight of femur link increased by 31 g (see Fig. 2c)
5	loosen(t)	26 994	Tibia link freely moving regardless of tibia servo position
6	rubber(t)	27 277	Tibia joint load increased with rubber band (see Fig. 2d)
7	rubber(f)	27 273	Femur joint load increased with rubber band (see Fig. 2d)
8	rubber(f,t)	27 493	Merge of rubber(t) and rubber(f) setups (see Fig. 2f)
9	collision	100	Different obstacles placed in the pathway of the leg to induce leg collisions

The datasets 1 and 2–8 have been collected using 2000 and 1000 randomly chosen target points within the leg's operational space, respectively, and interpolating the path between the targets with the maximum allowed step size of 0.4 mm, which is transferred into the joint coordinates using inverse kinematics. The path in joint coordinates is then executed in the open-loop by commanding the leg servo motors with the desired joint angles. For all the datasets, the desired joint angle θ_{des} and the real (measured) joint angle θ_{real} were collected from the daisy-chained leg servo motors at the highest possible sampling rate of 100 Hz. The Ordinary Least Squares method is used for training *linear regressor* and *polynomial regressor*, whereas the *ReLU regressor* has been trained using backpropagation.

The performance of the learned regressors is studied in five benchmarks focused on: (1) model precision, (2) model generalization to the cases with the

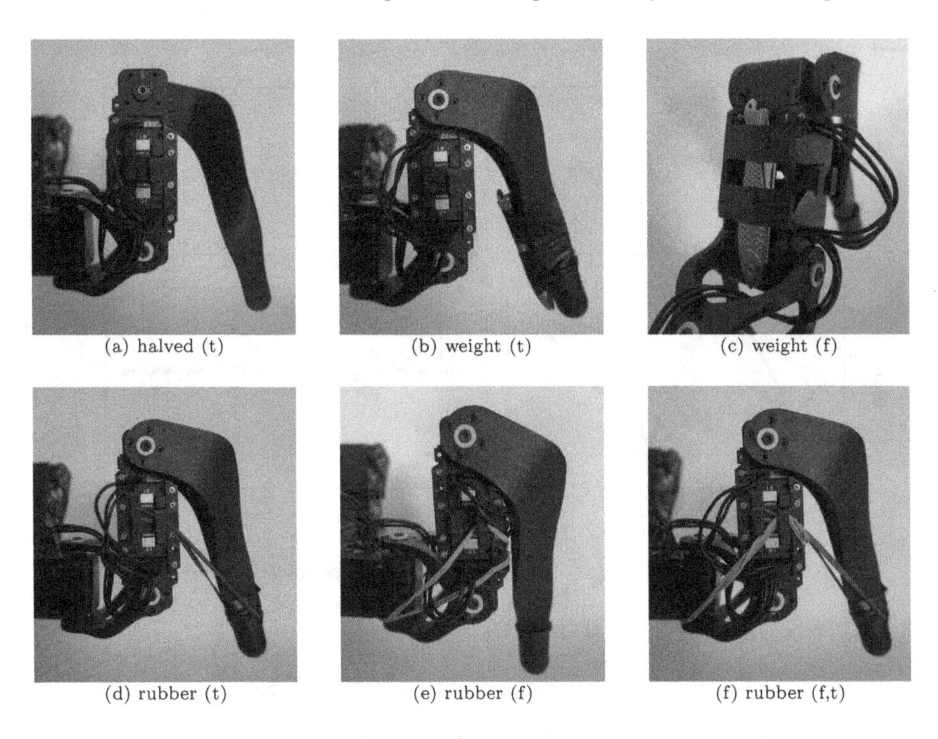

Fig. 2. Leg modifications to simulate non-stationarities and alter leg parameters.

induced non-stationarities, (3) size of the training set, (4) computational requirements, and (5) the final deployment in the leg contact detection scenario. In each benchmark, the trained regressors are requested to process the collected time-series testing data per individual sample. The testing error is calculated as the difference $\theta_{\text{err}} = |\theta_{\text{est}} - \theta_{\text{real}}|$ between the one-step look ahead regressor prediction and the corresponding real measured error. The cumulative mean absolute error (MAE) is then used to report the results.

Table 4. Cumulative mean absolute prediction error

Method	**Cumulative Mean Absolute Prediction Error** [rad]
Baseline dynamic model	0.0170
Linear regressor	0.0069
Polynomial regressor	0.0068
ReLU regressor	0.0075

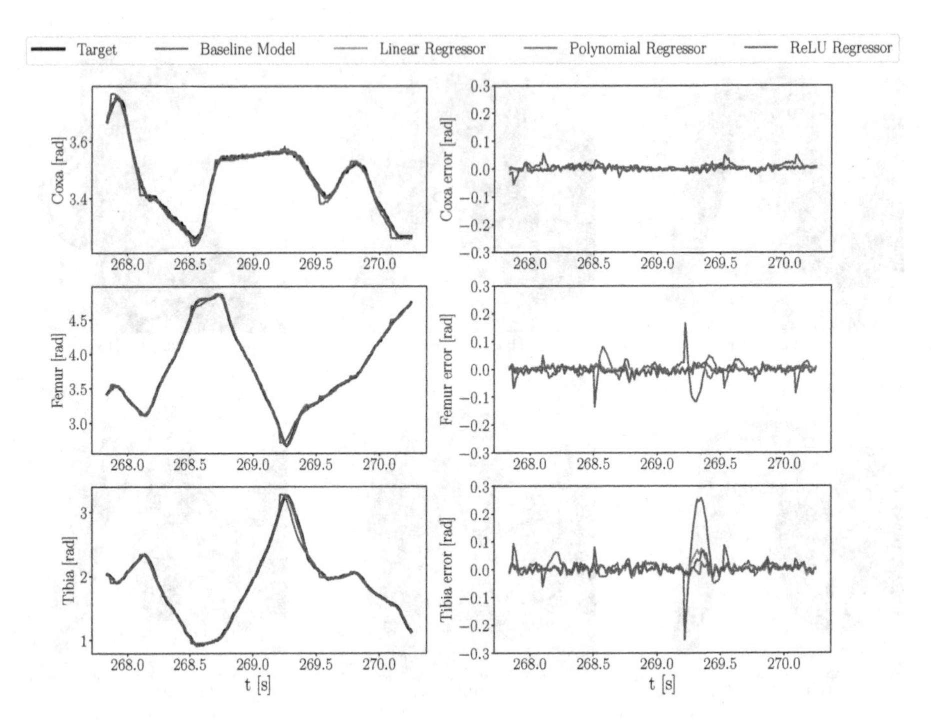

Fig. 3. Example of the estimated leg trajectory in joint angles (left column) and the corresponding prediction accuracy calculated as $\theta_{\text{err}} = |\theta_{\text{est}} - \theta_{\text{real}}|$ (right column) for considered regressors. In the presented example, the leg follows a random trajectory. The MAE of θ_{err} is used to report the results.

Model Precision has been studied on the regressors learned on the vanilla dataset and compared to the base analytical model. The vanilla dataset has been divided into training and test data with a 0.5:0.5 ratio. The cumulative mean absolute errors are depicted in Table 4, and an example of the estimate positions and the prediction error is shown in Fig. 3. The results indicate that considered ML approaches cope better with the leg position estimation than the baseline model [6].

Generalization ability has been examined using regressors learned using the vanilla dataset that has been then utilized for prediction using the datasets 2 to 8 collected on a modified leg mimicking parameter changes. For each scenario, the cumulative mean absolute error over all three servo motors has been computed to examine how regressors generalize leg dynamics and handle changes in its parameters. The results presented in Fig. 4 indicate that the ML-based approaches perform better compared to the baseline model.

Size of the Training Set influences the quality of the prediction. Besides, a new dataset can be collected when the model becomes inaccurate during the deployment, and the regressor can be retrained in an online learning fashion. Learning

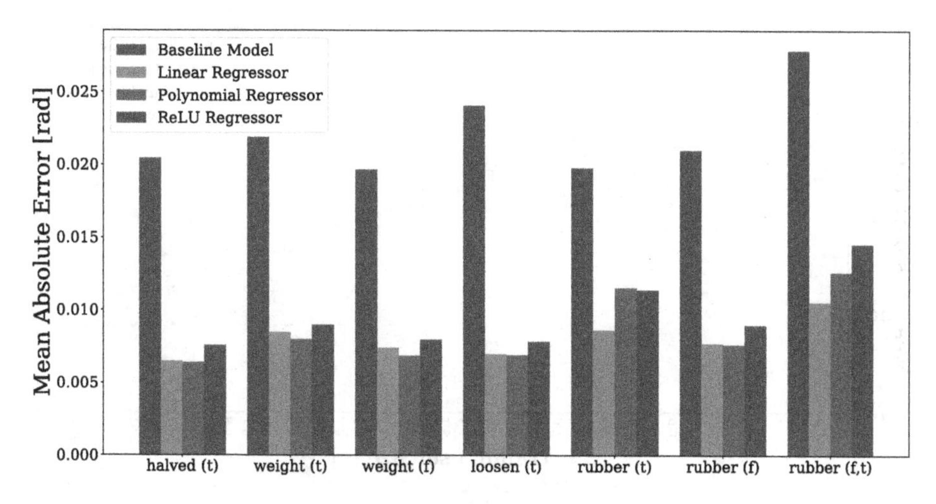

Fig. 4. Mean absolute prediction error of the regressors learned using the vanilla dataset in scenarios with differently modified leg morphology.

from a relatively small batch of data is desirable to enable relearning from data collected in the field. We examine the mean prediction accuracy based on the size of the training set. Since the servo motor joint angle is periodically read at the rate $\Delta t = 10\,\text{ms}$, it is possible to directly compute how long it takes to collect a dataset with a particular number of samples. Hence, the size of the vanilla dataset has been utilized to create a sequence of logarithmically increasing time intervals of training data corresponding to the period 0.1 s to 30 s. For each such time interval of m samples, a random starting point has been selected within the range $[0, n-m]$, where the n is the number of samples in the dataset. Following m samples have been selected from the vanilla dataset to learn the regressors initialized at random. Ten independent trials have been performed to examine the cumulative mean absolute prediction error. The five-number summary shows the minimum value, lower quartile, median value, upper quartile, and maximum value. The cumulative error per trial is computed using prediction error for all three servo motors of the leg. The influence of prediction error on the training set size is shown in Fig. 5.

The reported results suggest that the size of the training set required to surpass the baseline model by the learned regressor significantly depends on the particular regressor as both the error and its variance decrease with the size of the training set. The ReLU regressor seems to be unsuitable for online learning because a competitive performance with the baseline model is achieved with the considerably large training dataset, which is likely caused by the size of the hidden layer. On the other hand, for the linear and polynomial regressors, it takes only a few seconds of the collected data to surpass the laboriously crafted baseline dynamic model [6].

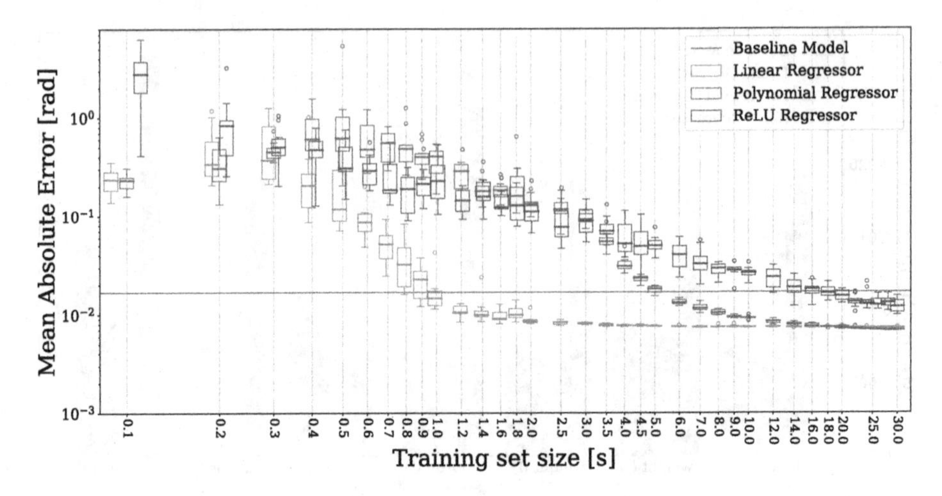

Fig. 5. Cumulative mean absolute prediction error for training set of different size. The shown five-number summary is computed from ten independent trials. Note both axes are in the logarithmic scale.

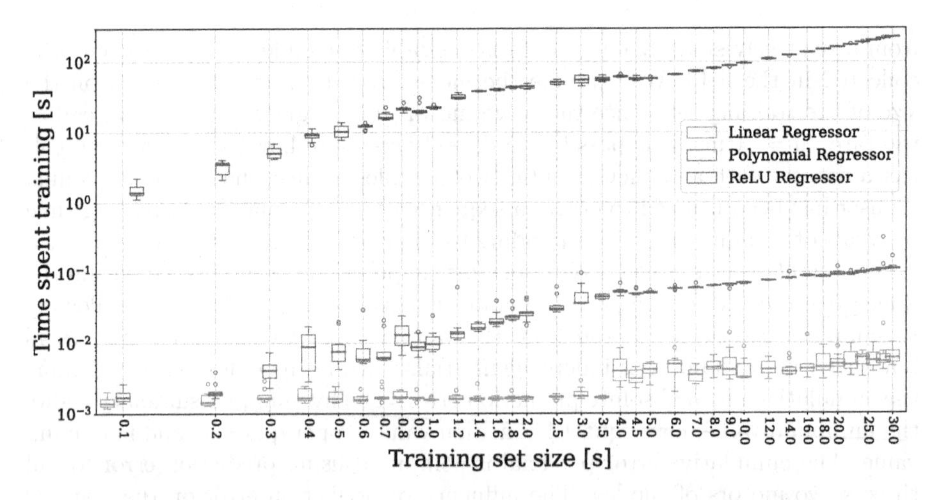

Fig. 6. Required computational time to train the examined regressors based on the size of the training set. The real computational time is shown as the five-number summary.

Computational Requirements are essential when the method is deployed onboard of the walking robots as computational-demanding methods increase power consumption and decrease the operational time. The required computational time for learning the changed-parameter leg dynamics increases the training time and slows down the average robot speed in online learning. The time spent in prediction might increase the gait control period and thus also decrease the robot speed. The training depends on the size of the training set. Therefore, the required computational time for regressors training has been examined using the vanilla dataset with logarithmically increasing time intervals of the training data starting at 0.1 s to 30 s. The plot of the five-number summary of the required computational time is depicted in Fig. 6. The mean required computational time for prediction using the baseline model and learned regressors is listed in Table 5.

Table 5. Mean required computational time for position prediction

Method	**Prediction time** [μs]
Baseline dynamic model [6]	6.7
Linear regressor	**0.5**
Polynomial regressor	13.8
ReLU regressor	9.5

The results indicate that real computational requirements are insignificant even for relatively large input data in the linear and polynomial regressors. The ReLU regressor is about several orders of magnitude more demanding because of the underlying backpropagation.

Contact Detection represents a practical use case of the position prediction that enables the legged robot to negotiate the terrain. In this setup, the leg follows a circular trajectory with the diameter 10 cm, regularly sampled to 100 data points, with period 1 s. The trajectory has been performed in six trials. During the first trial, denoted $\mathcal{T}_1$, the leg followed the trajectory freely without any collision. The collected data has been then used for the detection of leg contact with an obstacle. The contact is detected whenever the prediction error $\theta_{\text{err}} = |\theta_{\text{est}} - \theta_{\text{real}}|$ is above the predefined threshold value $e_{\text{thld}} = 0.052$ rad. An obstacle has been placed into the leg trajectory for all other trials causing the leg to collide at different trajectory parts. For the trials $\mathcal{T}_2$, $\mathcal{T}_3$, and $\mathcal{T}_4$, only the foot-tip has been in contact with the obstacle. For $\mathcal{T}_5$ and $\mathcal{T}_6$, the collision occurred with the femur link. The course of the position error θ_{err} shown up to the collision detection is visualized in Fig. 7.

The presented results suggest that the linear and polynomial regressors provide similar performance to the baseline dynamics model. In the descending part of the trajectory, these regressors predict the collisions using a few samples of the baseline model. The linear regressor reports the collision sooner than the baseline model. During the ascending phase of the circular movement, the errors of the regressors' prediction exceed the threshold a few samples late than the baseline. On the other hand, the ReLU regressor failed in all scenarios, which is most likely because the vanilla dataset size is not large enough to train the ReLU model with its 100 neurons in the hidden layer properly, albeit the main prediction error is lower than the baseline model, as shown in Table 4 and Fig. 5. As the comparison in Table 4 is based on the mean absolute error, it may cover up erroneous behavior that will only become apparent in the collision detection experiment.

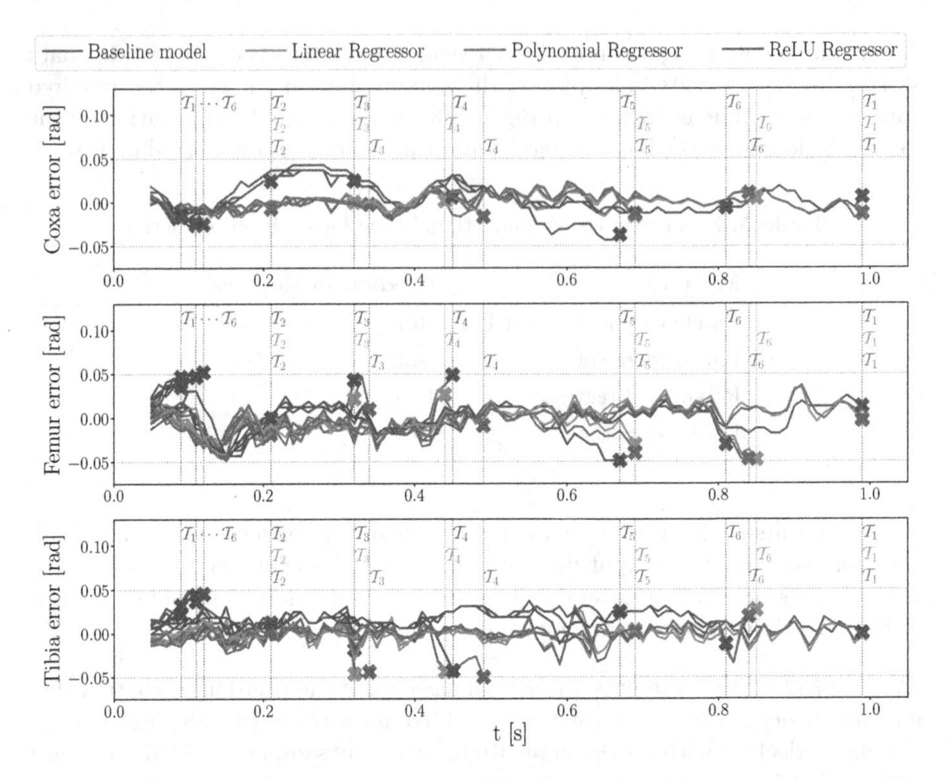

Fig. 7. Plots of the prediction accuracy calculated as $\theta_{\text{err}} = |\theta_{\text{est}} - \theta_{\text{real}}|$ of each leg's joints for the particular trials shown up to the collision detection using the threshold value $e_{\text{thld}} = 0.052\,\text{rad}$. The first trial $\mathcal{T}_1$ is an obstacle-free trajectory. An obstacle has been placed at a different part of the trajectory in the five other trials $\mathcal{T}_2, \ldots, \mathcal{T}_6$. The annotated vertical lines represent the contact of the corresponding regressor and trial with the respective color-coding.

5 Conclusion

Three learning-based approaches for inverse dynamics model learning of hexapod walking robot have been examined and compared with the baseline analytical dynamic model. Based on the reported results from five evaluation scenarios, the performance of the learned models is competitive to the baseline model, which requires laborious identification of the proper values of the model parameters. The learned models achieved higher precision than the baseline approach, and all the regressors demonstrate generalization to changes in the leg properties. The linear and polynomial regressors further show satisfactory performance for the practical deployment in the collision detection scenario. As our future work, we plan to deploy the regressors for online learning in real-life environments.

Acknowledgment. The presented work has been supported under the OP VVV funded project CZ.02.1.01/0.0/0.0/16_019/0000765 "Research Center for Informatics" and the Czech Science Foundation (GAČR) under research Project 18-18858S.

References

1. Aoi, S., Manoonpong, P., Ambe, Y., Matsuno, F., Wörgötter, F.: Adaptive control strategies for interlimb coordination in legged robots: a review. Front. Neurorobot. **11**, 39 (2017). https://doi.org/10.3389/fnbot.2017.00039
2. Bledt, G., Wensing, P.M., Ingersoll, S., Kim, S.: Contact model fusion for event-based locomotion in unstructured terrains. In: IEEE International Conference on Robotics and Automation (ICRA), pp. 4399–4406 (2018). https://doi.org/10.1109/ICRA.2018.8460904
3. Bloesch, M., Gehring, C., Fankhauser, P., Hutter, M., Hoepflinger, M.A., Siegwart, R.: State estimation for legged robots on unstable and slippery terrain. In: IEEE/RSJ International Conference on Intelligent Robots and Systems (IROS), pp. 6058–6064 (2013). https://doi.org/10.1109/IROS.2013.6697236
4. Camurri, M., Fallon, M., Bazeille, S., Radulescu, A., Barasuol, V., Caldwell, D.G., Semini, C.: Probabilistic contact estimation and impact detection for state estimation of quadruped robots. Robot. Autom. Lett. **2**(2), 1023–1030 (2017). https://doi.org/10.1109/LRA.2017.2652491
5. Chung, H.Y., Hou, C.C., Hsu, S.Y.: A CPG-inspired controller for a hexapod robot with adaptive walking. In: Automatic Control Conference (CACS), pp. 117–121. IEEE (2014). https://doi.org/10.1109/CACS.2014.7097173
6. Faigl, J., Čížek, P.: Adaptive locomotion control of hexapod walking robot for traversing rough terrains with position feedback only. Robot. Auton. Syst. **116**, 136–147 (2019). https://doi.org/10.1016/j.robot.2019.03.008
7. Focchi, M., Barasuol, V., Havoutis, I., Buchli, J., Semini, C., Caldwell, D.G.: Local reflex generation for obstacle negotiation in quadrupedal locomotion. Nature-Inspired Mobile Robotics, pp. 443–450 (2013). https://doi.org/10.1142/9789814525534_0056
8. Focchi, M., del Prete, A., Havoutis, I., Featherstone, R., Caldwell, D.G., Semini, C.: High-slope terrain locomotion for torque-controlled quadruped robots. Auton. Robots **41**(1), 259–272 (2016). https://doi.org/10.1007/s10514-016-9573-1
9. Haddadin, S., De Luca, A., Albu-Schaffer, A.: Robot collisions: a survey on detection, isolation, and identification. IEEE Trans. Robot. **33**(6), 1292–1312 (2017). https://doi.org/10.1109/TRO.2017.2723903
10. Hwangbo, J., Bellicoso, C.D., Fankhauser, P., Hutter, M.: Probabilistic foot contact estimation by fusing information from dynamics and differential/forward kinematics. In: IEEE/RSJ International Conference on Intelligent Robots and Systems (IROS), pp. 3872–3878 (2016). https://doi.org/10.1109/IROS.2016.7759570
11. Kolvenbach, H., Wisth, D., Buchanan, R., Valsecchi, G., Grandia, R., Fallon, M., Hutter, M.: Towards autonomous inspection of concrete deterioration in sewers with legged robots. J. Field Robot. (2020). https://doi.org/10.1002/rob.21964
12. Lubbe, E., Withey, D., Uren, K.R.: State estimation for a hexapod robot. In: IEEE/RSJ International Conference on Intelligent Robots and Systems (IROS), pp. 6286–6291 (2015). https://doi.org/10.1109/IROS.2015.7354274
13. Pedregosa, F., et al.: Scikit-learn: machine learning in Python. J. Mach. Learn. Res. **12**, 2825–2830 (2011). https://doi.org/10.5555/1953048.2078195

14. Polydoros, A.S., Nalpantidis, L., Krüger, V.: Real-time deep learning of robotic manipulator inverse dynamics. In: IEEE/RSJ International Conference on Intelligent Robots and Systems (IROS), pp. 3442–3448 (2015). https://doi.org/10.1109/IROS.2015.7353857
15. Raibert, M., Blankespoor, K., Nelson, G., Playter, R.: BigDog, the rough-terrain quadruped robot. IFAC Proc. Vol. **41**(2), 10822–10825 (2008). https://doi.org/10.3182/20080706-5-KR-1001.01833
16. Rouček, T., et al.: DARPA subterranean challenge: multi-robotic exploration of underground environments. In: Mazal, J., Fagiolini, A., Vasik, P. (eds.) MESAS 2019. LNCS, vol. 11995, pp. 274–290. Springer, Cham (2020). https://doi.org/10.1007/978-3-030-43890-6_22
17. Siciliano, B., Sciavicco, L., Villani, L., Oriolo, G.: Robotics: Modelling, Planning and Control. Springer Science & Business Media, London (2010). https://doi.org/10.1007/978-1-84628-642-1
18. Sigaud, O., Salaun, C., Padois, V.: On-line regression algorithms for learning mechanical models of robots: a survey. Robot. Auton. Syst. **59**(12), 1115–1129 (2011). https://doi.org/10.1016/j.robot.2011.07.006
19. Tedeschi, F., Carbone, G.: Design issues for hexapod walking robots. Robotics **3**(2), 181–206 (2014). https://doi.org/10.3390/robotics3020181
20. Tokui, S., et al.: Chainer: A deep learning framework for accelerating the research cycle. In: 25th ACM SIGKDD International Conference on Knowledge Discovery and Data Mining, pp. 2002–2011 (2019). https://doi.org/10.1145/3292500.3330756

Multi-UAV Mission Efficiency: First Results in an Agent-Based Simulation

Julian Seethaler(✉), Michael Strohal, and Peter Stütz

Institute of Flight Systems, University of the Bundeswehr Munich, 85577 Neubiberg, Germany
{julian.seethaler,michael.strohal,peter.stuetz}@unibw.de

Abstract. To assess the mission effectiveness and efficiency of future airborne systems-of-systems with autonomous components, appropriate performance models and metrics are required. In a first attempt, such metrics were derived systematically by subject-matter experts (SMEs) decomposing a specific mission and its tasks hierarchically, until measurable criteria were reached. Weights of the criteria were obtained through the Fuzzy Analytic Hierarchy Process (FAHP), using linguistic variables for the pairwise comparison of criteria on all decomposition levels.

This work demonstrates determination of such metrics in a multi-agent 2D simulation. The results are aggregated according to their respective weight, generating the mission run's total evaluation.

An air-to-ground operation by multiple unmanned aerial vehicles (UAVs) with pop-up threats present was chosen as an example mission. Deploying different force packages, their UAV models having different characteristics, in the simulation yields distinct aggregated evaluation outcomes. These signify differences in efficiency and thus suitability for the selected mission.

For investigation of result robustness local sensitivity analysis is used. For validation it will be required to have the SMEs compare their assessments of the conducted missions to the generated results.

Keywords: Agent-based simulation · AHP · Metrics · Systems effectiveness · Systems of systems · UAV

1 Introduction

Future airborne systems will consist of multiple interacting aerial vehicles with a certain degree of interdependence, e.g. in manned-unmanned teaming (MUM-T) with unmanned aerial vehicles (UAV), "mosaic warfare" [1], or even fully autonomous swarms with emergent behavior. But also currently missions are already conducted by simpler composites of cooperating aircraft, such as wingman configurations. These complex systems are called *systems of systems* (SoS) because the composite force is made up of individual distributed interacting and interdependent systems, i.e. the single aircraft.

Having reliable and meaningful measures of quality (MoQ) is relevant in (military) operations research (OR) [2], requirements engineering, platform development,

J. Mazal et al. (Eds.): MESAS 2020, LNCS 12619, pp. 169–188, 2021.
https://doi.org/10.1007/978-3-030-70740-8_11

for objective comparisons in aircraft procurement and deployment decisions, but also as utility functions for artificial intelligence (AI) and machine learning [3]. Generally, they are needed whenever one wants to judge or assess a system and its effectiveness or success. As effectiveness for e.g. reconnaissance missions is expected to benefit significantly from cooperation of multiple UAVs [4], relevant figures of merit are required especially in the context of systems of systems. Additionally, mission-objective based discovery of the best configurations and compositions of collaborating and cooperative autonomous systems is an ongoing effort [5], for which again appropriate MoQs are needed to determine and quantify what "better" or "best" actually means.

For objectively and reliably assessing systems of systems, only considering traditional single-aircraft measures of performance (MoP) such as turn or climb rates is not sufficient. Quantifying how effectively an aircraft has fulfilled its task(s), however, is possible for single aircraft and disaggregated systems alike. A system of systems' function and thereby effectiveness relies massively and non-linearly on the respective interactions and cooperation of its elements. In reality, their behavior also directly and indirectly depends on the dynamic mission environment, such as opposing forces' actions and interactions with the distributed subsystems, that can impact their respective contribution to the mission tasks. A prominent example is the sharing of sensor data via a communications channel, which could be partially or fully jammed.

Detailed mission-based evaluation methods have been developed for single aircraft [6, 7]. However, the involved manifold interdependences within composites of aircraft make it necessary to judge outcomes only rather than (technical) input parameters, because from these it cannot reliably be foreseen whether the desired effects will be achieved adequately. The primary intent behind the use of a single aircraft or a distributed system in a mission, after all, is to achieve a certain outcome or result.

Therefore, we propose a framework for comprehensive, transparent and traceable quantitative assessment of systems of systems in terms of effectiveness and efficiency, that is based on mission results. Here, by mission efficiency we mean the considered systems' effectiveness versus effort and cost in the mission-specific context. The overall effectiveness and efficiency of a system of systems can then be gained by aggregating the mission-based results for all relevant scenarios.

The method can be used and is demonstrated here in the context of constructive simulation, which can provide insights of predictive nature for systems under development, but in principle also allows for assessment after real-life missions. Due to the focus on mission outcomes it is equally valid for any number of own (blue) and opposing (red) systems. This paper shows how the method is applied to a specific mission in a multi-agent simulation.

2 Concept

The overall idea is depicted in Fig. 1: The left side illustrates the many cases covered by the agent-based simulation, on the manned-unmanned spectrum in one dimension and the homogeneity-heterogeneity spectrum of the considered force packages in the other dimension. By employing agent-based simulation, all configurations and combinations, such as "one vs. Many" or "many vs. Many", can be assessed.

The right part of Fig. 1 shows the process, how from the specific mission(s) multi-criteria decision analysis (MCDA) methods can be used by stakeholders and subject-matter experts (SMEs) to derive metrics. In aggregated form these yield the total quality integral (or sum) J over all agents. The result of the quality integral numerically represents the figure of merit of the executed mission and thus the effectiveness of the system in this run. The relative importance of the metrics is described mathematically by weightings that are determined from the expert knowledge of the participants during the metric derivation process prior to all measurements.

Finally, the central closed loop in Fig. 1 stands for the iterative use of Monte Carlo (MC) sampling of the same mission vignette to gain a stable average value of J., indicating the system's performance rather than possible outliers e.g. due to random circumstances. The J value in turn can be used for optimization of the mission runs in further simulation replications by employing different mission plans and/or changing the system of systems' configuration by altering the number of agents or changing technical parameters of the entities such as equipment. Multi-dimensional optimization can be achieved using Pareto Fronts [8].

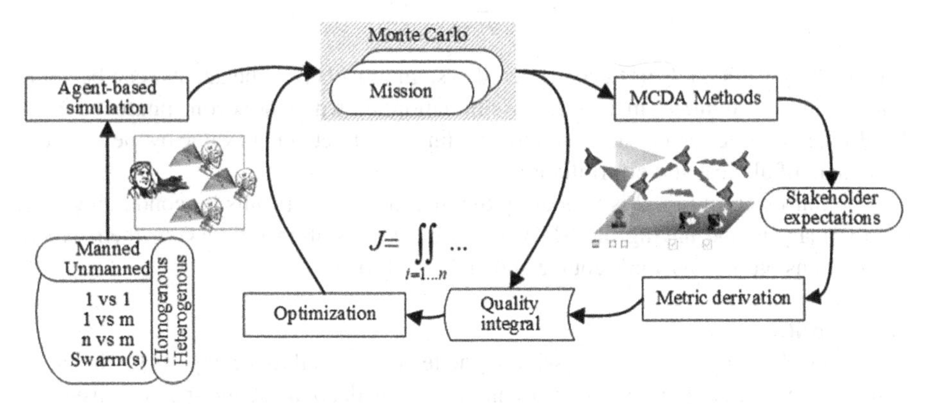

Fig. 1. General concept: derivation and use of metrics in systems-of-systems simulation

By aggregating the effectiveness values of one considered cooperative system over all relevant scenarios, ideally with optimized mission plans, a thorough quantified *capability* assessment can be achieved. Also termed *ability*, this is a numerical indication of how well the system is able to perform the required tasks.

In addition to capability, *reliability* and *availability* determine a system's overall effectiveness [9]. These are equivalent to asking the question "are all relevant components available and functionally operational for the mission?" After all, a perfectly capable unit is still useless when it is not available at the required theater. This question can be illuminated by a stochastic model (e.g. as seen in [5]) in a wider MC context including maintenance, that should also take supply of fuel and other expendables into account.

2.1 AHP-Based Metric Derivation

In order to find what to measure in the first place, a systematic metric derivation process has been conducted [10] by SMEs in the fields of aircraft assessment and mission result analysis, as suggested in the concept (Fig. 1). The upper part of Fig. 2 illustrates the structure of the metric derivation process in more detail: For a specific scenario, brainwriting and moderated group discussion is used to obtain the hierarchically decomposed, measurable mission objectives and cost criteria. Then the Fuzzy Analytic Hierarchy Process (Fuzzy AHP or FAHP) is used iteratively in conjunction with group discussions to yield the criteria weightings.

The quality integral J is the accumulated result. Equation (1) generally gives this overall measure of effectiveness – in accordance to the method of *simple additive weighting* (SAW) – as a sum for all N elementary criteria over their specific weight multiplied by their respective normalized numerical assessment value j and a sign $\sigma = \pm 1$ for benefit (positive) or effort/cost (negative):

$$J = \sum_{k=1}^{N} \sigma_k w_k j_k \tag{1}$$

The resulting value of J for the considered system models (i.e. the agents) can be computed from measurements in the mission simulation, shown as the second phase in Fig. 2. An efficiency value can be calculated by dividing the effectiveness value by the absolute of the sum of all cost and effort items.

The third step in Fig. 2 is validating the method and its results by conducting sensitivity analysis and having the SMEs compare the resulting rating(s) to their expert observations on the mission's course and end result(s).

Decomposition

After the well-defined, specific mission vignette is presented to the experts by the moderator in some detail, the first step is to hierarchically decompose the specific mission's objectives and its tasks into a tree-like multi-level structure, where the end nodes without any children are called *leaves* or *elementary criteria.* These elementary criteria should be measurables or easily derived from measurements in the simulation or a real-life mission in the end. The SMEs at first decompose the mission independently from another in a session of *brainwriting* [11] to avoid inadvertently influencing each other via group dynamics and to retain their thoughts and personal expertise for later. Afterwards, in the moderated *group discussion stage* a consensus must be established about all criteria and their structure. Criteria and metrics should be brought forward by the SMEs themselves, however it is possible that some are suggested by the moderator and then critically reviewed by the experts. Also, the optimization goals (maximization, minimization, satisfaction, or fixation) for all criteria must be determined at this stage.

Weighting via Fuzzy AHP

The importance of all found criteria is indicated by their respective weights w. These can be obtained by applying the *Analytic Hierarchy Process* (AHP) [12], where criteria are ordered and weighted by pairwise comparison. In each comparison the strength of

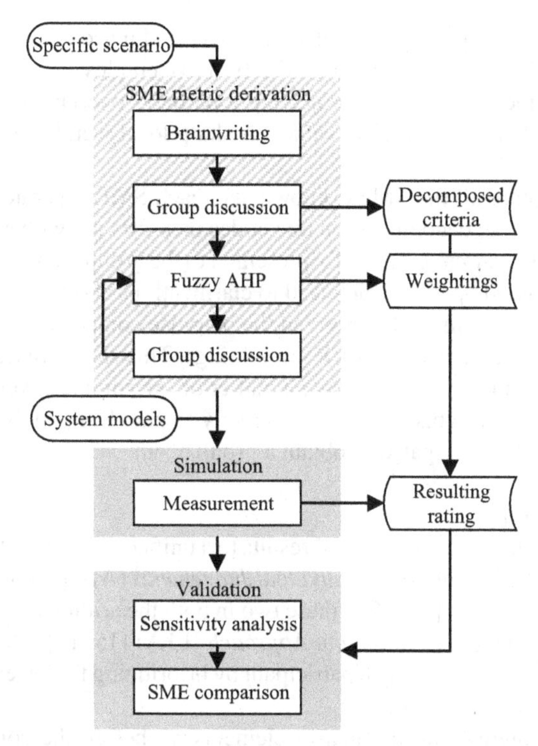

Fig. 2. Process flow diagram of SME metric derivation, simulation and validation, as proposed by us in [10]

dominance a_{ij} of one item i over another criterion j is expressed on an integer scale. All comparisons at each decomposition node yield a square n-by-n comparison matrix A with diagonal of ones ($a_{ii} = 1$) and reciprocal off-diagonal elements $a_{ij} = 1/a_{ji}$.

By calculating and normalizing the principal eigenvector, i.e. the eigenvector $(w_1,\ldots,w_n)$ belonging to the largest eigenvalue $\lambda_{\max}$ of this matrix as seen in Eq. 2, one derives the local (i.e. relative to criteria at the same parent node) weights w_i of the considered criteria. Global weights in the decomposition structure are computed by multiplication of a node's local weight with all its parent nodes' local weights.

$$A\overrightarrow{w} = \lambda_{\max}\overrightarrow{w}, \quad \overrightarrow{w} = (w_1, \ldots, w_n) \tag{2}$$

$$CI = \frac{\lambda_{\max} - n}{n - 1} \tag{3}$$

The principal eigenvalue related to the dimension of the matrix gives the consistency index (CI) as shown in Eq. 3. The consistency ratio (CR) is then given by comparing it to the average consistency index RI of random matrices of the same size as $CR = CI/RI$. CR should not exceed 0.1 to ensure avoidance of circular dominance logic [13]. If at any node of the hierarchical decomposition structure a SME produces a comparison matrix with too high a CR, another iteration of the pairwise comparison at this node is required for this SME.

For more accurate representation of subjectivity and inaccuracy in human decision making, we used the variant Fuzzy AHP (FAHP) [14], employing seven linguistic variables to compare the criteria, each projecting a partially overlapping triangular fuzzy set, instead of single fixed numerical values. A sharp local weight value is obtained by defuzzification.

In order to capture all the SMEs' knowledge, each SME separately completed the Fuzzy AHP stage for all decomposition tree nodes in a software with graphic user interface (GUI). The results were graphically presented to the group along with the respective CR, and then discussed openly, which lead to clarifications about some criteria's meanings [10]. Subsequently, all SMEs repeated the pairwise comparison process to achieve more consistent (significantly lower CR) weightings. This can be repeated iteratively if necessary, as shown in Fig. 2, and is only a matter of few minutes. After ensuring sufficient consistency, the weights can be used directly to gain SME-individual assessment results, or they can be aggregated to obtain a group result.

Weight Aggregation

For aggregation of the individual SMEs' results to a unified group result several methods were evaluated: The *Aggregation of Individual Judgements* (AIJ) [15], the *Aggregation of Individual Priorities* (AIP) [15, 16] – these two in both the arithmetic and the geometric mean variants –, and the *Loss Function Approach* (LFA) [15, 17]. The latter explicitly accounts for the consistency of each participant by prioritizing for lower CR, but appears the least transparent to the SMEs.

AIJ relies on aggregating the matrix elements a_{ij} before the computation of the weights. In AIP on the other hand, first the local weights w_i are computed, then those are averaged over all participants by taking the arithmetic or geometric mean.

Geometric-mean AIP was chosen as preferred method, because only this alternative fulfills all relevant rationality axioms [15], and is also most intuitive and thus transparent and relatable to the SMEs.

SME Survey

We conducted a survey with the SMEs to evaluate the metric derivation process. The majority rated it positively: it was perceived as better structured and including relevant criteria much more completely than previous approaches, but also as more transparent and traceable. Additionally, the surveyed SMEs confirmed the relevance of new assessment approaches over single-aircraft parameters when considering interaction-heavy composites of aircraft in mission contexts [10].

2.2 From Measurement to Metric

Now that the mission-specific elementary criteria and their respective weights have been derived, it must be determined how to quantitatively and objectively measure their degree of fulfillment. This is achieved by defining metrics, represented by lower-case j in Eq. 1. Important examples are elaborated in the following paragraphs.

Critically for commensurability, each metric needs normalization, otherwise comparison of different criteria and attaching weights would not make any sense, not being on the same scale. Measurement and collection of required data should be the same in simulation or reality, although abstractions and simplifications might occur.

Cost

Some simulation results are rather straightforward to quantify in a metric at first glance, such as cost. The total cost of the mission run of all cost categories (fuel, spent armament etc.), e.g. in Dollars or Euros, can be simply added up. In our simplified model maintenance cost is calculated on a per-flight-hour basis, but the conceptual approach also supports higher-fidelity cost calculations. For normalization, the maximally possible total cost of the mission can be used.

Furthermore, any personnel loss is deemed unacceptable, so a value of infinity is attached to it.

Experience

On the other hand, a metric for the gain of experience for human operators, AI, planning and maintenance personnel is not easily derived. Very little research on quantification of experience, experience gain, and experience-induced rise in productivity can be found, as it is dependent on the individual and highly task-specific [18]. Still, the amount of time spent doing a certain task is the main indicator of experience.

Thus, we suggest quantifying the benefit from gained experience based on the time(s) spent during the considered mission. A function mapping years of experience to productivity for Air Force personnel was proposed by R.T. Roth [19], which we write as:

$$P = \begin{cases} \sqrt{1 + \left(\frac{t\,[years]}{20} - 1\right)^2} & for\ t < 20 \\ 1 + 0.04 \cdot (t[years] - 20) & for\ t \geq 20 \end{cases} \tag{4}$$

Here the experience time t is in years. The productivity P is zero at zero years of experience and reaches one ("full productivity") after 20 years. In between that, the productivity behaves like a quarter circle, meaning less/slower growth with more experience. After the first 20 years, Roth supposes that the productivity increases linearly [19].

In this model, productivity gain obviously depends on the previously existing amount of experience, which consequently is required as an input value. To find the gain in productivity caused by the increase of experience due to the mission's execution, we convert the time spent on an activity to typical year-equivalent, i.e. the fraction of the time in a state during or because of the conducted mission versus the total time in this state in a typical year in the service.

Threat

Time logging is also important for metrics of the cost/effort type: observability of the considered system and portion of the mission spent under threat or in danger.

The basic idea is to add up all times, when the system can be detected or attacked respectively. Detectability can be assumed if a type of signature such as acoustic, radar or infrared (IR) exceeds a set threshold, which itself can be a function of time. The threshold can be derived by setting an upper limit for the allowable detection probability.

For Radar cross-section (RCS) this can be based on signal-to-noise ratio (S/N). In air-to-air situations angular relations play a major role, because relative positioning determines the applicability of several types of missiles [20].

For simplification, summing the times while an aircraft is inside a certain radius or distance from the origin of the threat yields a first approximation of this metric. In many cases, the maximum will be given by the total simulated mission time. The minimum is zero, meaning no portion of the mission time was spent under any threat.

The 'physical' effect(s) on the cooperative systems or its parts if or when it actually is attacked is considered separately – mainly in terms of resulting lower positive mission outcomes, but also by higher cost for maintenance, spare parts, etc.

Effect Chain

In our survey [10] SMEs agreed that often the main desired effect of an aerial mission is to impede some red force's capability or functionality for a defined minimal time, e.g. transportation. In the mission-specific metric derivation it is paramount to find how that capability of the opponent can be defined and how its impediment can be measured.

For e.g. a convoy, the number of immobilized ground vehicles can be counted. For an installation or compound however, functionalities can be more abstract and therefore harder to quantify.

The benefit of partial fulfillment can be better represented by defining utility and preference functions for the criterion. These map the degree of fulfillment to respective utility, for example destroying the first target out of five can have significantly higher relevance and thus larger increase in benefit than destroying the fifth out of the same number.

2.3 Aggregation of Results

Having derived the proper metrics and their individual weights and obtained the actual measurement data (example given in Sect. 3), several MCDA methods can be used to aggregate these intermediate results into an overall measure of performance for the whole decomposed mission tree structure [21].

The most straightforward is *Simple Additive Weighting* (SAW), where normalized elementary criteria metric values are multiplied by their attached weight and then collectively added up as in Eq. (1). The resulting number is, very intuitively, interpreted as degree of compliance towards the mission's objectives. "100%", the absolute best case, would mean having all desired effects in full, at no cost. In SAW the result is calculated separately for all considered alternatives, and then can be used for ranking. SAW is limited however, as it cannot represent e.g. diminishing returns in benefit or interdependent (sub-)criteria.

Other methods rely on comparison of all considered alternatives, so they can only be meaningfully applied when comparing alternative, i.e. different, mission runs. Differences here can be external, i.e. in the mission environment, neutral or opposing forces, or internal, i.e. regarding the considered system of systems itself.

Utility Analysis (UA) attaches a point value to every criteria value, signifying its utility. However, this method often is not seen as suitable for its weaknesses: criteria interdependence cannot be taken into account, its linear transformation functions are deemed unrealistic, and small differences in the criteria evaluations are overemphasized by UA [21].

Consequently, *outranking* methods, which also account for conflicting or contradicting information, are preferred [22]: The *Technique for Order Preference by Similarity to Ideal Solution* (TOPSIS) relies on calculating Euclidian distances to the ideal (best) and worst alternatives, and from these the relative closeness to the ideal alternative, which is used to rank all compared options [21, 23].

The *Preference Ranking Organisation Method for Enrichment Evaluations* (PROMETHEE) [22] makes use of an individual preference function for every criterion. These functions represent how strongly some difference at the criterion leads to a (dis)advantage, and can, in principle, be of arbitrary shape. Outranking relations are calculated by paired comparison; the rank order is derived from that by comparing how often an alternative dominates versus how often it is dominated. A variant of PROMETHEE also allows for two or more alternatives being ranked as equal (indifferent) [21, 22]. PROMETHEE is the most complete and competent out of these methods and even can be appropriately explained to stakeholders/SMEs rather easily.

2.4 Analysis

After obtaining an assessment result on all the mission's decomposition nodes by any of the MCDA methods mentioned in Sect. 2.3, they must be visualized. *Spider plots* using relative percentages emphasize the differences between the considered alternatives (the compared mission runs).

Local and possibly global *sensitivity analysis* should be conducted to gain insight on criticality of parameters and the robustness of the assessment and ranking against insecurities or variations of the measurement values [7]. Importantly, local sensitivity analysis of the weights also shows how much results depend on the a-priori assumptions and preferences (choice of weights) made by the SMEs. Plotting the resulting spread of J quality values versus a criterion's weight illustrates if and at what weight a change of rank occurs – i.e. one alternative becomes rated higher than another because of the weight modification at the criterion. This allows the SMEs to directly see the impact of their prioritization of the criteria.

3 Experiments in a Simulation Environment

Having completed the metric derivation process for an example mission, a simulation testbed for the overall concept was implemented to demonstrate and explore its utility and usability.

3.1 Goal

For the prototype simulation environment, the primary goal was to deliver just enough complexity to gather the relevant data for the derived metrics, but not high real-world fidelity of any simulated systems. Furthermore, we strived to depict significant interactions, and to capture and enable (seemingly) emergent behavior in the simulated autonomous system of systems. In order to retain full control and optimal insights about the mechanisms, adapting an off-the-shelf simulation environment was dismissed. The simulation is rather highly abstracted and quite simplified, with the intention of accommodating fast execution, in order to eventually enable parameter studies and multidimensional sampling. The future goal is to have many Monte Carlo (MC) runs, i.e. many replications on the same initial state, as needed to generate assessments for systems of systems over diverse mission vignettes.

Thus, the simulation was to be multi-agent based and focused on interaction. Additionally, all relevant measurements had to be implemented. The minimalist graphics output of a two-dimensional (2D) "god's view"-type only serves as an overview of the current state, mainly as a debugging tool. In the future it will also be used for SME validation (as indicated in the lower portion of Fig. 2).

3.2 Scenario and System of Systems

The specific mission vignette being considered as an example is that of an air-to-ground operation: finding and effecting on a small number of ground targets in a defined, previously unreconnoitered area of operations, where surface-to-air missile (SAM) pop-up threats are present. We had the criteria for this scenario decomposed by SMEs in [10].

The ground targets, which have no own anti-air capability, move independently based on predefined waypoints and have a small radius of detectability. SAM sites are active from the initial simulation time or are activated at a given later point, i.e. "pop-up".

The airborne system of systems to be considered here is a swarm-like homogenous composite of autonomous UAV agents. All agents have the same sensing capabilities regarding the ground targets and SAM sites, and the same type and number of armaments. UAVs will try to evade a detected SAM site. Collisions between UAVs need to be avoided and they should direct each other to a target if an entity has detected one. Thus, for information interchange they rely on a common communications (comms) channel, to which they broadcast their position, velocity, and the detection of a ground target at each time step.

3.3 Simulation Design

The simulation of the mission scenario and the system of systems outlined above was implemented in MATLAB [24], in object-oriented style. A multi-agent approach [25] is ideal, as all entities (blue and red forces) are to act on their own. This way, interaction and interdependence is introduced in the required non-deterministic way. The experiment is designed in a 2D space, further simplification is achieved by omitting terrain or atmospheric effects.

The UAV-agents' movement with limited turn rate and velocity is based on a fourth-order Runge-Kutta (RK4) solver; fuel consumption is assumed only to be dependent on the UAV's speed. The UAV agents independently pick random target coordinates *(x,y)* from a uniform distribution in the given area of operation. Once an agent has reached those (to a given distance delta), it picks a new target location at random. This behavior is interrupted when a ground target or a SAM site is detected, or another UAV gets too close. When a UAV has detected a ground target, it broadcasts its location to the other agents, which will then pick their next target location in the vicinity.

The UAVs' AI (implemented in a "controller" class) uses a *Behavior Tree* (BT) [26] approach for decision making, which was translated to a finite-state machine (FSM) in *Stateflow*. Stateflow is a logic tool that is provided with MATLAB and allows for graphical overview of state transitions (chart), and live inspection during the active run highlighting the current state. Additionally, there is a data logger object for each agent. An agent's armament is modelled in another sub-object, with each UAV carrying a certain number of missiles. If a missile is fired at a target, a hit (with a defined damage value) is instantaneously determined probabilistically from a random distribution.

Uniform random sampling also is used for modelling the loss rate in communications reception, impacting the UAVs' interaction. Further uncertainty is introduced when the agent's controller obtains the UAV's location by global positioning system (GPS), which is modelled with random imprecision.

The waypoint-following ground targets are each their own independent agent. Their health points (HP) are diminished by the damage value of missile hits, and they are declared incapacitated at zero HP. Each target has a defined radius where it can be detected, while at smaller distance it can be attacked by an UAV. The detection by an UAV decreases non-linearly in between these radii. Above a set threshold for the UAV's sensing system, the agent will communicate having encountered a target to the other entities and stay close to this position.

SAM sites, too, have a detectability radius, but also a maximum distance for sensing UAVs themselves. Inside a lower radius their agent logic will attack a detected UAV. Time inside the detection radius of an active SAM location is counted as time under threat for the respective UAV. One hit on an UAV is assumed to fully disable it. Each SAM site has a maximum number of targets it can affect at the same time step. Again, hits are determined stochastically based on inverse transform sampling of uniform random numbers, as SAM hit probability decreases with distance. All random number generators (RNGs) use constant seeds (unique to each entity) to allow for reproducibility in testing, but seeding will be time-based for future high-replication MC runs.

For the assessment part, an extensive class was implemented, with classes for the MCDA methods mentioned in Sect. 2.3. At any state of the simulation all logged data can be accumulated and applied to the predefined metrics. For example, experience gain from mission times is converted to productivity gain via Eq. (4). The results are combined to the mission decomposition and weighting data, which is structured in a tree class, for aggregation to the partial (at all nodes) and overall assessment results. Furthermore, the whole simulation state can be exported for later import to facilitate comparative assessment of different mission runs.

Figure 3 depicts the flow of the simulation process: In the beginning, the environment (SAM sites, ground targets, operational area limits) is set along with initial values for all agents. Then in the simulation time loop, first all UAV agents are simulated: their sensing of ground targets and SAM sites, and reception of messages from other UAVs, their decision process in a Stateflow chart, their armament's effect if any at this time, their movement, followed by transmitting a message to the communications channel about their state and position. Finally, all relevant data is saved to the data logger object.

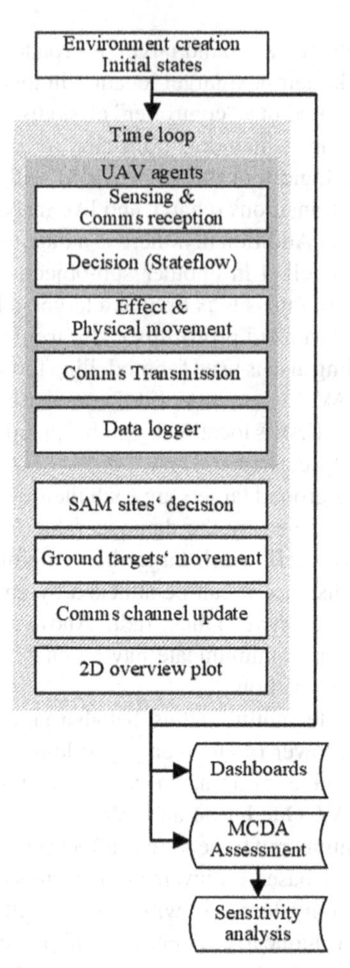

Fig. 3. Schematic simulation structure overview (runtime consecutive arrows are not shown in the time loop for compactness.)

Afterwards, the behavior of the SAM sites is calculated, then the ground targets are moved along their waypoints, and the UAVs' communications channel is updated. Lastly, the 2D overview is plotted.

Graphical Presentation

This plot of the current simulation state is the main graphical output during the simulation, for an example see Fig. 4. Additionally, at any time the user can have the gathered data plotted in concise synoptic visualizations: Dashboards for single agents, an aggregated dashboard for the whole simulation, and a cost dashboard for all UAVs. These dashboards also serve for SME validation of the generated mission assessment results.

An aggregated dashboard depicting data from another simulation run is partially shown in Fig. 5. The dashboard summarizes the current state of the mission: pie charts show how many of the UAVs are alive and how many were successfully launched, how many of the ground targets have been disabled, how many SAM sites are intact and how many are active. A bar charts adds up the various measured times for each UAV: total simulated time, airtime, time inside a SAM site and detectable by a SAM site, and finally the time detecting the "probability cloud" of a ground target. Further charts are concerned with all UAVs' speed over time, missiles spent over time, missiles spent and hit totals, evasive maneuvers over time and totals, fuel left over time for each UAV and the current fuel status (used vs. Left fuel) for all UAVs.

After generating a mission assessment, at every parent node of the criteria decomposition a spider plot (also known as *radar chart*, for an example see Fig. 6) can be generated. These plots show the performance assessment of all children nodes, each criterion with its own axis. One alternative is normalized to 100% at each axis to highlight the relative differences.

Furthermore, sensitivity analysis results are given graphically; for an example see Fig. 7. On the y-axis of these plots one can see the total assessment result J, e.g. compliance percentage for SAW aggregation. The x-axis shows the weight of the selected criterion node. Varying the node weight between 0 and 100% and the assessment data (intensities) of the children nodes by $\pm 5\%$ gives a cone-shaped graph for each alternative, showing how strongly the overall assessment correlates with this specific node. Currently used values, also employed as start values, are indicated by dashed lines. If alternatives cross, it can be observed at which weight rank changes will occur.

Run Variants

From our first experimental application three different runs are listed in Table 1: We compared a reference run to one with a significant environmental change (no second SAM site popping up), and one with a change to the system of systems, i.e. to a property of the UAVs, by having lowered the communications reliability by only 0.5% points. In reality this could be caused e.g. by a different receiver or antenna subsystem.

All three runs were executed with six homogenous UAV agents with three missiles each (one could be fired at each time step), launched consecutively 30 time units apart at the same position. There were three ground targets, two of them on the move. Each ground target required two hits to be fully disabled. Two stationary SAM sites were present, one being active from the start of the simulation. The air-to-ground missile hit probability was set at 80%. All parameters not mentioned in Table 1, including all random seeds, were kept identical over the three runs. The execution of each run was stopped after the same simulation time and the assessments were made according to the previously elaborated process.

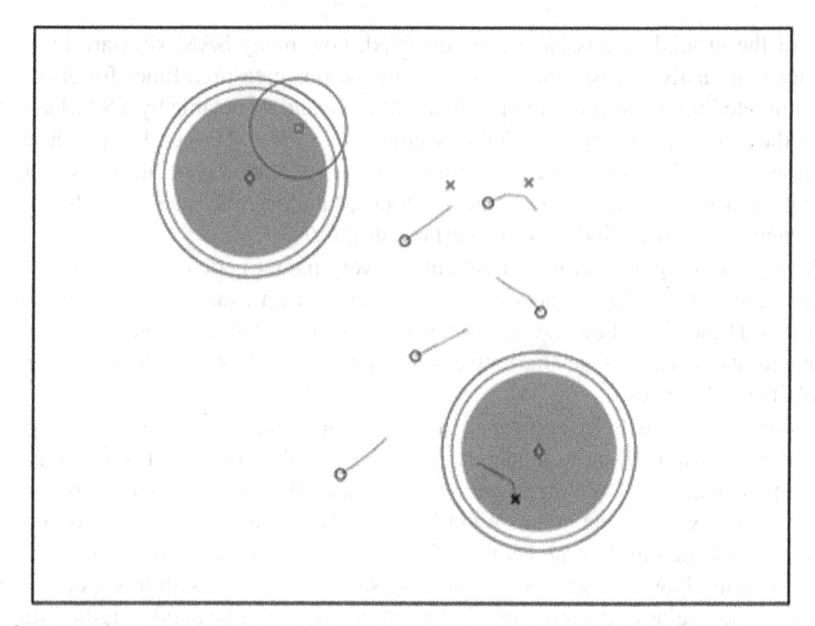

Fig. 4. Example (image section) of a 2D overview plot of a simulation state in the chosen mission: Active SAM sites are indicated in red with rhombic centers, live UAV agents by black circular markers, live ground targets by violet square markers. The disabled units are marked by an x in the respective color. Trails represent respective former positions; circles indicate detection distances of the SAM sites and ground targets.

Table 1. Simulated example mission runs: differences in setup

#	Pop-up SAM site	Comms reception probability
1	After 50 time units	99.5%
2	None	99.5%
3	After 50 time units	99.0%

4 First Results

The mission results for these three runs of the same mission first differ in terms of cost: In variants 1 and 3 the SAM site popping up after a pre-set time disables one UAV, which means the whole unit cost must be added to the cost total. Secondly, in variation 3, due to the lower communications reliability, one less ground target is disabled, which means a significantly lower effect rating. These main outcomes are listed in Table 2, which also gives the compliance percentages for the whole mission according to the SAW method.

As expected, the total mission rating is significantly lower if the main intention of disabling ground targets is fulfilled to a lesser degree, as seen in case 3. Case 2 benefits from having lost no UAV – entailing lower cost, but also more gained experience by the

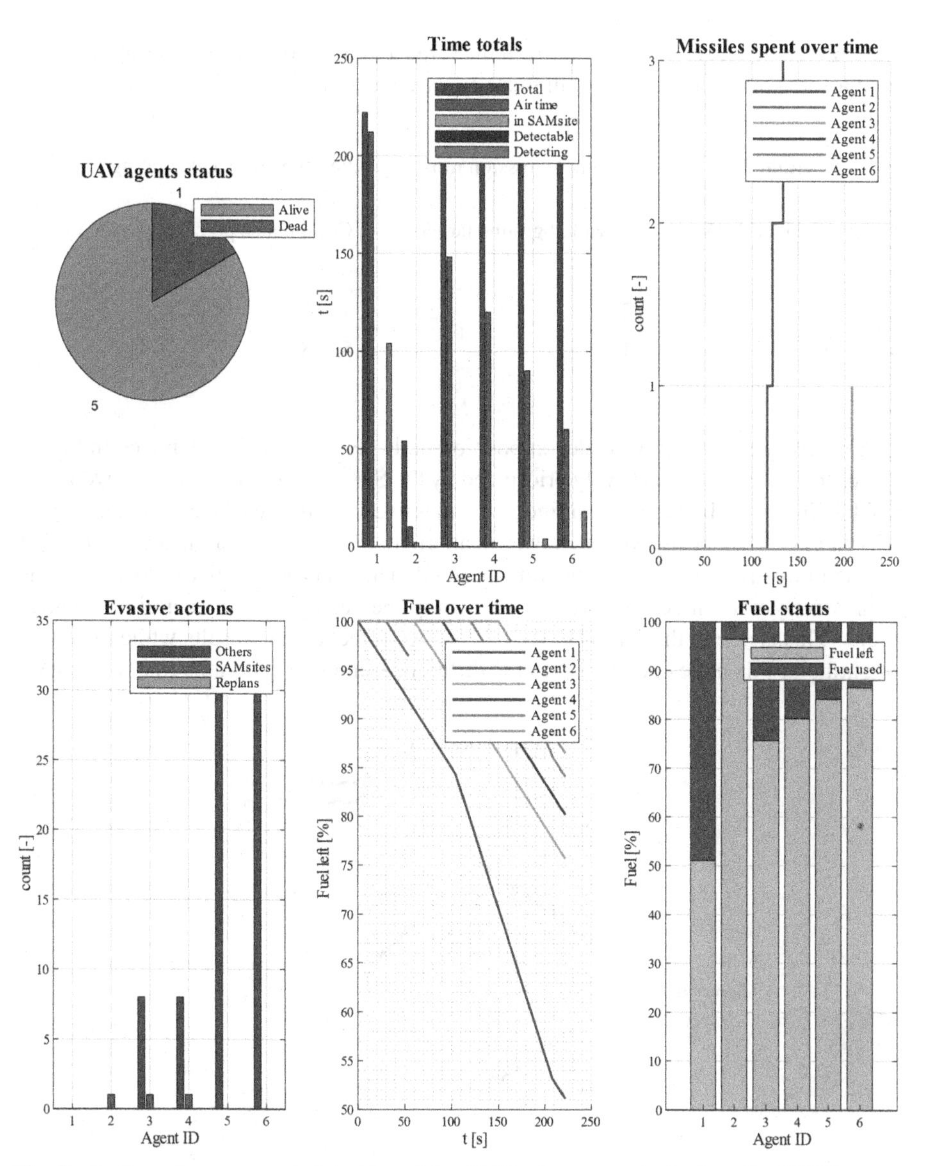

Fig. 5. Section of an aggregated dashboard for SMEs: (upper row) pie chart for UAVs' status, time totals, missiles spent over time; (lower row) cumulated evasive actions, fuel left over time, fuel status left vs. used.

maintenance crew. Case 1 is rated rather closely to case 2: the main objective has been completed at the same degree, since the swarm was robust against the loss of an agent.

One of the generated spider plots can be seen in Fig. 6 as an example. It shows the productivity gain by the amount of experience generated in the mission by the time-based generalization as seen in Sect. 2.2, which omits possible learnings from the loss of a

UAV. At this node, runs 1 and 3 perform the same and thus completely overlap, whereas run 2 affords the maintenance team more experience as all six UAVs survive.

Table 2. Simulated example mission runs: outcomes and compliance rating

#	Disabled UAVs	Disabled ground targets	Compliance percentage (SAW)
1	1	2	68.02%
2	0	2	72.37%
3	1	1	50.18%

Sensitivity analysis for the decomposition node "effect chain" is depicted in Fig. 7. This criterion has been massively prioritized by the SMEs [10] at a normalized weight of 56.64% (indicated by the vertical line), so it massively influences the compliance rating of its parent node. For any weight, however, alternative 3 is clearly dominated by the other two, as in this run only one target was disabled. The other alternatives' cones overlap progressively with increasing weight, indicating increasingly close overall assessment results. This is due to these runs having the same "effect chain" results while increased weighting of this node deemphasizes the differences at other criterions (mainly cost).

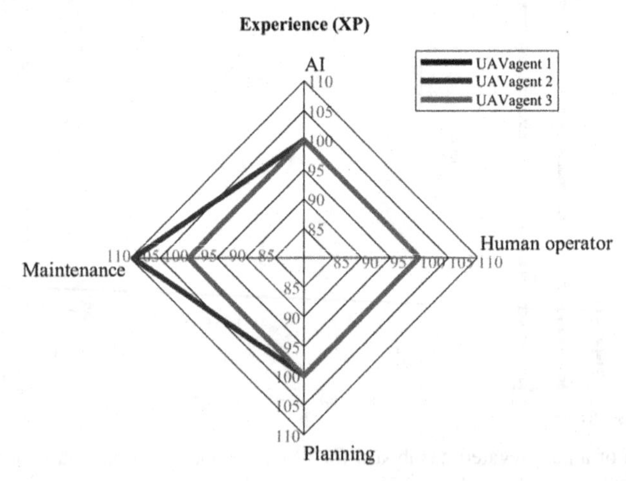

Fig. 6. Spider plot for comparison of alternatives at the "experience" node; alternative 1 at this criteria node is normalized to 100%, alternative 3 performs the same as alternative 1 at this criterion and thus they completely overlap. Alternative 2 yields 10% more maintenance productivity gain through experience.

Overall, the total mission outcomes are relatively close to each other, because only small variations were introduced. However, the mission results differ and thus, as expected and required, the quantified mission assessment shows differences accordingly. These differences allow the considered alternatives to be ranked: the UAV type with less reliable communications (alternative 3) clearly is significantly less adequate for the simulated mission vignette. The impact of this system-immanent change is significantly larger than the effect of a change in the mission environment in this first example.

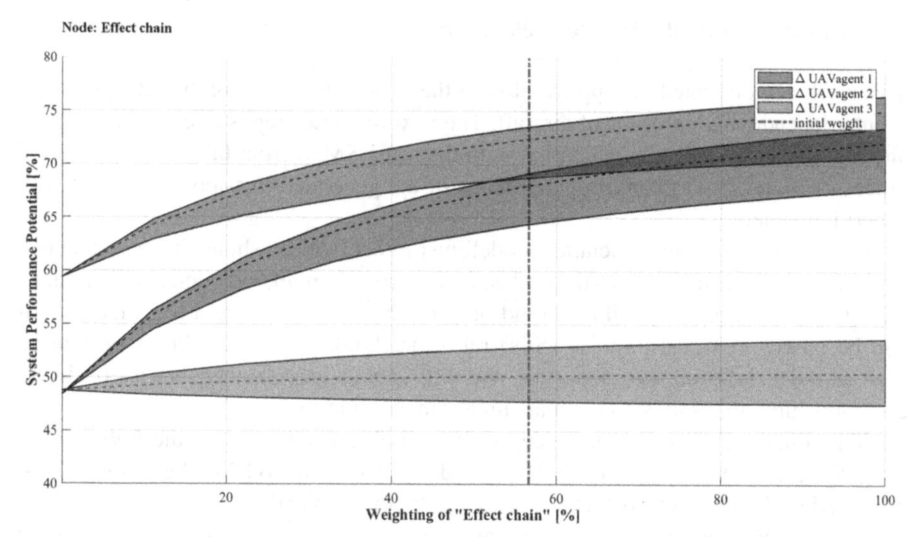

Fig. 7. Local sensitivity analysis at the node "effect chain": weight of the node – initially determined by the SMEs at 56.64% of the mission as indicated by the vertical line – on the x-axis, compliance percentage on the y-axis. The variance cones are derived from variation of the input values by ±5%, the dashed lines indicate the value based on the initially calculated outcome.

5 Discussion

The experimental results demonstrate that the methodic concept can indeed be applied to assess mission results and thereby to compare the performance of airborne system of systems. A negative difference in subsystem capability is represented by a lower mission effectiveness score according to its impact on the mission execution in the given example.

Even in this rather low-fidelity simulation environment, small differences in the system of systems – especially regarding interaction and interdependence, as highlighted by the communications reliability – lead to noticeable changes in the mission outcomes. Those then can be seen impacting the specific criterion node and the total mission ratings. Furthermore, the simulation already enables parameter studies and multidimensional sampling in the presented swarm-type scenario.

Overall, these first results are positively indicating viability for the overall concept of assessing a composite force of multiple aircraft based on mission outcomes. However, this calls for a rather large sampling space, not just over multiple relevant mission vignettes, but also over several plan and execution variants. This huge sampling space necessitates systematic stratification on the one hand, and on the other hand underlines the importance of a careful choice of the considered mission vignettes. The mission scenarios not only directly affect computation time, but of course immediately impact the system assessment.

6 Conclusion and Future Research

This work demonstrated the applicability of the approach to mission-based assessment of interdependent composites of aircraft. The next required step is to close the loop, as already indicated in Fig. 2: validation with the same SMEs from the metric derivation process is necessary to confirm congruence of the generated mission assessment with expert judgement.

In future research, more detailed modelling (higher fidelity) should be implemented, including more realistic flight dynamics, and more differentiated "detectability time" measurement. Sensor capability should not only be determined by a fixed radius, but e.g. be based on signal-to-noise (S/N) ratio thresholds for several kinds of sensors such as radar, infrared, and acoustics. We will also suggest further metrics on lower decomposition nodes to accommodate higher model fidelity.

Regarding agent AI, proper Behavior Trees (BTs), using C++ and the *Robot Operating System* (ROS) framework [27], instead of FSMs with BT-like behavior will be evaluated for better inter-process and inter-agent communication.

Finally, MC sampling over various mission scenarios will be conducted to gain a total assessment of an aerial system of systems, entailing an exploration and discussion of appropriate sampling stratification methods and Design of Experiments (DoE).

References

1. Grayson, T.: Mosaic Warfare, 27 July 2018. https://www.darpa.mil/attachments/STO-Mosaic-Distro-A.pdf. Accessed 6 August 2020
2. Jaiswal, N.K.: Military Operations Research. Quantitative Decision Making. International Series in Operations Research & Management Science, vol. 5. Kluwer Academic Publishers, Boston (1997)
3. Ernest, N.D.: Genetic fuzzy trees for intelligent control of unmanned combat aerial vehicles, University of Cincinnati (2015)
4. Stodola, P., Drozd, J., Mazal, J., Hodický, J., Procházka, D.: Cooperative unmanned aerial system reconnaissance in a complex urban environment and uneven terrain. Sensors (Basel, Switzerland) (2019). https://doi.org/10.3390/s19173754
5. Hodicky, J., Prochazka, D.: Modelling and simulation paradigms to support autonomous system operationalization. In: Mazal, J., Fagiolini, A., Vasik, P. (eds.) MESAS 2019. LNCS, vol. 11995, pp. 361–371. Springer, Cham (2020). https://doi.org/10.1007/978-3-030-43890-6_29
6. Morawietz, S., Strohal, M., Stütz, P.: A mission-based approach for the holistic evaluation of aerial platforms: implementation and proof of concept. In: 18th AIAA/ISSMO Multidisciplinary Analysis and Optimization Conference 2017, Denver, Colorado, USA, 5–9 June 2017: held at the AIAA Aviation Forum 2017. 18th AIAA/ISSMO Multidisciplinary Analysis and Optimization Conference, Denver, Colorado. Curran Associates Inc., Red Hook (2017). https://doi.org/10.2514/6.2017-4152

7. Morawietz, S., Strohal, M., Stütz, P.: A decision support system for the mission-based evaluation of aerial platforms: advancements and final validation results. In: 18th AIAA Aviation Technology, Integration, and Operations Conference 2018. Atlanta, Georgia, USA, 25–29 June 2018: held at the AIAA Aviation Forum 2018. 2018 Aviation Technology, Integration, and Operations Conference, Atlanta, Georgia. Curran Associates Inc., Red Hook (2018). https://doi.org/10.2514/6.2018-3975
8. Ruf, C., Zwick, M., Morawietz, S., Stütz, P.: Enhancing automated aerial reconnaissance onboard UAVs using sensor data processing-characteristics and pareto front optimization. In: AIAA Scitech 2019 Forum. AIAA Scitech 2019 Forum, San Diego, California. American Institute of Aeronautics and Astronautics, Reston, Virginia (2019). https://doi.org/10.2514/6.2019-1541
9. Habayeb, A.R.: Systems Effectiveness. Elsevier Science, Burlington (1987)
10. Seethaler, J., Strohal, M., Stütz, P.: Finding metrics for combat aircraft mission efficiency: an AHP-based approach. In: Deutscher Luft- und Raumfahrtkongress 2020 (in publication). Deutsche Gesellschaft für Luft- und Raumfahrt - Lilienthal-Oberth e.V., Bonn (2020). https://doi.org/10.25967/530013
11. vanGundy, A.B.: Brainwriting for new product ideas: an alternative to brainstorming. J. Consum. Market. **1**, 67–74 (1983)
12. Saaty, R.W.: The analytic hierarchy process—what it is and how it is used. Math. Modelling (1987). https://doi.org/10.1016/0270-0255(87)90473-8
13. Mu, E., Pereyra-Rojas, M.: Practical Decision Making. An Introduction to the Analytic Hierarchy Process (AHP) Using Super Decisions v2. Springer, Cham (2017). https://doi.org/10.1007/978-3-319-33861-3
14. Buscher, U., Wels, A., Franke, R.: Kritische Analyse der Eignung des Fuzzy-AHP zur Lieferantenauswahl. In: Bogaschewsky, R., Eßig, M., Lasch, R., Stölzle, W. (eds.) Supply Management Research. Aktuelle Forschungsergebnisse 2010; [Tagungsband des wissenschaftlichen Symposiums Supply Management; Advanced studies in supply management, Bd. 3], vol. 140, 1st edn., pp. 27–60. Gabler, Wiesbaden (2010). https://doi.org/10.1007/978-3-8349-8847-8_2
15. Ossadnik, W., Schinke, S., Kaspar, R.H.: Group aggregation techniques for analytic hierarchy process and analytic network process: a comparative analysis. Group Decis. Negot. **25**(2), 421–457 (2015). https://doi.org/10.1007/s10726-015-9448-4
16. Forman, E., Peniwati, K.: Aggregating individual judgments and priorities with the analytic hierarchy process. Eur. J. Oper. Res. (1998). https://doi.org/10.1016/S0377-2217(97)00244-0
17. Cho, Y.-G., Cho, K.-T.: A loss function approach to group preference aggregation in the AHP. Comput. Oper. Res. (2008). https://doi.org/10.1016/j.cor.2006.04.008
18. Schriver, A.T., Morrow, D.G., Wickens, C.D., Talleur, D.A.: Expertise differences in attentional strategies related to pilot decision making. Hum. Factors (2008). https://doi.org/10.1518/001872008X374974
19. Roth, R.T.: Quantifying Experience in the Value of Human Capital, ADA186585. US Air Force Academy, Colorado Springs (1987). https://apps.dtic.mil/sti/pdfs/ADA186585.pdf. Accessed 9 July 2020
20. Kelly, M.J.: Performance measurement during simulated air-to-air combat. Hum. Factors **30**(4), 495–506 (1988)
21. Morawietz, S.: Konzipierung und Umsetzung eines Unterstützungssystems zur vergleichenden Bewertung von Luftfahrzeugen. Dissertation, Universität der Bundeswehr München (2018)
22. Geldermann, J., Lerche, N.: Leitfaden zur Anwendung von Methoden der multikriteriellen Entscheidungsunterstützung. Methode: PROMETHEE, Georg-August-Universität (2014). www.uni-goettingen.de/en/sh/57609.htmlde/document/download/285813337d59201d34806cfc48dae518.pdf/MCDA-Leitfaden-PROMETHEE.pdf. Accessed 24 Jan 2020

23. Hwang, C.-L., Yoon, K.: Multiple Attribute Decision Making. Methods and Applications: A State the Art Survey. Lecture Notes in Economics and Mathematical Systems, vol. 186. Springer, Berlin (1981). https://doi.org/10.1007/978-3-642-48318-9
24. The Mathworks Inc.: MATLAB. Version 2019b, Natick, Massachusetts (2019)
25. Zhu, X., Liu, Z., Yang, J.: Model of collaborative UAV swarm toward coordination and control mechanisms study. Procedia Comput. Sci. (2015). https://doi.org/10.1016/j.procs.2015.05.274
26. Colledanchise, M., Ögren, P.: Behavior Trees in Robotics and AI: An Introduction. Artificial Intelligence and Robotics Series 2018. Chapman & Hall/CRC (2017–2020)
27. Colledanchise, M., Faconti, D.: BehaviorTree.CPP (2014–2020). https://github.com/BehaviorTree/BehaviorTree.CPP. Accessed 2 Aug 2020

The GRASP Metaheuristic for the Electric Vehicle Routing Problem

David Woller(✉), Viktor Kozák, and Miroslav Kulich

Czech Institute of Informatics, Robotics, and Cybernetics, Czech Technical University in Prague, Jugoslávských partyzánů 1580/3 160 00 Praha 6, Prague, Czech Republic
{wolledav,viktor.kozak,kulich}@cvut.cz,
http://imr.ciirc.cvut.cz

Abstract. The Electric Vehicle Routing Problem (EVRP) is a recently formulated combination of the Capacitated Vehicle Routing Problem (CVRP) and the Green Vehicle Routing Problem (GVRP). The goal is to satisfy all customers' demands while considering the vehicles' load capacity and limited driving range. All vehicles start from one central depot and can recharge during operation at multiple charging stations. The EVRP reflects the recent introduction of electric vehicles into fleets of delivery companies and represents a general formulation of numerous more specific VRP variants. This paper presents a newly proposed approach based on Greedy Randomized Adaptive Search Procedure (GRASP) scheme addressing the EVRP and documents its performance on a recently created dataset.GRASP is a neighborhood-oriented metaheuristic performing repeated randomized construction of a valid solution, which is subsequently further improved in a local search phase. The implemented metaheuristic improves multiple best-known solutions and sets a benchmark on some previously unsolved instances.

Keywords: Electric vehicle routing problem · Greedy randomized adaptive search · Combinatorial optimization

1 Introduction

This paper addresses the Electric Vehicle Routing Problem (EVRP) recently formulated in [14]. The EVRP is a challenging $\mathcal{NP}$-hard combinatorial optimization problem. It can be viewed as a combination of two variants of the classical Vehicle Routing Problem (VRP) - the Capacitated Vehicle Routing Problem (CVRP) and the Green Vehicle Routing Problem (GVRP). In the VRP, the goal is to minimize the total distance traveled by a fleet of vehicles/agents, while visiting each customer exactly once. In the CVRP, the customers are assigned an integer-valued positive demand, and the vehicles have limited carrying capacity. Thus, an additional constraint of satisfying all customers' demands while respecting the limited vehicle capacity is added to the VRP. Concerning the GVRP, the terminology is not completely settled, but the common idea among different formulations aims at minimizing the environmental impact, typically by taking into

J. Mazal et al. (Eds.): MESAS 2020, LNCS 12619, pp. 189–205, 2021.
https://doi.org/10.1007/978-3-030-70740-8_12

consideration the limited driving range of alternative fuel-powered vehicles and the possibility of refueling at rarely available Alternative Fuel Stations (AFSs). The EVRP has the same objective as the VRP while incorporating the additional constraints from CVRP and GVRP. It is sometimes alternatively named CGVRP, while the name is EVRP often used for other variants of the GVRP (e.g., with considering the non-linear time characteristic of the recharging process or the influence of carried load on the energy consumption).

A method based on the Greedy Randomized Adaptive Search Procedure (GRASP) metaheuristic, together with the preliminary results, is presented in the paper. A novel dataset was introduced in [13], and it is the first publicly available dataset of EVRP instances. A subset of this dataset is used in the competition [12]. This paper is the first to give results to the whole dataset and thus represents a valuable benchmark for future researchers. As the competition results were not known at the time of writing, the results are compared only to the organizers-provided best-known values for the smallest instances.

1.1 Related Works

Various approaches were successfully applied to the numerous variants of the VRP, and many of these can also be adapted to the EVRP formulation solved. For example, a recent survey [5] focused only on the variants of EVRP presented a total number of 79 papers. Most of these consider additional constraints intended to reflect real-world conditions, such as using heterogeneous or mixed vehicle fleet, hybrid vehicles, allowing partial recharging or battery swapping, considering different charging technologies and the non-linearity of charging function, dynamic traffic conditions, customer time windows, and others. The problem formulation introduced in [14] stands out, as it leaves out all but the most fundamental constraints on battery and load capacity. Thus, addressing it might produce a universal method adjustable for more specific variants.

According to [5], the most commonly applied metaheuristics are Adaptive Large Neighborhood Search (ALNS), Genetic Algorithms (GA), Large Neighborhood Search (LNS), Tabu Search (TS), Iterative Local Search (ILS) and Variable Neighborhood Search (VNS). The GRASP metaheuristic deployed in this paper is, therefore, not a commonly used one. However, it follows similar principles as other neighborhood-oriented metaheuristics, such as TS, ILS, or VNS. Exact methods such as Dynamic Programming or various Branch-and-Bound/Branch-and-Cut techniques are frequently used as well, but these are generally not suitable for solving larger instances in a reasonable time.

According to [14], the EVRP variant solved was first formulated in [9]. So far, only a few papers are dealing with this problem, and for each one of them, the exact formulation slightly varies. The first one is [18], which additionally limits the maximum number of routes. It presents a solution method based on the Ant Colony System (ACS) algorithm. The second one is [15], which considers the maximum total delivery time. It presents a Simulated Annealing (SA) algorithm operating with four different local search operators (swap, insert, insert AFS, and delete AFS). Then, [17] proposes a novel construction method and a memetic algorithm consisting of a local search and an evolutionary algorithm. Similarly to [15], the local search combines the Variable Neighborhood Search (VNS) with the Simulated Annealing acceptance criterion. The operators used in the local search are 2-opt, swap, insert, and inverse. Unlike the previous approaches, the proposed algorithm is memetic, thus maintains a whole set of solutions. The organizers of [12] themselves also presented a solution method together with the new dataset in [13]. Similarly to [18], they employ an Ant Colony Optimization metaheuristic. However, the problem formulation in [13] differs from [14] and the previously mentioned methods, as it evaluates the energy consumption as a function of the current load. Thus, their results are not directly comparable.

2 Methods

2.1 Problem Formulation

The EVRP can be described as follows: given a fleet of EVs, the goal is to find a route for each EV, such that the following requirements are met. The EVs must start and end at the central depot and serve a set of customers. The objective is to minimize the total distance traveled, while each customer is visited exactly once, for every EV route the total demand of customers does not exceed the EV's maximal carrying capacity and the EV's battery charge level does not fall below zero at any time. All EVs begin and end at the depot, EVs always leave the AFS fully charged (or fully charged an loaded, in case of the depot), and the AFSs (including the depot) can be visited multiple times by any EV. An example of a solved EVRP instance is shown in Fig. 1. Here, the depot, the AFSs, and the customers are represented by a red circle, black squares, and blue circles, respectively. The blue line represents the planned EVRP tour.

The EVRP mathematical formulation as introduced in [14] follows.

$$\min \sum_{i \in V, j \in V, i \neq j} w_{ij} x_{ij}, \tag{1}$$

s.t.

$$\sum_{j \in V, i \neq j} x_{ij} = 1, \forall i \in I, \tag{2}$$

$$\sum_{j \in V, i \neq j} x_{ij} \leq 1, \forall i \in F', \tag{3}$$

$$\sum_{j \in V, i \neq j} x_{ij} - \sum_{j \in V, i \neq j} x_{ji} = 0, \forall i \in V, \tag{4}$$

$$u_j \leq u_i - b_i x_{ij} + C(1 - x_{ij}), \forall i \in V, \forall j \in V, i \neq j, \tag{5}$$

$$0 \leq u_i \leq C, \forall i \in V, \tag{6}$$

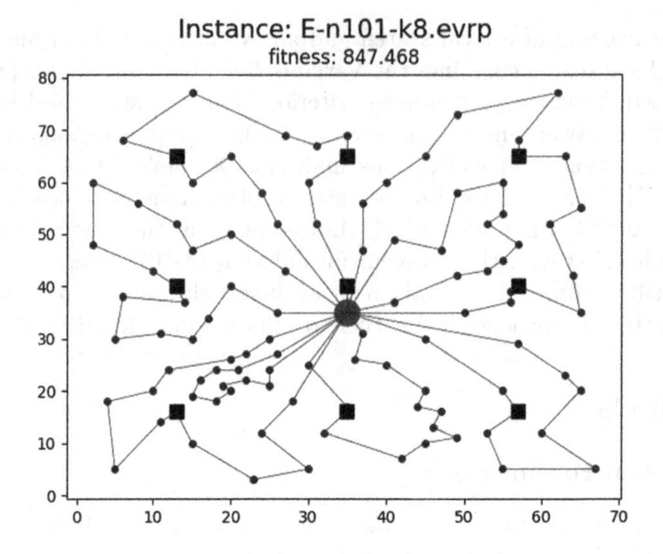

Fig. 1. Solved EVRP instance

$$y_j \leq y_i - h w_{ij} x_{ij} + Q(1 - x_{ij}), \forall i \in I, \forall j \in V, i \neq j, \tag{7}$$

$$y_j \leq Q - h w_{ij} x_{ij}, \forall i \in F' \cup D, \forall j \in V, i \neq j, \tag{8}$$

$$0 \leq y_i \leq Q, \forall i \in V, \tag{9}$$

$$x_{ij} \in \{0, 1\}, \forall j \in V, i \neq j, \tag{10}$$

where $V = \{D \cup I \cup F'\}$ is a set of nodes. Set I denotes the set of customers, set F' denotes set of β_i node copies of each AFS $i \in F$ (i.e., $|F'| = \sum_{i \in F} \beta_i$) and D denotes the central depot. Lets also define $E = \{(i, j) | i, j \in V, i \neq j\}$ as a set of edges in the fully connected weighted graph $G = (V, E)$. Then, x_{ij} is a binary decision variable corresponding to usage of the edge from node $i \in V$ to node $j \in V$ and w_{ij} is the weight of this edge. Variables u_i and y_i denote, respectively, the remaining carrying capacity and remaining battery charge flevel of an EV on its arrival at node $i \in V$. Finally, the constant h denotes the consumption rate of the EVs, C denotes their maximal carrying capacity, Q the maximal battery charge level, and b_i the demand of each customer $i \in I$.

For the purposes of formal components description, let's also define an EVRP tour T as a sequence of nodes $T = \{v_0, v_1, ..., v_{n-1}\}$, where v_i is a customer, a depot or an AFS and n is the length of the tour T. Finally, let

$$w(T) = \sum_{i=0}^{n-2} w_{i,i+1} \tag{11}$$

be the weight of the whole tour T.

2.2 GRASP Metaheuristic

GRASP is a well-established metaheuristic first introduced in [6] in 1989. Since then, it was successfully applied to numerous operations research problems, such as routing, covering and partition, location, minimum Steiner tree, optimization in graphs, assignment, and scheduling [7]. Its industrial applications include fields such as manufacturing, transportation, telecommunications, graph and map drawing, power systems, computational biology, or VLSI.

GRASP is a multi-start metaheuristic suitable for computing near-optimal solutions of combinatorial optimization problems. It is described in Algorithm 1. At the beginning, the best found tour T^* is initialized as empty and its weight $w(T^*)$ is set to infinity (lines 1–2). Then, the following process is repeated until a stop condition is met. A tour T visiting all customers is built from scratch, using a greedy randomized construction (line 4). If T is not a valid EVRP tour (e.g. constraints on load or battery capacity are not satisfied), tour T is fixed by a repair procedure (line 5–6). After that, the tour T is improved by a local search procedure (line 7), where a local minimum is found. The best tour found overall T^* is then updated, if T yields better weight (line 8–9). In this application, the stop condition is defined by a maximal number of GRASP iterations $MaxIters$.

Algorithm 1: Greedy Randomized Adaptive Search (GRASP)

Input: max. number of iterations $MaxIters$
Output: best tour found T^*

```
1  T* ← ∅
2  w(T*) ← ∞
3  for i = 1 to MaxIters do
4  |   T ← greedyRandomizedConstruction()
5  |   if !isValid(T) then
6  |   |   T ← repair(T)
7  |   T ← localSearch(T)
8  |   if w(T) < w(T*) then
9  |   |   T* ← T
10 return T*
```

2.3 Construction

In each iteration of the GRASP metaheuristic, a new valid EVRP tour is to be constructed. This tour serves as a starting point to the subsequent local search. According to the GRASP philosophy, a commonly used Nearest Neighbor (NN) heuristic was utilized for the greedy randomized construction. Due to the additional constraints imposed by the EVRP formulation, the NN construction can produce an invalid EVRP tour. Therefore, a repair procedure is needed. A novel procedure called Separate Sequential Fixing (SSF) was designed for this purpose.

Nearest Neighbor (NN) Construction. [10] is a commonly used algorithm to find an approximate solution to the Travelling Salesman Problem (TSP). The EVRP is equivalent to the TSP if the battery and load constraints are omitted. As these constraints cannot be easily incorporated into an iterative construction, it is convenient to determine only the order of the customers with the NN construction. The construction is described in Algorithm 2. The input to the algorithm is a set of all customers I, the depot D, and a set of edges E_{ID} in the fully connected weighted graph $G_{ID} = (D \cup I, E_{ID})$. Note that the AFSs F' and the corresponding edges are not considered in this phase. The output is then a TSP tour, which starts and ends at D and visits all customers. At the very beginning, the depot is added to T (line 1). Then, a first customer to be visited is randomly selected (line 3). After that, the remaining customers are greedily added to the tour one by one. Each time, the customer which is closest to the previously added customer is selected (line 5–8). Finally, the depot is added again and the tour is closed (line 9).

Algorithm 2: NN construction

Input: graph $G_{ID} = (D \cup I, E_{ID})$
Output: tour visiting all customers T_{TSP}

```
1  T_TSP.append(D)
2  Mark all customers in I as unvisited
3  Randomly select c ∈ I
4  T_TSP.append(c), mark c as visited
5  while all customers not visited do
6  |   Find the shortest edge from c to an unvisited c' ∈ I
7  |   T_TSP.append(c'), mark c' as visited
8  |   c ← c'
9  T_TSP.append(D)
10 return T
```

Separate Sequential Fixing (SSF) Repair Procedure is a newly proposed method designed to repair such an EVRP tour, where the constraints on battery or load capacity are not met. It consists of two phases, in which the constraints violations are fixed separately. If the following two assumptions are satisfied, the SSF procedure guarantees to produce a valid EVRP tour. First, the graph of all AFS and the depot must be connected. Second, each customer must be reachable from at least one AFS or depot.

The first phase of SSF is described in Algorithm 3. It takes a TSP tour over all of the customers T_{TSP} as an input and outputs a valid CVRP tour T_{CVRP}. All nodes are sequentially copied from the T_{TSP} to the T_{CVRP}, and the current vehicle load $CurLoad$ is held. If the current load is not sufficient for satisfying the next customer, the depot is added to the T_{CVRP} first.

Algorithm 3: SSF repair procedure - phase 1

```
    Input: tour visiting all customers T_TSP
    Output: valid CVRP tour T_CVRP
 1  T_CVRP ← ∅
 2  T_CVRP.append(T_TSP.popFront())
 3  CurLoad ← MaxLoad
 4  for Next in T_TSP do
 5  |   if CurLoad ≥ demand(Next) then
 6  |   |   T_CVRP.append(Next)
 7  |   |   CurLoad ← CurLoad − demand(Next)
 8  |   else
 9  |   |   T_CVRP.append(D)
10  |   |   T_CVRP.append(Next)
11  |   |   CurLoad ← MaxLoad − demand(Next)
12  return T_CVRP
```

The second phase is described in Algorithm 4. It takes a T_{CVRP} as an input and outputs a valid EVRP tour T_{EVRP}. This time, the nodes are sequentially copied from the T_{CVRP} to the T_{EVRP}. Initially, the depot is added to the T_{EVRP} and the current battery level $CurBattery$ is set to maximum (line 2–3). Then, the following loop is performed for all of the remaining nodes in the T_{CVRP}. The last node already added to the T_{EVRP} is denoted as $Current$ (line 5). The next node to be added is denoted as $Next$, $NextBattery$ is the potential battery level in the $Next$ node (line 6–9), and $NextAFS$ is the AFS, which is closest to the $Next$ node (line 10). If the $Next$ node is directly reachable from $Current$ and the vehicle will not get stuck in it, $Next$ is added to T_{EVRP} (line 11–15). Otherwise, the $CurrentAFS$ node, which is the closest AFS to $Current$, is determined (line 17). Then, a sequence of AFSs from $CurrentAFS$ to $NextAFS$ is added to T_{EVRP} (line 18). This sequence is obtained as the shortest path on a graph of all AFSs and a depot. Due to the condition about not getting stuck (line 13) and the two SSF assumptions, $CurrentAFS$ is always reachable from $Current$. After $NextAFS$, $Next$ can be added as well, and the loop is repeated until the T_{CVRP} is not empty.

Algorithm 4: SSF repair procedure - phase 2

```
   Input: valid CVRP tour T_CVRP
   Output: valid EVRP tour T_EVRP
 1 T_EVRP ← ∅
 2 T_EVRP.append(T_CVRP.popFront())
 3 CurBattery ← MaxBattery
 4 for Next in T_CVRP do
 5     Current ← T_EVRP.back()
 6     if isAFS(Next) then
 7         NextBattery ← MaxBattery
 8     else
 9         NextBattery ← CurBattery − getConsumption(Current, Next)
10     NextAFS ← getClosestAFS(Next)
11     Reachable ← CurBattery ≥ getConsumption(Current, Next)
12     Stuck ← NextBattery < getConsumption(Next, NextAFS)
13     if Reachable & !Stuck then
14         T_EVRP.append(Next)
15         CurBattery ← NextBattery
16     else
17         CurAFS ← getClosestAFS(Current)
18         T_EVRP.append(getPath(CurAFS, NextAFS))
19         T_EVRP.append(Next)
20         CurBattery ← MaxBattery − getConsumption(T_EVRP.back(), Next)
21 return T_EVRP
```

2.4 Local Search

This section provides a detailed description of the local search that is performed within the GRASP metaheuristic described in Sect. 2.2. The local search uses several local search operators, corresponding to different neighborhoods of an EVRP tour. The application of these operators is controlled by a simple heuristic. For this purpose, the Variable Neighborhood Descent (VND) and its randomized variant (RVND) were selected [3].

(Randomized) Variable Neighborhood Descent - (R)VND is a heuristic commonly used as a local search routine in other metaheuristics. It has a deterministic variant (VND) and a stochastic one (RVND). Both variants are described in Algorithm 5.

The input is a valid EVRP tour T and a sequence of local search operators $\mathcal{N}$, corresponding to different neighborhoods in the search space. The output is a potentially improved valid EVRP tour T. Both of the heuristic variants perform the local search sequentially in the neighborhoods in $\mathcal{N}$. In the case of the RVND, the sequence of the neighborhoods is randomly shuffled first (line 2), whereas, in the VND, the order remains fixed. Then, the heuristic attempts to improve the current tour T in the i-th neighborhood $\mathcal{N}_i$ according to the best improvement scenario (line 4). This corresponds to searching the local optimum in $\mathcal{N}_i(T)$. Each time an improvement is made, the local search is restarted and T is updated accordingly (line 5–7). The VND then starts again from the first neighborhood in $\mathcal{N}$, while the RVND randomly reshuffles the neighborhoods first (line 8). The algorithm terminates when no improvement is achieved in any of the neighborhoods.

Algorithm 5: (Rand.) Variable Neighborhood Descent - (R)VND

Input: valid EVRP tour T, neighborhoods $\mathcal{N}$
Output: potentially improved valid EVRP tour T

```
1  i ← 1
2  Randomly shuffle N // RVND only
3  while i ≤ |N| do
4      T' ← argmin_{T̃ ∈ N_i(T)} w(T̃)
5      if w(T') < w(T) then
6          T ← T'
7          i ← 1
8          Randomly shuffle N // RVND only
9      else
10         i ← i + 1
11 return T
```

Local Search Operators. A description of individual local search operators corresponding to different neighborhoods follows. Several operators commonly used in problems, where the solution can be encoded as a permutation (e.g., the TSP), were adapted for the EVRP. These operators are the 2-opt [4] and 2-string, which is a generalized version of numerous other commonly used operators.

An essential part of a neighborhood-oriented local search is efficient cost update computation. As both the 2-string derived operators and the 2-opt take two input parameters, the time complexity of exploring the whole neighborhood is $\mathcal{O}(n^2)$, where $n = |V|$. A naive approach is to apply every possible combination of parameters, create a modified tour $\tilde{T}$, and determine its weight $w(\tilde{T})$ in order to discover the most improving move. However, evaluating $w(\tilde{T})$ is a $\mathcal{O}(n)$ operation, thus the time complexity of the local search in each neighborhood would become $\mathcal{O}(n^3)$. This could be prevented by deriving $\mathcal{O}(1)$ cost update functions δ, which can be expressed as a difference between the sum of removed edges weights and the sum of newly added edges weights. Thus, a positive value of the cost update corresponds to an improvement in fitness and vice versa.

As the operators can produce an invalid EVRP tour, each local optimum candidate $\tilde{T} \in \mathcal{N}_i(T)$ is determined and checked for validity before acceptance as T'.

2-Opt is an operator commonly used in many variants of classical planning problems such as TSP or VRP. It takes a pair of indices i, j, and a tour T as an input and returns a modified tour $\tilde{T}$, where the sequence of nodes from i-th to j-th index is reversed. It must hold, that $i < j$, $i \geq 0$ and $j < n$.

The cost update function δ_{2-opt} can be evaluated as

$$\delta_{2-opt} = w_{i-1,i} + w_{j,j+1} - w_{i-1,j} - w_{i,j+1}, \tag{12}$$

where the indices are expressed w.r.t. to the tour T.

2-String and Its Variants is a generalized version of several commonly used operators, which can be obtained by fixing some of the 2-string parameters. The 2-string operator takes five parameters: a tour T, a pair of indices i, j valid w.r.t. to T, and a pair of non-negative integers X, Y. It returns a modified tour $\tilde{T}$, where the sequence of X nodes following after the i-th node in T is swapped with the sequence of Y nodes following after the j-th node. It must hold, that $i \geq 0$, $j \geq i + X$ and $j + Y \leq n - 1$. The following operators can be derived by fixing the values of X and Y:

- 1-point: $X = 0, Y = 1$
- 2-point: $X = 1, Y = 1$
- 3-point: $X = 1, Y = 2$
- or-opt2: $X = 0, Y = 2$
- or-opt3: $X = 0, Y = 3$
- or-opt4: $X = 0, Y = 4$
- or-opt5: $X = 0, Y = 5$

When performing the local search, the complementary variants of these operators (e.g., 1-point with $X = 1, Y = 0$) are considered as well.

The cost update function $\delta_{2-string}$ can be evaluated as

$$\delta_{2-string} = cut_1 + cut_2 + cut_3 + cut_4 - add_1 - add_2 - add_3 - add_4, \tag{13}$$

where cut_1 corresponds to the edge weight after i-th node in T, cut_2 to the edge after $i + X$, cut_3 to the edge after j and cut_4 to the edge after $j + Y$. Then, add_1 is the weight of the edge added after the index i-th node in T, add_2 of the edge added after the reinserted block of Y nodes, add_3 of the edge added after j and add_4 of the edge added after the reinserted block of the X nodes. For some combinations of the parameters, some of these values evaluate to zero, which must be carefully treated in the implementation. For example, if $X \neq 0$, then $cut_2 = w_{i+X,i+X+1}$, otherwise $cut_2 = 0$.

3 Results and Discussion

This section documents the parameters tuning and evaluates the performance of the proposed GRASP metaheuristic on the dataset introduced in [13]. Note that the CGVRP formulation used in [13] slightly differs from the EVRP formulation used in this paper and [14]. In [13], the energy consumption depends on the traveled distance and the current vehicle load, whereas in [14], it depends only on the traveled distance. Thus, the best-known values provided in [13] are not relevant when using the more general formulation from [14].

The stop condition of the metaheuristic is also adopted from [14]. An individual run terminates after $25000n$ fitness evaluations of a tour T, where $n = |V|$ is the actual problem size, that is, the total number of unique nodes. A single call to a distance matrix is counted as $1/n$ of an evaluation. Consistently with [14], each problem instance is solved 20 times, with random seeds ranging from 1 to 20. The experiments were carried out on a computer with an Intel Core i7-8700 3.20GHz processor and 32 GB RAM.

The rest of this section is structured as follows. Section 3.1 provides detailed information about the used dataset [13]. The process of tuning GRASP parameters on a subset of the dataset is described in Sect. 3.2. The final results of the tuned metaheuristic on the whole dataset are given in Sect. 3.3.

3.1 Dataset Description

The dataset consists of 4 types of instances, denoted as E, F, M, and X. These EVRP instances were created from already existing CVRP instances by a procedure described in [13], which adds a minimum number of charging stations such that all the customers are reachable from at least one charging station. The E instances are generated from the CVRP benchmark set from [1], the M instances from [2], the F instances from [8] and the X instances from [16]. The E and F instances are small to medium-sized, as they contain between 21 and 134 customers. The M instances are medium-sized (100 to 199 customers), and the X instances are large (143 to 1000 customers). Only instances X and E are addressed in [12]. Some parameters used in [12] instances (e.g., energy consumption rate and maximal energy capacity) slightly differ from the values used in [13]. Here, values from [12] are used when solving the X and E instances.

3.2 Parameters Tuning

The proposed GRASP metaheuristic was implemented in C++ and tuned on the E and X instances from [12]. Three settings of the local search were addressed: randomization of the local search (VND or RVND), neighborhood descent strategy (best improvement - BI or first improvement - FI), and selection of the best performing subset of the operators. The individual setups' names encode the parameter settings. For example, setup rvnd_BI_ls:255 stands for RVND, best improvement, and operators subset no. 255 (which is 11111111 in binary representation and corresponds to using all eight operators). As it was not feasible to

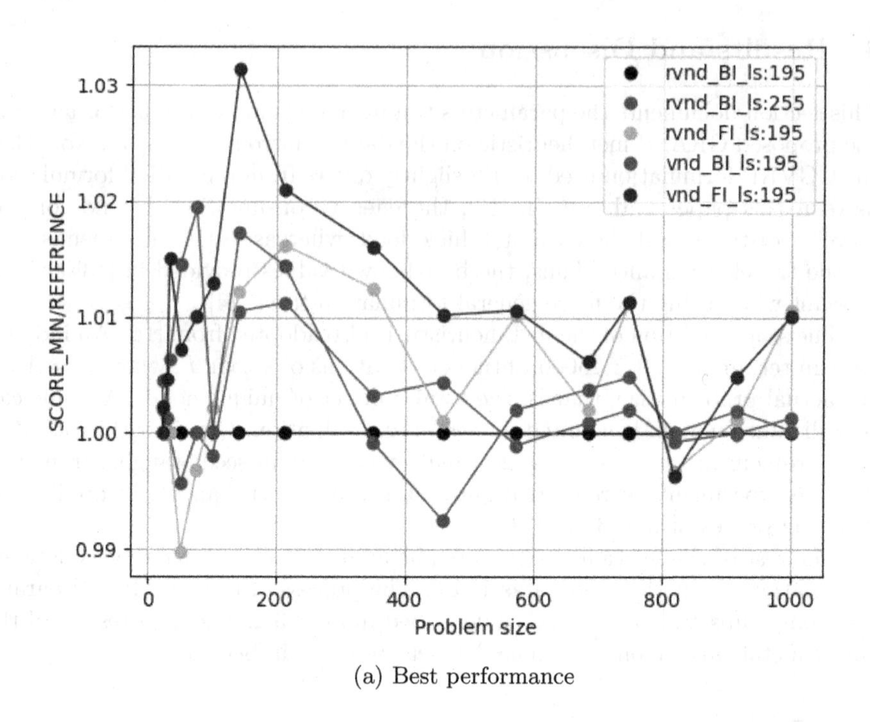

(a) Best performance

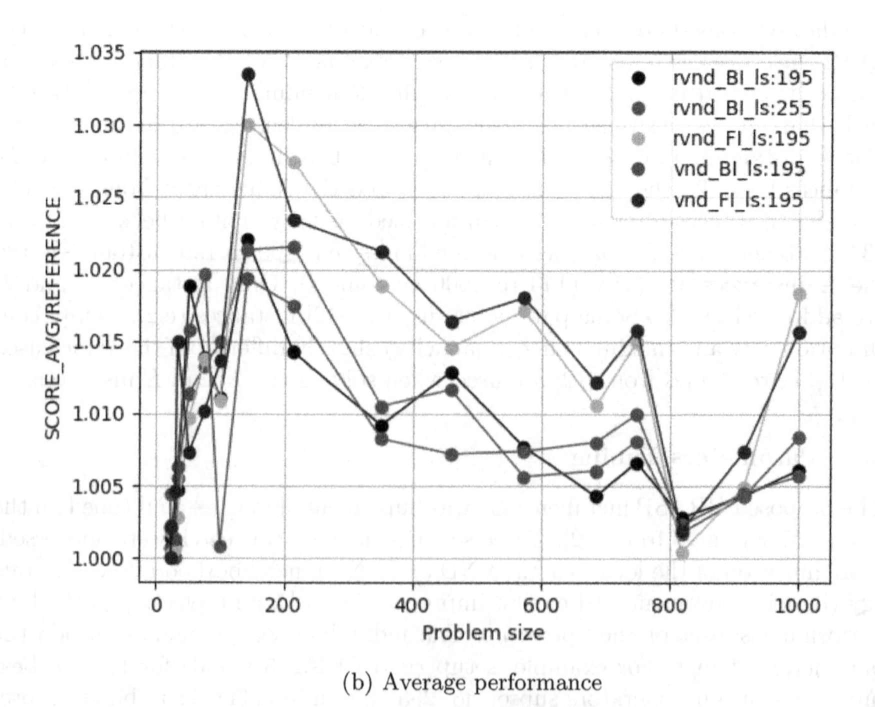

(b) Average performance

Fig. 2. Parameters tuning

tune all three parameters simultaneously, the process was split into two phases. First, all 255 possible subsets out the eight operators described in Sect. 2.4 were tested. In this phase, the remaining two parameters were set to RVND and BI. Second, all four combinations of the two remaining parameters were tested simultaneously, while the already selected subset of operators was fixed.

In the first phase, the method rvnd_BI_ls:195 performed best, as it yielded the lowest best score most frequently. The operators used in this setup are 2-opt, 1-point, 2-point, 3-point. Interestingly, no or-opt operator is included. The results of the second phase are presented in Fig. 2 and in Table 1. Fig. 2a displays the best performance of the individual setups, Fig. 2b the average performance and Table 1 provides counts of achieving the lowest (=best) score for each setup. The results are plotted relative to the best score achieved by the setup rvnd_BI_ls:195, which serves as a reference. It turns out that rvnd_BI_ls:195 is also best in terms of achieving the lowest best score (9 times out of 17 instances) in the second phase. However, vnd_BI_ls:195 is slightly better in terms of achieving the lowest average score. The first improvement descent strategy is generally rather unsuccessful. Based on these results, the setup rvnd_BI_ls:195 is selected as the final method. As can be seen in Fig. 2, the differences among individual setups are minor. When averaged across all instances, the worst setup vnd_FI_ls:195 is worse by 1% in terms of the best score and only by 0.5% in terms of the average score than the reference setup.

Table 1. Parameters tuning

GRASP setup	Lowest avg - cnt	lowest best - cnt
rvnd_BI_ls:195	4	9
rvnd_BI_ls:255	3	7
rvnd_FI_ls:195	2	6
vnd_FI_ls:195	3	2
vnd_BI_ls:195	5	2

3.3 Final Results

This section documents the performance of the tuned GRASP metaheuristic on all the available instances from the dataset [13]. The results on the E instances are presented in Table 2. The authors of [14] provided fitness values of the best-known solutions for these instances in [12]. These values are shown in the right-most column. The implemented GRASP metaheuristic found better solutions for 5 out of the 7 instances - the improved values are displayed in bold font in Table 2. The current best-known solution scores are given in the column marked as BKS. The best score obtained by the GRASP metaheuristic is at most by 1.2% worse than the BKS (instance E-n22-k4). On the other hand, the GRASP metaheuristic in some cases improved the previous best score by as much as 6% (instances E-n51-k5 and E-n101-k8).

The results on the X, F and M instances are given in Tables 3, 4 and 5 respectively. As no other solution values were known at the time of writing, the presented scores are intended as an initial benchmark for future research.

Table 2. GRASP results on competition E instances

instance	best	mean ± stdev	worst	t_{avg}(s)	BKS	prev. best
E-n22-k4	389.32	389.89 ± 0.41	390.19	0.09	384.68	384.68
E-n23-k3	**571.95**	572.36 ± 0.56	573.13	0.10	**571.95**	573.13
E-n30-k3	512.19	512.67 ± 0.31	512.88	0.13	511.25	511.25
E-n33-k4	**841.08**	845.06 ± 1.56	846.83	0.15	**841.08**	869.89
E-n51-k5	**536.09**	546.21 ± 5.32	562.32	0.30	**536.09**	570.17
E-n76-k7	**701.63**	711.36 ± 5.27	721.21	0.58	**701.63**	723.36
E-n101-k8	**847.47**	856.86 ± 6.90	871.10	0.96	**847.47**	899.88

Table 3. GRASP results on competition X instances

instance	best	mean ± stdev	worst	t_{avg}(s)
X-n143-k7	16460.80	16823.00 ± 157.00	17071.90	1.79
X-n214-k11	11575.60	11740.70 ± 80.41	11881.90	4.76
X-n351-k40	27521.20	27775.30 ± 111.99	28019.90	27.26
X-n459-k26	25929.20	26263.30 ± 134.66	26527.40	30.75
X-n573-k30	52584.50	52990.90 ± 246.79	53591.00	52.28
X-n685-k75	72481.60	72792.70 ± 189.53	73206.10	111.70
X-n749-k98	82187.30	82733.40 ± 213.21	83170.40	245.03
X-n819-k171	166500.00	166970.00 ± 211.84	167370.00	492.28
X-n916-k207	345777.00	347269.00 ± 654.93	348764.00	1108.73
X-n1001-k43	77636.20	78111.20 ± 315.31	78914.00	191.70

Table 4. GRASP results on F instances

instance	best	mean ± stdev	worst	t_{avg}(s)
F-n49-k4-s4	729.97	731.02 ± 0.88	732.57	0.21
F-n80-k4-s8	247.80	248.00 ± 0.45	249.21	0.49
F-n140-k5-s5	1177.97	1179.61 ± 1.96	1186.70	1.55

Table 5. GRASP results on M instances

instance	best	mean ± stdev	worst	t_{avg}(s)
M-n110-k10-s9	829.00	829.29 ± 0.41	830.11	1.04
M-n126-k7-s5	1066.00	1068.05 ± 1.40	1070.10	1.36
M-n163-k12-s12	1068.60	1081.42 ± 8.24	1106.36	2.31
M-n212-k16-s12	1359.55	1377.25 ± 9.82	1395.23	4.45

4 Conclusion

This paper addresses the recently formulated Electric Vehicle Routing Problem and presents the GRASP metaheuristic for finding high-quality solutions in a reasonable time. The performance is tested on the recently proposed dataset [13] of CGVRP instances, which is also the first one publicly available. For the instances with best-known solution values, the GRASP metaheuristic proves to be competitive, as it improves the best-known solution for 5 out of 7 instances. The rest of the instances with no previous solution values is solved as well, and the results are presented for future reference. The main strength of the implemented GRASP metaheuristic lies in efficient local search, which is sped up by using constant-time cost update functions. The key component when applying the GRASP to the EVRP is a newly proposed robust repair procedure called Separate Sequential Fixing (SSF).

Concerning future work, extracting problem-specific information will be tested. For example, all of the implemented local search operators do not consider the currently unused AFSs, as they are inspired by methods for TSP-like problems and explore only some permutation-based neighborhood of the current solution. Also, the initial NN construction is only distance-based, and meeting the battery and load constraints is left entirely for the repair procedure. Adopting informed methods for initial construction, such as those discussed in [11], might prove beneficial, especially for the larger instances where the quality of the initial solution is crucial. Besides that, it is important to compare the metaheuristic with other methods addressing similar problem formulations, such as [18] or [17]. These were tested on a dataset that is not publicly available and was not obtained at the time of writing.

Acknowledgements. This work has been supported by the European Union's Horizon 2020 research and innovation program under grant agreement No 688117. The work of David Woller and Viktor Kozák has also been supported by the Grant Agency of the Czech Technical University in Prague, grant No. SGS18/206/OHK3/3T/37.

References

1. Christofides, N., Eilon, S.: An algorithm for the vehicle-dispatching problem. J. Oper. Res. Soc. **20**(3), 309–318 (1969). https://doi.org/10.1057/jors.1969.75

2. Christofides, N., Mingozzi, A., Toth, P.: Exact algorithms for the vehicle routing problem, based on spanning tree and shortest path relaxations. Math. Program. **20**(1), 255–282 (1981). https://doi.org/10.1007/BF01589353
3. Duarte, A., Sánchez-Oro, J., Mladenović, N., Todosijević, R.: Variable neighborhood descent. In: Martí, R., Pardalos, P.M., Resende, M.G.C. (eds.) Handbook of Heuristics, pp. 341–367. Springer, Cham (2018). https://doi.org/10.1007/978-3-319-07124-4_9
4. Englert, M., Röglin, H., Vöcking, B.: Worst case and probabilistic analysis of the 2-Opt algorithm for the TSP. Algorithmica **68**(1), 190–264 (2013). https://doi.org/10.1007/s00453-013-9801-4
5. Erdelic, T., Carić, T., Lalla-Ruiz, E.: A survey on the electric vehicle routing problem: variants and solution approaches. J. Adv. Transp. **2019**, 48 (2019). https://doi.org/10.1155/2019/5075671
6. Feo, T.A., Resende, M.G.: A probabilistic heuristic for a computationally difficult set covering problem. Oper. Res. Lett. **8**(2), 67–71 (1989). https://doi.org/10.1016/0167-6377(89)90002-3
7. Festa, P., Resende, M.G.C.: GRASP. In: Martí, R., Pardalos, P., Resende, M. (eds.) Handbook of Heuristics. Springer, Cham (2018). https://doi.org/10.1007/978-3-319-07124-4_23
8. Fisher, M.L.: Optimal solution of vehicle routing problems using minimum K-trees. Oper. Res. **42**(4), 626–642 (1994). https://doi.org/10.1287/opre.42.4.626. https://www.jstor.org/stable/171617
9. Goncalves, F., Cardoso, S., Relvas, S.: Optimization of distribution network using electric vehicles: A VRP problem. Technical report. University of Lisbon (2011)
10. Gutin, G., Yeo, A., Zverovich, A.: Traveling salesman should not be greedy: Domination analysis of greedy-type heuristics for the TSP. Discrete Appl. Math. **117**(1–3), 81–86 (2002). https://doi.org/10.1016/S0166-218X(01)00195-0
11. Kozák, V., Woller, D., Kulich, M.: Initial solution constructors for capacitated green vehicle routing problem. In: Modelling and Simulation for Autonomous Systems (MESAS) **2020** (2020)
12. Mavrovouniotis, M.: CEC-12 Competition on Electric Vehicle Routing Problem (2020). https://mavrovouniotis.github.io/EVRPcompetition2020/. Accessed 23 Nov 2020
13. Mavrovouniotis, M., Menelaou, C., Timotheou, S., Ellinas, G.: A benchmark test suite for the electric capacitated vehicle routing problem. In: 2020 IEEE Congress on Evolutionary Computation (CEC), pp. 1–8 (2020). https://doi.org/10.1109/CEC48606.2020.9185753
14. Mavrovouniotis, M., Menelaou, C., Timotheou, S., Panayiotou, C., Ellinas, G., Polycarpou, M.: Benchmark Set for the IEEE WCCI-2020 Competition on Evolutionary Computation for the Electric Vehicle Routing Problem. Technical report, KIOS Research and Innovation Center of Excellence, Department of Electrical and Computer Engineering, University of Cyprus, Nicosia, Cyprus (2020). https://mavrovouniotis.github.io/EVRPcompetition2020/TR-EVRP-Competition.pdf
15. Normasari, N.M.E., Yu, V.F., Bachtiyar, C.: Sukoyo: A simulated annealing heuristic for the capacitated green vehicle routing problem. Mathematical Problems in Engineering **2019** (2019). https://doi.org/10.1155/2019/2358258
16. Uchoa, E., Pecin, D., Pessoa, A., Poggi, M., Vidal, T., Subramanian, A.: New benchmark instances for the capacitated vehicle routing problem. Eur. J. Oper. Res. **257**, 845–858 (2016). https://doi.org/10.1016/j.ejor.2016.08.012

17. Wang, L., Lu, J.: A memetic algorithm with competition for the capacitated green vehicle routing problem. IEEE/CAA J. Autom. Sinica **6**(2), 516–526 (2019). https://doi.org/10.1109/JAS.2019.1911405
18. Zhang, S., Gajpal, Y., Appadoo, S.S.: A meta-heuristic for capacitated green vehicle routing problem. Ann. Oper. Res. **269**(1–2), 753–771 (2018). https://doi.org/10.1007/s10479-017-2567-3

Stereo Camera Simulation in Blender

Tomáš Pivoňka(✉) and Libor Přeučil

Czech Institute of Informatics, Robotics and Cybernetics, Czech Technical University in Prague, Jugoslávských partyzánů 1580/3, 160 00 Prague, Czech Republic
{Tomas.Pivonka,Libor.Preucil}@cvut.cz

Abstract. The development and implementation of computer vision methods require an appropriate data-set of images for testing. Capturing real images is time-consuming, and it can be difficult in some applications. Furthermore, a precise measurement of distances and geometric transformations between camera positions for particular images is not always possible. In these cases, synthetic data can be beneficially used. Besides, they allow precisely setting camera positions and simulating various types of camera movements. The method for generation of stereo camera data-set is presented in this paper. It is built on the 3D creation suite Blender. The method is designed for visual odometry and 3D reconstruction testing, and it simulates a stereo camera movement over the captured 3D model. The images are directly inputted to the tested system. Together with the known ground truth of camera positions, they allow testing particular steps of the system. The proposed method was used for testing 3D reconstruction of a camera-based car undercarriages scanner.

Keywords: Stereo vision · Camera simulation · Blender

1 Introduction

3D reconstruction of a scene is one of the basic tasks in computer vision. Although a 3D model can be computed from a set of monocular camera images only, its scale is ambiguous. Determining an exact scale requires more information about a reconstructed object or relative positions of camera frames. Stereo camera satisfies the requirements because it is composed of two monocular cameras with known mutual positions. It returns a precise 3D model from each frame. Stereo cameras are also used for visual odometry or SLAM (simultaneous localization and mapping).

The presented work is a part of project KASSANDRA, whose main aim is to develop a camera-based scanner of car undercarriages. The scanner serves for a security inspection of cars. It contains two stereo pairs of cameras. It is placed under a road level and captures undercarriage while a car passes over it. Because it is impossible to capture the whole undercarriage in one frame, particular 3D models reconstructed from stereo images are stitched together. Visual odometry is used for computing individual models' mutual position.

J. Mazal et al. (Eds.): MESAS 2020, LNCS 12619, pp. 206–216, 2021.
https://doi.org/10.1007/978-3-030-70740-8_13

The main principles of computer vision are based on analytic geometry. The formulas contain a lot of coefficients, which represent inner camera parameters or mutual position of cameras in a stereo pair. Because it is not usually possible to determine coefficients values directly, they are gained from camera calibration. The correctness of all coefficients and transformations used in computations is required because even one error leads to a whole system's failure or considerably influences a result.

For evaluation and debugging of systems similar to the presented one, it is necessary to have testing data with known ground truth. Especially for a visual odometry evaluation, the accurate ground truth of transformations between particular images is crucial. It is difficult to measure a camera's precise motion while capturing real data. For basic evaluation of the method, a synthetic data-set can be beneficially used because all camera positions are known.

The method presented in this paper serves to generate synthetic stereo images, which simulate stereo camera motion. The system is implemented in advanced 3D creation suite Blender [3]. It allows to create or display 3D models, render images with virtual cameras, and it supports python scripts. Python is used to generate stereo images of an assigned model. The main advantage of synthetic data is the precise determination of all parameters and transformations. In addition, the images with corresponding calibration parameters can be directly processed by the tested system and used for debugging all of the steps. The method's limitation is the simplified model of cameras and artificial images, which do not include all real conditions. It also can not debug errors caused by wrong camera calibration.

The paper is structured into five chapters. In Sect. 2, there is introduced Blender software, the pinhole camera model, stereo vision, April camera calibration, and there are presented works related to a camera simulation. The proposed method is described in Sect. 3. A synthetic data-set generated by the method and its usage for 3D reconstruction testing are proposed in Sect. 4. Finally, the method and its results are discussed in Sect. 5.

2 State of the Art

2.1 Blender and Other Simulation Tools

Blender is the open-source 3D creation suite [3], which offers tools for the whole 3D creation pipeline. It covers modeling, rendering, simulation of objects' movements and interactions, and making animations. The software is also used by PC game developers.

Besides an advanced graphical interface, there is API for Python scripting. It serves to create new tools or automate processes. In this work, it is used for simulation of camera movement along the desired trajectory and capturing images.

There are several 3D creation suits with similar functionality as Blender. They are Autodesk 3DS Max [2] or Pixologic ZBrush [16], but both programs are paid.

Further, stereo cameras can also be simulated in robotic simulators like Gazebo [7], CoppeliaSim [4], or Microsoft Air-Sim [1]. Gazebo and CoppeliaSim also offer essential tools for scene modeling, but more complex objects have to be created in more advanced editors and imported as 3D models. Simulation software based on Gazebo presented in [8] uses Blender to create a model of an environment. One of the proposed method's main advantages is that Blender is the only software required for processing the whole pipeline from creating a 3D model to camera simulation.

2.2 Pinhole Camera

Pinhole camera model [9] describes transformation from a general world coordinate system to an image coordinate system. It represents central projection of a point in front of a camera to a sensor. It is the projection of a 3D point to 2D point in image coordinates. The model is simplified because it does not include any camera distortion. Distortion is usually removed by the undistortion of the captured image based on camera distortion coefficients.

The formula representing the transformation from world spatial coordinates to image coordinates is shown in Eq. (1), and the same formula with particular coordinates and parameters is presented in Eq. (2). $\vec{u}$ represents an image point with coordinates u, v. $\vec{x}$ is a point in 3D with coordinates X, Y, Z. Both $\vec{u}$ and $\vec{x}$ are vectors in homogeneous spaces. $\mathbf{K}$ is a camera matrix with horizontal and vertical focal lengths f_u, f_v and coordinates of principal point in the image c_u, c_v. Rotation matrix $\mathbf{R}$ and vector $\vec{t}$ represents an orientation and position of a camera in the world space.

$$\lambda \cdot \vec{u} = \mathbf{K} \cdot \left[\mathbf{R} \mid \vec{t}\right] \cdot \vec{X} \tag{1}$$

$$\begin{bmatrix} \lambda \cdot u \\ \lambda \cdot v \\ \lambda \end{bmatrix} = \begin{bmatrix} f_u & 0 & c_u \\ 0 & f_v & c_v \\ 0 & 0 & 1 \end{bmatrix} \cdot \left[\mathbf{R} \mid \vec{t}\right] \cdot \begin{bmatrix} X \\ Y \\ Z \\ 1 \end{bmatrix} \tag{2}$$

In this work, units of spatial and image coordinates are meters and pixels. Further, focal lengths f_u and f_v are considered to be equal. A camera parameters in Blender are a focal length f_{mm} in millimeters, a width w_{mm} (or a height) of a chip in millimeters, and vertical and horizontal camera resolution w_{px}, h_{px} in pixels. The relationship between desired focal length f and Blender parameters is expressed in Eq. (3).

$$f = \frac{w_{px} \cdot f_{mm}}{w_{mm}} \tag{3}$$

2.3 April Camera Calibration

Camera calibration serves to determine camera distortion coefficients, a focal length, and a principal point position. During the calibration, a printed pattern is captured by a camera from different view positions. Based on the pattern detection in images, all coefficients are computed to minimize the error of the camera. The output of the calibration is usually text files with computed coefficients and camera parameters.

April camera calibration [15] is used for calibration of real cameras in the car undercarriages' scanner [5]. Therefore, it is also assumed to be used with a real camera in this work, and simulation parameters are adjusted to it. Further, a calibration model with three coefficients for radial distortion and two coefficients for tangential distortion is assumed. Because there is no distortion in simulation, all these coefficients are equal to zero. Even they are not used in the simulation, a tested system can require them. April calibration allows multiple camera calibration, which returns the mutual position of cameras in a stereo pair.

Each camera calibrated by the April camera calibration is determined by intrinsic and extrinsic parameters. The intrinsic ones are horizontal and vertical focal length, principal point coordinates, and distortion coefficients in a selected format. The extrinsic parameters describe a camera position in a world coordinate system set as a local coordinate system of the first calibrated camera. The camera position is determined by three rotation angles (Roll, Pitch, Yaw system) and x, y, z coordinates of a translation vector. All six parameters represent transformation from a global coordinate system to a system of a camera.

2.4 Camera Simulation in Blender

One of the Blender's possible applications in mobile robotics is simulating an environment for mobile robots. The system presented in [10] combines Blender with Matlab. Matlab serves to simulate robot dynamics and to perform a visual SLAM. Blender simulates a camera mounted on a robot and captures a simulated environment from positions computed in Matlab.

The possibility of testing visual SLAM systems on data-sets from Blender is proposed in [14]. Simulating many environments with various properties allows systematic analysis of SLAM methods and their mutual comparison. The proposed system is used for testing of sparse visual odometry algorithm on three synthetic data-sets of cities.

Structure-from-motion (SfM) task is 3D object reconstruction from different camera positions. The comparison of many SfM systems is presented in [11]. Because creating a data-set of real objects with precise ground truth requires expensive specialized hardware, and it is time-demanding, the authors used Blender to extend the data-set.

Synthetic image data can be beneficially used for neural network training. The supervised teaching requires a lot of data with known ground truth. For example, real data-sets for segmentation usually have to be created manually. By contrast, synthetic data-sets give ground truth for segmentation automatically. In [6], the Blender is used to simulate a camera of an unmanned aerial vehicle for neural network training. Besides, the presented method allows to generate large urban models automatically.

The presented method and all systems described in this chapter use Blender to simulate regular cameras. In [13], there is shown, that Blender can also be used for modeling plenoptic cameras, which are more complicated optical systems. They have a special layer of micro-lenses between a sensor and the main lens. The primary motivation was the possibility to study plenoptic cameras without the necessity to have expensive hardware.

3 Stereo Camera Simulation System

3.1 3D Model

To create stereo images that can be directly processed by the tested system, a proper 3D model similar to a real environment is necessary. Its required properties depend on a tested method. In general, most of the methods in computer vision fail in low textured environments.

The 3D model can be directly designed in Blender or loaded from a file. The appearance of a model is further influenced by lighting. Blender offers several types of light sources, which can be added to a scene. For essential evaluation of implemented algorithms and parameters, a too detailed model may cause undesirable failures. Using of uniform lighting and diffusive materials only is sufficient.

The final scene for stereo images' rendering contains a 3D model placed on the desired position in a global coordinate system, light sources, and two cameras. Whereas model and lighting positions are fixed, camera positions are set by capturing software. Because all scene-setting is done in Blender graphical user interface, a user has full control over scene appearance.

3.2 Stereo Camera Simulation

In the proposed system, a stereo camera is represented by two camera objects with a fixed mutual position. Because Blender simulates cameras' hardware properties, it is necessary to set focal length, size of a sensor, and camera resolution. All mentioned parameters determine a camera matrix. The pinhole camera model with no shift of optical center is used.

Cameras' mutual position is determined by a transformation from the first to second camera coordinates. It is represented by a vector of translation and the Roll-Pitch-Yaw system of Euler angles. The angles can be directly converted to the rotation matrix and vice versa.

In the system, the first camera's position is set in global coordinates and directly determines a second camera position. The shift from first to second camera position is represented as a 6 DoF (degrees of freedom) rigid body motion. It corresponds to the transformation of coordinates from second to first camera coordinates, which is an inverse assigned transformation representing the mutual position of cameras. While most computer vision methods use camera coordinates with z-axis oriented towards a captured object and y-axis corresponding to y-axis in image coordinates, in Blender camera local coordinates, z-axis and y-axis are oriented oppositely. The transformation between these two types of coordinates is rotation around the x-axis by 180°.

The equations (4) and (5) represent a conversion from the assigned mutual position represented by vector $\vec{t}_{12}$ and rotation matrix $\mathbf{R_{12}}$ to rigid body motion from first to second camera $\vec{t}$, $\mathbf{R}$. $\mathbf{R_X}$ is a rotation matrix of the rotation around the x-axis by 90°. The final transformation is in local coordinates of the first camera.

$$\mathbf{R} = \mathbf{R_X} \cdot \mathbf{R_{12}^T} \cdot \mathbf{R_X} \tag{4}$$

$$\vec{t} = -\mathbf{R_X} \cdot \mathbf{R_{12}^T} \cdot \vec{t}_{12} \tag{5}$$

Computing of second camera position $\vec{t}_{C2}$, $\mathbf{R_{C2}}$ in global coordinate system is shown in equations (6), (7). First camera global position is represented by $\vec{t}_{C1}$, $\mathbf{R_{C1}}$.

$$\mathbf{R_{C2}} = \mathbf{R_{C1}} \cdot \mathbf{R} \tag{6}$$

$$\vec{t}_{C2} = \vec{t}_{C1} + \mathbf{R_{C1}} \cdot \vec{t} \tag{7}$$

3.3 Capturing Software

The pipeline of the proposed system is depicted in Fig. 1. In the first step, the 3D model is created or loaded, and light sources and two cameras are set. The rendering of images and setting of cameras positions is done by python script automatically. Other inputs to the program are mutual positions of cameras, an initial position of the first camera, and a relative shift of the camera between stereo images. The position and the shift are represented by a vector of translation and Euler angles.

At the beginning of the script, positions of cameras are set. Rendering of images and camera movement is performed in a loop. In each step, images of both cameras are rendered and stored. Afterward, the first camera position is shifted according to the shift from the input represented by a rotation matrix $\mathbf{R_m}$ and a translation vector $\vec{t}_m$. The computation of the new position of first camera in step i is expressed in equations (8), (9).

$$\mathbf{R_{C1}^{i+1}} = \mathbf{R_{C1}^{i}} \cdot \mathbf{R_m} \tag{8}$$

$$\vec{t}_{C1}^{\,i+1} = \vec{t}_{C1}^{\,i} + \mathbf{R_{C1}^{i}} \cdot \vec{t}_m \tag{9}$$

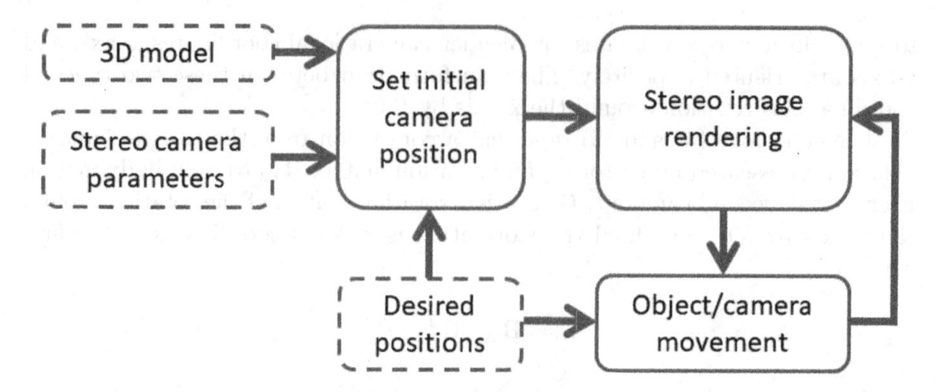

Fig. 1. A pipeline of the presented method for synthetic data rendering.

4 Experiments

4.1 Synthetic Data

The proposed method serves for testing visual odometry and the 3D reconstruction method of car undercarriages scanner [5]. The scanner uses a system of mirrors, and a stereo image is captured only by one camera. It reduces the necessary depth of a scanner under a road level. On the other hand, a camera center is projected behind a mirror, and it is not possible to determine the precise position of a camera and to measure undercarriage motion relative to the camera.

The proposed system was used to generate testing data for evaluation and debugging software. A simple 3D model with similar properties to a car undercarriage was created. It consists of a planar rectangle with size 5 × 3 m, three cuboids (one beveled), and four hemispheres, as shown in Fig. 2. The frontal faces of cuboids were textured by an image with a lot of visual features to ensure the stability of feature-based visual odometry. Hemispheres were added for testing 3D reconstruction.

Two cameras of a stereo pair were placed similarly to positions from camera calibration of a real prototype. The resolution of cameras was 1920 x 530 px identically to real cameras. Cameras in the simulator are pinhole cameras with an optical center in the middle of a sensor. Radial distortion and optical center shift, which is significant for the mirror system because of taking both images by one camera, was not simulated. Therefore, the mutual position can not be the same as in original camera calibration. An example of a rendered stereo image is in Fig. 3.

4.2 Data Evaluation

To evaluate the correctness of rendered data, they were tested in a Meshroom software [12]. The test was performed to verify camera positions and 3D model

Fig. 2. A 3D model created for an evaluation of visual odometry and 3D reconstruction methods for a car undercarriages scanner.

properties. Meshroom is a free software for structure from motion, and it reconstructs a 3D model from a proposed set of images. Because it does not work directly with stereo cameras, the final model is ambiguous on a scale. It does not distinguish between cameras, and it works with data as belonging from one camera only. The test was performed only for a part of the whole model. The result of a 3D reconstruction is shown in Fig. 4. There is depicted a textured 3D model together with particular camera positions. The movement of a stereo camera was only 1D translation above a model. It is possible to observe that all camera positions were reconstructed correctly (with right orientation to the model and positions on a line), and the final model also fits well with the original. Errors at the borders of the model are caused by a lack of information for reconstruction, and they are not related to synthetic data.

4.3 Testing of 3D Reconstruction System

The data-set presented in Sect. 4.1 was successfully used for debugging 3D undercarriage scanner software [5]. It uses stereo reconstruction for building 3D models of captured parts of an undercarriage, which are connected together based on visual odometry. The main advantages of synthetic data testing were a precise simulation of various motions and known properties of a reconstructed model. The model created by the system on the synthetic data is in Fig. 5.

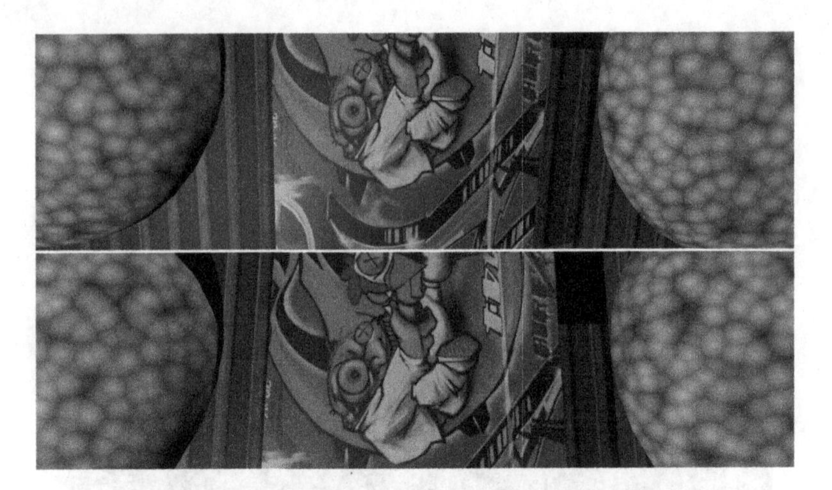

Fig. 3. An example of a synthetic stereo image from the simulator.

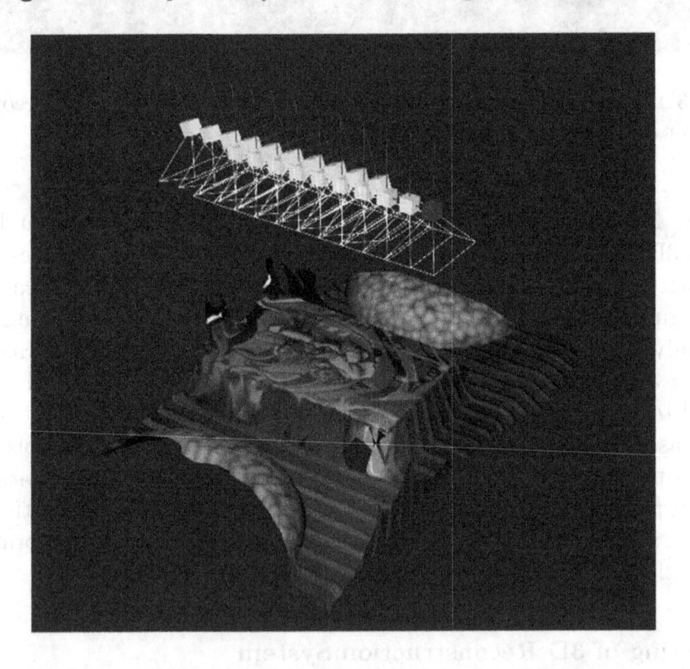

Fig. 4. The rendered data-set evaluation by a structure from motion software.

Fig. 5. A model reconstructed by the tested undercarriage scanner software.

5 Conclusions

The method for generating a data-set of synthetic stereo images in Blender was presented. It serves to the testing of computer vision systems. The main advantages of the approach are an easy creation of a well-suited 3D model and the known ground truth of all transformations, which is difficult to arrange in real conditions. The data-set with reference values can be used for the evaluation and debugging of computer vision systems. Generated data can be directly put to the system's input. Besides, knowledge of all transformations and model properties can serve to check intermediate results.

The proposed method was used to generate a synthetic data-set for vehicle undercarriages scanner. The well-textured model, with a similar shape to undercarriages, was created. The system generates a data-set with similar properties to the system with mirrors. Synthetic data were used mainly for simulating more complex trajectories and evaluation of 3D reconstruction.

The capturing in real conditions is still necessary for evaluating the performance and stability of the method, but it has higher demands on time, people, and material resources. While for the evaluation of basic functionality and correctness of tested methods, the synthetic data are sufficient.

Acknowledgements. The project is solved in collaboration with VOP CZ, s.p. and it is supported by the Ministry of the Interior of the Czech Republic under the project number VI20172020080.

References

1. Air-sim. https://microsoft.github.io/AirSim/

2. Autodesk 3ds max. https://www.autodesk.com/products/3ds-max
3. Blender. https://www.blender.org/
4. Coppeliasim. https://www.coppeliarobotics.com/
5. Dörfler, M., Pivoňka, T., Košnar, K., Přeučil, L.: Application of surface reconstruction for car undercarriage inspection. In: 2020 3rd International Conference on Intelligent Robotic and Control Engineering (IRCE), pp. 47–51 (2020). https://doi.org/10.1109/IRCE50905.2020.9199251
6. Gao, Q., Shen, X., Niu, W.: Large-scale synthetic urban dataset for aerial scene understanding. IEEE Access **8**, 42131–42140 (2020). https://doi.org/10.1109/ACCESS.2020.2976686
7. gazebo. http://gazebosim.org/
8. Giubilato, R., Masili, A., Chiodini, S., Pertile, M., Debei, S.: Simulation framework for mobile robots in planetary-like environments. In: 2020 IEEE 7th International Workshop on Metrology for AeroSpace (MetroAeroSpace), pp. 594–599 (2020). https://doi.org/10.1109/MetroAeroSpace48742.2020.9160154
9. Hartley, R.I., Zisserman, A.: Multiple View Geometry in Computer Vision, 2nd edn. Cambridge University Press, Cambridge (2004). ISBN: 0521540518
10. Lima, M.V.P., Bastos, V.B., Kurka, P.R.G., Araujo, D.C.: vslam experiments in a custom simulated environment. In: 2015 International Conference on Indoor Positioning and Indoor Navigation (IPIN), pp. 1–7 (2015). https://doi.org/10.1109/IPIN.2015.7346757
11. Marelli, D., Bianco, S., Celona, L., Ciocca, G.: A blender plug-in for comparing structure from motion pipelines. In: 2018 IEEE 8th International Conference on Consumer Electronics - Berlin (ICCE-Berlin), pp. 1–5 (2018). https://doi.org/10.1109/ICCE-Berlin.2018.8576196
12. Meshroom. https://alicevision.org/#meshroom
13. Michels, T., Petersen, A., Palmieri, L., Koch, R.: Simulation of plenoptic cameras. In: 2018–3DTV-Conference: The True Vision - Capture, Transmission and Display of 3D Video (3DTV-CON), pp. 1–4 (2018). https://doi.org/10.1109/3DTV.2018.8478432
14. Particke, F., Kalisz, A., Hofmann, C., Hiller, M., Bey, H., Thielecke, J.: Systematic analysis of direct sparse odometry. In: 2018 Digital Image Computing: Techniques and Applications (DICTA), pp. 1–6 (2018). https://doi.org/10.1109/DICTA.2018.8615807
15. Richardson, A., Strom, J., Olson, E.: AprilCal: assisted and repeatable camera calibration. In: 2013 IEEE/RSJ International Conference on Intelligent Robots and Systems, pp. 1814–1821 (2013). https://doi.org/10.1109/IROS.2013.6696595
16. Zbrush2021. http://pixologic.com/features/about-zbrush.php

Vision-Based Localization for Multi-rotor Aerial Vehicle in Outdoor Scenarios

Jan Bayer(✉) and Jan Faigl

Faculty of Electrical Engineering, Czech Technical University in Prague, Technicka 2, 166 27 Prague, Czech Republic
{bayerja1,faiglj}@fel.cvut.cz
https://comrob.fel.cvut.cz

Abstract. In this paper, we report on the experimental evaluation of the embedded visual localization system, the Intel RealSense T265, deployed on a multi-rotor unmanned aerial vehicle. The performed evaluation is targeted to examine the limits of the localization system and discover its weak points. The system has been deployed in outdoor rural scenarios at altitudes up to 20 m. The Absolute trajectory error measures the accuracy of the localization with the reference provided by the differential GPS with centimeter precision. Besides, the localization performance is compared to the state-of-the-art feature-based visual localization ORB-SLAM2 utilizing the Intel RealSense D435 depth camera. In both types of experimental scenarios, with the teleoperated and autonomous vehicle, the identified weak point of the system is a translation drift. However, taking into account all experimental trials, both examined localization systems provide competitive results.

1 Introduction

A precise and reliable localization system is necessary for many robotics and related applications, including mapping, augmented reality, and fully autonomous deployments of mobile robots. In this work, we consider the mobile robotics domain with the primary focus on localization systems for Unmanned Aerial Vehicles (UAVs) in applications such as autonomous navigation, exploration, or perimeter monitoring, where a full 6 DOF localization is required. The existing localization solutions can be broadly divided into two classes. The first class contains systems that require external infrastructures, such as the Global Navigation Satellite System (GNSS) [13] or optical systems operating on the line of sight [6,7,17]. The great advantage of these localization systems is the limited accumulation of localization errors, even in large-scale scenarios. However, these methods are limited by the needed infrastructure, particularly not suitable for large scale indoor environments without prior preparations or scenarios with many obstacles shading signal or line of sight. On the other hand, systems of the second class rely only on sensors mounted on the robot. These methods are much more suitable for unknown environments, where external infrastructure cannot be prepared in advance. Such localization systems mainly use Light Detection and Ranging (LiDAR) sensors [22] and different types of cameras [11].

J. Mazal et al. (Eds.): MESAS 2020, LNCS 12619, pp. 217–228, 2021.
https://doi.org/10.1007/978-3-030-70740-8_14

Fig. 1. The unmanned aerial vehicle DJI Matrice 600, which was deployed during the experimental evaluation.

In recent years, sensors for visual localization have been introduced, including the Intel RealSense T265 tracking camera [15] that is referred to as the T265 for short in the rest of the paper. The T265 provides image data, and it is also capable of processing the data by its embedded visual processing unit Movidius Myriad 2.0, capable of doing all the computations needed for visual localization. The power effectiveness is a great advantage of embedded solutions for the localization since the visual localization algorithm might run on dedicated computational resources of the camera itself, saving the main onboard computational resources of the UAV. The saved computational power can be thus utilized for navigation and real-time mapping tasks, or even more sophisticated tasks like autonomous exploration. The features and benefits of embedded solutions based on the T265 and recent successful deployments on the ground legged walking robots [4,16] motivate us to examine its performance for localization of small UAV, see Fig. 1. Thus, we evaluate the T265 as a vision-based localization system that uses affordable off-the-shelf lightweight cameras. Besides, we compare the performance to the state-of-the-art localization method ORB-SLAM2 [20].

The rest of the paper is organized as follows. Principles and related evaluations of the visual localization methods are presented in Sect. 2. The evaluation method utilized for the evaluation of the visual localization is overviewed in Sect. 3. The resulting localization precision measured during the deployments is reported in Sect. 4. The concluding remarks on the achieved localization precision and limits of the embedded localization system are presented in Sect. 5.

2 Related Work

Two recent compact sensors for vision-based localization are available on the market nowadays: the ZED mini [27] and Intel RealSense T265 [15], shown in Fig. 2a. Both of them are passive fisheye stereo cameras. ZED mini primary targets augmented reality applications, and it requires external computational power for GPU-based acceleration of the visual localization algorithms. On the other hand, the T265 runs all the computations onboard using a power-efficient visual processing unit, which can save the computational and power resources such, e.g., as reported in [4]. The T265 is thus a suitable choice for embedded localization for small robotic vehicles such as multi-rotor UAVs.

(a) Intel RealSense T265. (b) Intel RealSense D435.

Fig. 2. Cameras used for the visual localization.

The parameters of the T265 are promising, but the experimental evaluation is important for the verification of the localization performance in real-world scenarios because the real localization performance is affected by multiple factors that are related to the

- precision of the localization algorithm itself;
- properties of the environment;
- sensory equipment;
- computational resources (if limited);
- and motion of the platform with the sensors.

The experimental evaluations (to the best of the authors' knowledge) have been reported only for indoor conditions or carried by humans [1,2,5,18,23]. In these scenarios, the T265-based localization system is reported to with satisfactory performance. Thus, in the herein presented results, we aim to push the localization system to its limits. We deploy the system in real-world outdoor scenarios with the DJI Matrice 600 operating in a rural-like environment. The precision of the localization is measured as in [23] using the well-established metric of the Absolute Trajectory Error (ATE) [28].

Besides, a state-of-the-art vision-based localization method has been selected to provide a baseline solution for the selected scenarios using a traditional approach with cameras and data processing on dedicated standard computational resources. There are many different visual localization approaches mentioned in the literature. One of the differences between the methods is in the required sensory equipment. Techniques such as [9,10,21] use monocular cameras. Other approaches use stereo cameras [24,29] and RGB-D cameras [8,30]. The stereo and RGB-D cameras can be considered more sophisticated than monocular cameras regarding the sensory equipment, and their advantage is the reduced drift of the map scale. Besides, the map initialization is easier as bootstrapping is avoided [12]. Thus, we focus on the stereo and RGB-D localization methods capable of being used with various sensors.

A representative approach is the ORB-SLAM2 [20], a state-of-the-art publicly available localization method, reported to perform well on datasets [19,26,28]

Fig. 3. Examples of image features detected by ORB-SLAM2 [20].

and often used as the baseline for comparing different localization approaches [2]. Based on our previous work [3,4], we have chosen to use the feature-based localization method ORB-SLAM2 together with the RGB-D camera Intel RealSense D435 [14], shown in Fig. 2b. An example of the detected ORB features used by ORB-SLAM2 is in Fig. 3.

3 Evaluation Method for Localization Systems

In the literature, the localization systems are compared by the precision of trajectories obtained during an experiment. A comparison of the localization systems solely based on the trajectories allows us to compare black-box localization systems like the T265, where a detailed description of the particular algorithms is not provided. For such purposes, there is a well-established metric of the *Absolute Trajectory Error* (ATE) [28]. A ground truth trajectory and trajectories estimated by the localization systems under the examination are required to compute the ATE. Besides, it is also necessary to have the trajectory estimate and ground truth with poses that correspond to the same timestamps. Therefore, the trajectories have to be time-synchronized using interpolation or by finding the nearest neighbor [28] if the ground truth trajectory is provided with a different frequency than the trajectory estimate. In this work, the linear interpolation of the positions and *Linear Quaternion Interpolation* (LERP) of the orientation represented by the quaternions is utilized.

Once the trajectories are time-synchronized, they are processed by the ATE metrics. The ATE is defined in [28] by the equation

$$\mathbf{F}_i = \mathbf{Q}_i^{-1}\,\mathbf{S}\,\mathbf{P}_i, \tag{1}$$

where the matrices $\mathbf{Q}_i$ and $\mathbf{P}_i$ are SE(3) pose of the ground truth and estimated trajectory, respectively. The matrix $\mathbf{S}$ is a transformation between the coordinate frames of the ground truth and trajectory estimate. According to [28], the transformation is obtained by minimizing the squared distances between the corresponding positions of the trajectory estimate and the ground truth.

The average value is used to generate a statistical indicator from the error for the whole trajectory

$$\overline{\mathrm{ATE}_t} = \frac{1}{n}\sum_{i=1}^{n} \parallel trans(\mathbf{F}_i) \parallel, \tag{2}$$

where $trans()$ computes the size of the translation from the SE(3) matrix. In the results reported in Sect. 4, we assume only the translation errors because the ground truth is provided in 3 DOF only.

4 Results

The UAV has been experimentally deployed in the outdoor environment of the rural deployment scenario shown in Fig. 4. The UAV was flying in the experimental setup at different altitudes: 5, 10, 15, and 20 m driven manually by a human operator and autonomously by the DJI autopilot, to verify the ability of the localization system to work with the absence of close objects while being exposed to different motions. Two pairs of the Intel RealSense cameras were mounted on the UAV. The first pair is pointed to the front tilted by 45°, the second pair is oriented to the rear side and tilted as well, see Fig. 5.

Data from all the sensors were collected using the ROS middleware [25]. The localization from tracking cameras is captured directly during the experiments together with the RGB-D data. The localization provided by the ORB-SLAM2 was generated by processing the RGB-D data from the rosbag dataset captured during the experiment. The ROS rosbag captures data incoming from the sensors as provided by the sensors. It also enables processing the data at different speeds to simulate different computational power. Thus, the RGB-D data were processed by the ORB-SLAM2 at two different speeds. The first processing speed is to simulate online processing, and it is denoted *online*-speed. The second speed is denoted *half*-speed, and it corresponds to two times more computational time for processing the captured images than processing them in real-time. The ORB-SLAM2 was run on the regular computer with the Intel i5-5257U processor running at 2.7 GHz with 4 GB of memory. The Differential GPS provided the ground truth trajectory with a centimeter precision at 10 Hz.

4.1 Localization Error

The mean ATE for all experimental trials is summarized in Table 1. Each experimental trial contains several circular flights that are illustrated on Trial 1 in Fig. 6. The results show that the ORB-SLAM2 provides the best results with the front camera in Trial 1 and Trial 3. On the other hand, the T265 provided better results in Trial 2 and Trial 4. In the case the UAV was teleoperated manually, the angular velocity of the helicopter was always under 30° s^{-1}. In Trial 2,

Fig. 4. The deployment scenario in the rural-like environment.

Fig. 5. The sensory equipment used during the experiments attached to the UAV; the first pair of the Intel RealSense cameras is pointed to the front. The second pair is pointed to the rear side.

the UAV flew autonomously between preselected waypoints. On each waypoint, the UAV turned with an angular velocity above $30^\circ\ \mathrm{s}^{-1}$, which induced failure of the localization based on the ORB-SLAM2 in combination with the front D435 camera. Trial 2 is visualized in Fig. 7. It is possible to overcome the localization failure using more computational power, which can be observed for *half*-speed results. However, the drift at the corners of the trajectory is still very high; see Fig. 7b. Contrary, the effect of the high rotational speed on the localization quality is not observed for the T265.

The relatively small localization error provided by the T265 can be observed for Trial 4 shown in Fig. 8, where the UAV flew at altitudes of 10 m, 15 m, and 20 m. Based on the deformation of the trajectory estimated by the T265, shown

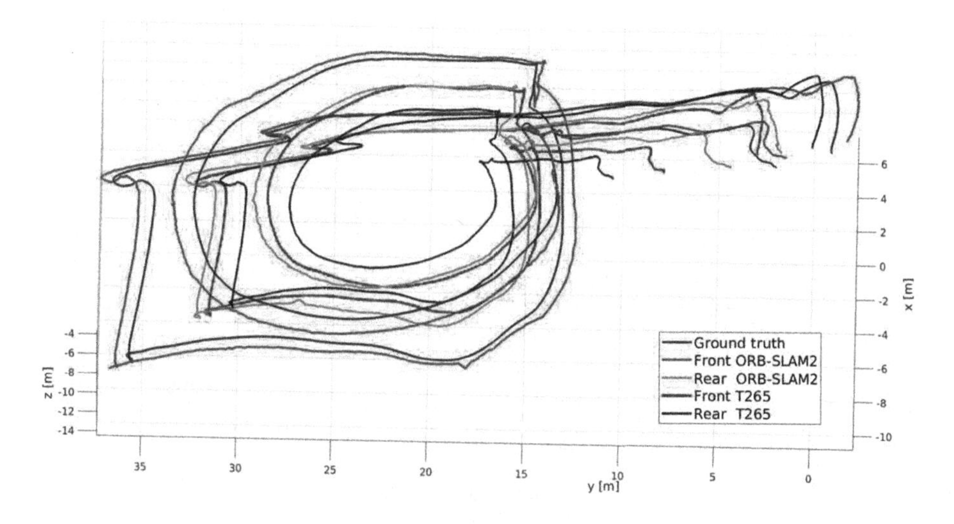

Fig. 6. Trial 1: two circular flights.

Table 1. The mean absolute trajectory error for front and rear setup in all four experimental trials.

Trial	Flight mode	Length [m]	T265		ORB-SLAM2			
					Online-speed		*Half*-speed	
			Front	Rear	Front	Rear	Front	Rear
Trial 1	Manual	174	3.01	4.01	0.75	2.41	**0.69**	2.38
Trial 2	Autonomous	170	**1.57**	4.63	Lost	4.12	7.12	4.58
Trial 3	Manual	596	3.84	6.37	1.92	4.09	**1.63**	3.07
Trial 4	Manual	1175	**0.91**	6.14	1.96	4.92	4.51	3.77

All results provided by the ORB-SLAM2 were evaluated on a computer with the Intel i5-5257U and 4 GB of memory. *Half*-speed means that the ORB-SLAM2 has twice more time to process the incoming data than in the *online* case.

in Fig. 7a and Fig. 8, it can be observed that the localization suffers mostly from the translation drift, not the orientation drift.

4.2 Remarks on Practical Usage of the Intel RealSense Cameras

The deployed sensor system contained the two T265 and two D435 cameras, connected to a single Intel NUC computer. Generally, the USB 3 bandwidth of various NUC computers is sufficient to stream data from more than two pairs of Intel RealSense cameras. However, the camera driver for Ubuntu with the ROS[1] requires a particular launch sequence to detect all the cameras correctly.

[1] Available at https://github.com/IntelRealSense/realsense-ros.

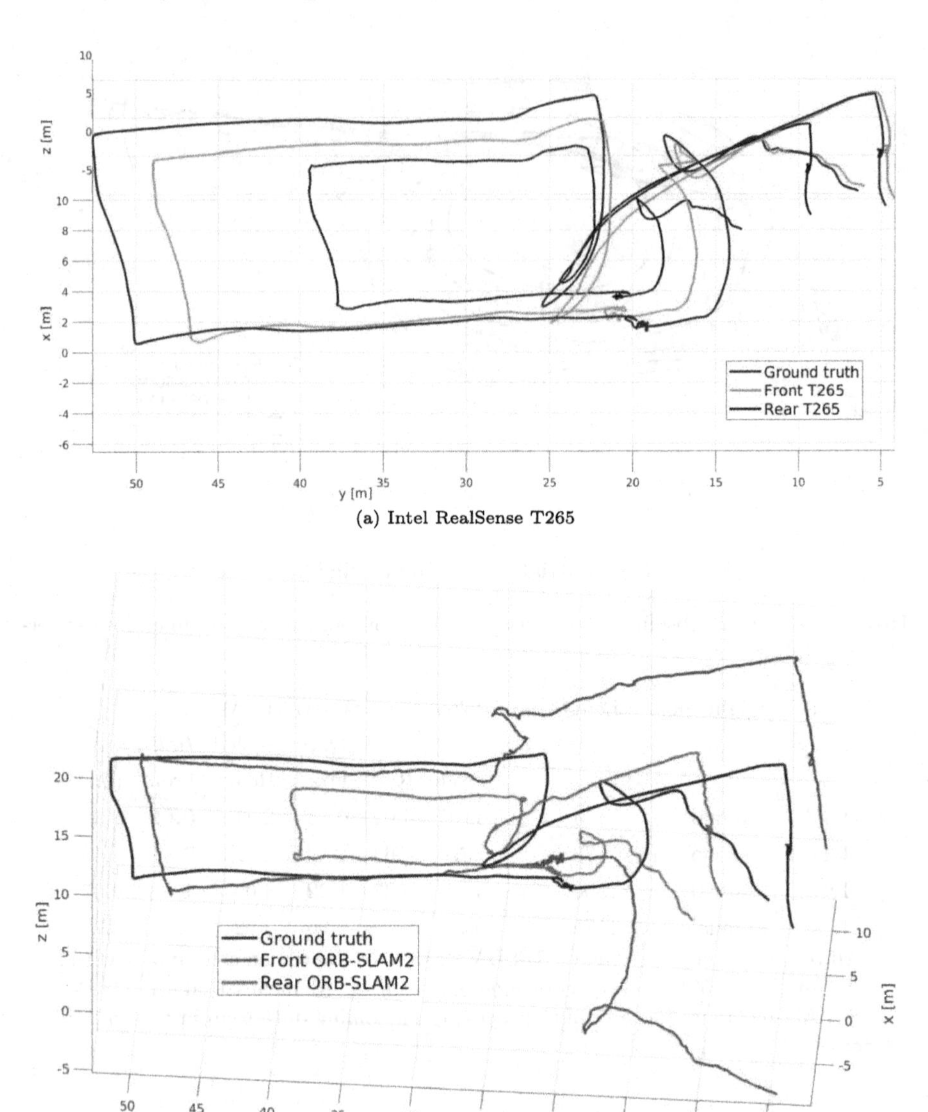

(a) Intel RealSense T265

(b) ORB-SLAM2

Fig. 7. Trial 2: UAV trajectories obtained during the autonomous flight.

The cameras have been therefore launched one-by-one in a fixed order with more than 20 s delay. Besides, resetting USB hubs before starting the launch sequence has been found necessary.

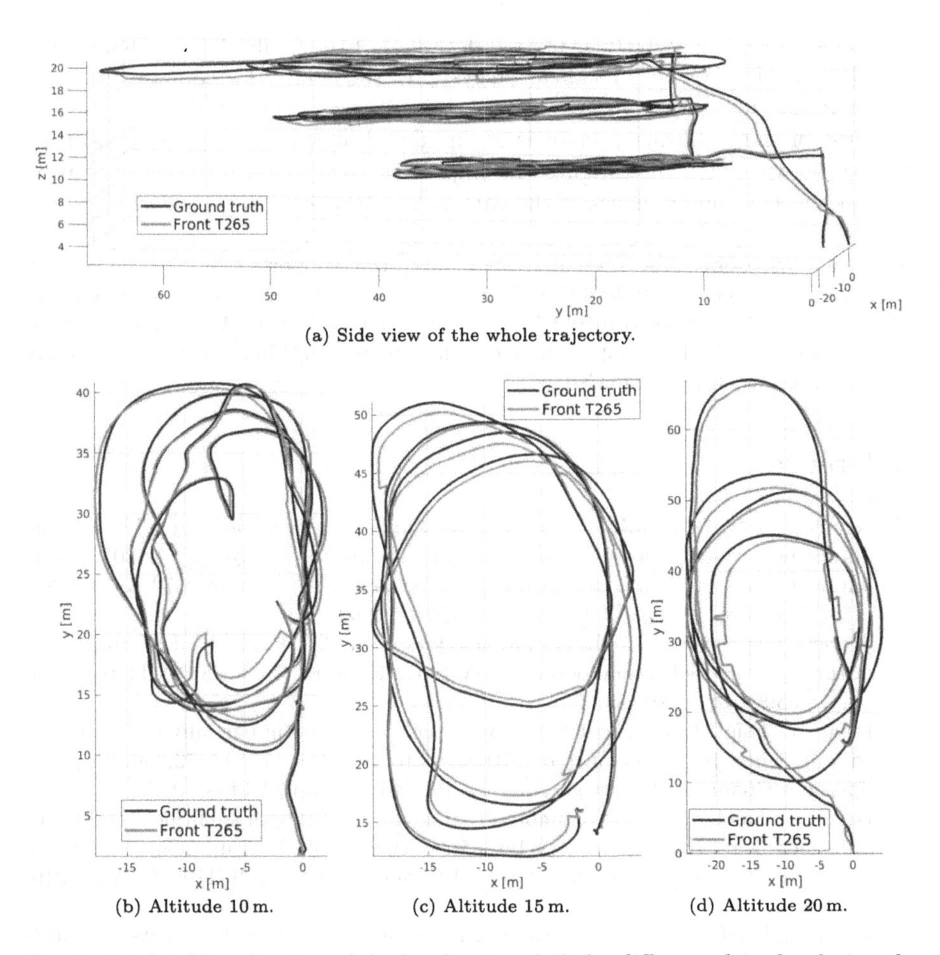

(a) Side view of the whole trajectory.

(b) Altitude 10 m.

(c) Altitude 15 m.

(d) Altitude 20 m.

Fig. 8. Trial 4: Visualization of the localization drift for different altitudes during the long UAV flight. The places where the T265 induced fast changes of the estimated pose are visible at the altitude of 20 m; this phenomenon is observed from 10 m altitude for the rear T265 camera.

5 Conclusion

The examined embedded localization system Intel RealSense T265 provided competitive results in the realistic outdoor deployment scenario to the state-of-the-art localization method ORB-SLAM2. In half of the experimental trials, the T265 performed even better than the ORB-SLAM2. Especially in the trial, where the UAV was controlled autonomously with an angular velocity above $30^\circ\ s^{-1}$ at corners of the trajectory. The effect of the high angular velocity of the UAV has not been observed for the T265, which makes it superior to the ORB-SLAM2 in applications where such motions are required. Moreover, the flight altitude effect on the localization performance was for the front T265 camera observed only at altitudes above 15 m. On the other hand, in half of the trials, T265 suffered from translation drift. For both localization systems, the front UAV cameras provided better results in nearly all cases.

Most of the localization error is accumulated at the start of the trajectory. Therefore, we aim to investigate the impact of the takeoff on the localization system and the consequences of the camera orientation in future work.

Acknowledgement. The presented work has been supported by the Technology Agency of the Czech Republic (TAČR) under research Project No. TH03010362 and under the OP VVV funded project CZ.02.1.01/0.0/0.0/16_019/0000765 "Research Center for Informatics". The support under grant No. SGS19/176/OHK3/3T/13 to Jan Bayer is also gratefully acknowledged.

References

1. Agarwal, A., Crouse, J.R., Johnson, E.N.: Evaluation of a commercially available autonomous visual inertial odometry solution for indoor navigation. In: 2020 International Conference on Unmanned Aircraft Systems (ICUAS), pp. 372–381 (2020). https://doi.org/10.1109/ICUAS48674.2020.9213962
2. Alapetite, A., Wang, Z., Hansen, J., Zajączkowski, M., Patalan, M.: Comparison of three off-the-shelf visual odometry systems. Robotics **9**, 56 (2020). https://doi.org/10.3390/robotics9030056
3. Bayer, J., Faigl, J.: Localization fusion for aerial vehicles in partially GNSS denied environments. In: Mazal, J. (ed.) MESAS 2018. LNCS, vol. 11472, pp. 251–262. Springer, Cham (2019). https://doi.org/10.1007/978-3-030-14984-0_20
4. Bayer, J., Faigl, J.: On autonomous spatial exploration with small hexapod walking robot using tracking camera Intel RealSense T265. In: European Conference on Mobile Robots (ECMR), pp. 1–6 (2019). https://doi.org/10.1109/ECMR.2019.8870968
5. Bayer, J., Faigl, J.: Handheld localization device for indoor environments. In: International Conference on Automation, Control and Robots (ICACR) (2020)
6. Collective of authors: vicon motion systems inc. https://www.vicon.com/. Accessed 1 Aug 2020
7. Collective of authors: leica geosystems ag @ONLINE. http://leica-geosystems.com/products/total-stations. Accessed 3 Mar 2020
8. Endres, F., Hess, J., Sturm, J., Cremers, D., Burgard, W.: 3-D mapping with an RGB-D camera. IEEE Trans. Robot. **30**(1), 177–187 (2014). https://doi.org/10.1109/TRO.2013.2279412
9. Engel, J., Koltun, V., Cremers, D.: Direct sparse odometry. IEEE Trans. Pattern Anal. Mach. Intell. **40**(3), 611–625 (2016). https://doi.org/10.1109/TPAMI.2017.2658577
10. Forster, C., Pizzoli, M., Scaramuzza, D.: SVO: fast semi-direct monocular visual odometry. In: IEEE International Conference on Robotics and Automation (ICRA), pp. 15–22 (2014). https://doi.org/10.1109/ICRA.2014.6906584
11. Fuentes-Pacheco, J., Ruiz-Ascencio, J., Rendón-Mancha, J.M.: Visual simultaneous localization and mapping: a survey. Artif. Intell. Rev. **43**(1), 55–81 (2015). https://doi.org/10.1007/s10462-012-9365-8

12. Gauglitz, S., Sweeney, C., Ventura, J., Turk, M., Höllerer, T.: Live tracking and mapping from both general and rotation-only camera motion. In: IEEE International Symposium on Mixed and Augmented Reality (ISMAR), pp. 13–22 (2012). https://doi.org/10.1109/ISMAR.2012.6402532
13. Hofmann-Wellenhof, B., Lichtenegger, H., Wasle, E.: GNSS - Global Navigation Satellite Systems: GPS, GLONASS, Galileo, and More. Springer, Vienna (2007). https://books.google.cz/books?id=Np7y43HU_m8C
14. Intel RealSense depth camera D435. https://www.intelrealsense.com/depth-camera-d435/. Accessed 4 Aug 2020
15. Intel RealSense tracking camera T265. https://www.intelrealsense.com/tracking-camera-t265/. Accessed 4 Aug 2020
16. Kim, D., et al.: Vision aided dynamic exploration of unstructured terrain with a small-scale quadruped robot. In: 2020 IEEE International Conference on Robotics and Automation (ICRA), pp. 2464–2470 (2020). https://doi.org/10.1109/ICRA40945.2020.9196777
17. Lightbody, P., Krajník, T., Hanheide, M.: A versatile high-performance visual fiducial marker detection system with scalable identity encoding. In: Proceedings of the Symposium on Applied Computing, pp. 276–282. ACM (2017)
18. Mahmoud, A., Atia, M.M.: Hybrid IMU-aided approach for optimized visual odometry. In: 2019 IEEE Global Conference on Signal and Information Processing (GlobalSIP), pp. 1–5 (2019). https://doi.org/10.1109/GlobalSIP45357.2019.8969460
19. Menze, M., Geiger, A.: Object scene flow for autonomous vehicles. In: IEEE Conference on Computer Vision and Pattern Recognition (CVPR), pp. 3061–3070 (2015). https://doi.org/10.1109/CVPR.2015.7298925
20. Mur-Artal, R., Tardós, J.D.: ORB-SLAM2: an open-source SLAM system for monocular, stereo, and RGB-D cameras. IEEE Trans. Robot. **33**(5), 1255–1262 (2017). https://doi.org/10.1109/TRO.2017.2705103
21. Mur-Artal, R., Montiel, J.M.M., Tardós, J.D.: ORB-SLAM: a versatile and accurate monocular SLAM system. IEEE Trans. Robot. **31**(5), 1147–1163 (2015). https://doi.org/10.1109/TRO.2015.2463671
22. Opromolla, R., Fasano, G., Rufino, G., Grassi, M., Savvaris, A.: LIDAR-inertial integration for UAV localization and mapping in complex environments. In: International Conference on Unmanned Aircraft Systems (ICUAS), pp. 444–457 (2016). https://doi.org/10.1109/ICUAS.2016.7502580
23. Ouerghi, S., Ragot, N., Boutteau, R., Savatier, X.: Comparative study of a commercial tracking camera and orb-slam2 for person localization. In: Proceedings of the 15th International Joint Conference on Computer Vision, Imaging and Computer Graphics Theory and Applications - Volume 4: VISAPP, pp. 357–364. INSTICC, SciTePress (2020). https://doi.org/10.5220/0008980703570364
24. Pire, T., Fischer, T., Castro, G., Cristóforis, P.D., Civera, J., Berlies, J.J.: S-PTAM: stereo parallel tracking and mapping. Robot. Auton. Syst. **93**, 27–42 (2017). http://www.sciencedirect.com/science/article/pii/S0921889015302955
25. Quigley, M., et al.: ROS: an open-source robot operating system. In: ICRA Workshop on Open Source Software (2009)
26. Smith, M., Baldwin, I., Churchill, W., Paul, R., Newman, P.: The new college vision and laser data set. Int. J. Robot. Res. **28**(5), 595–599 (2009). https://doi.org/10.1177/0278364909103911
27. ZED Mini. https://www.stereolabs.com/zed-mini/. Accessed 4 Aug 2020

28. Sturm, J., Engelhard, N., Endres, F., Burgard, W., Cremers, D.: A benchmark for the evaluation of RGB-D SLAM systems. In: IEEE International Conference on Intelligent Robots and Systems (IROS), pp. 573–580 (2012). https://doi.org/10.1109/IROS.2012.6385773
29. Usenko, V., Engel, J., Stückler, J., Cremers, D.: Direct visual-inertial odometry with stereo cameras. In: IEEE International Conference on Robotics and Automation (ICRA), pp. 1885–1892 (2016). https://doi.org/10.1109/ICRA.2016.7487335
30. Zhou, Y., Kneip, L., Li, H.: Semi-dense visual odometry for RGB-D cameras using approximate nearest neighbour fields. In: 2017 IEEE International Conference on Robotics and Automation (ICRA), pp. 6261–6268 (2017). https://doi.org/10.1109/ICRA.2017.7989742

AxS/AI in Context of Future Warfare and Security Environment

Model of Surveillance in Complex Environment Using a Swarm of Unmanned Aerial Vehicles

Petr Stodola[1](✉), Jan Drozd[2], and Jan Nohel[1]

[1] Department of Intelligence Support, University of Defence, Brno, Czech Republic
petr.stodola@unob.cz

[2] Department of Tactics, University of Defence, Brno, Czech Republic

Abstract. This paper examines the model of autonomous surveillance using a swarm of unmanned aerial vehicles with the simultaneous detection principle. This model enables to specify a number of sensors needed to detect an object of interest located in the area of interest; objects are detected only if scanned by the specified number of sensors simultaneously. The model plans deployment of individual vehicles in the swarm during the surveillance operation in a such a way that the surveillance is performed in the maximum quality; the quality is measured as a percentage of the area of interest that is covered during the operation. Furthermore, the surveillance is assumed to be conducted in the complex area of operations (including urban environments, build-up areas, or mountain environments with very uneven terrain) where occlusions caused by obstacles or terrain may occur often. For solution, the metaheuristic algorithm based on the simulated annealing is proposed. This algorithm deploys the number of waypoints, from which the monitoring is performed, maximizing the surveillance quality and taking the simultaneous detection principle into consideration. The algorithm is verified by a set of experiments based on typical surveillance scenarios.

Keywords: Surveillance · Unmanned aerial vehicles · Simultaneous detection · Urban environment · Metaheuristic algorithm · Experiments

1 Introduction

Contemporary armed conflicts are different than armed conflicts twenty years and more ago. One of the most significant characteristics of contemporary armed conflicts is changeable situation on the battlefield as well as countless information flow from different sources with different reliability. Moreover, most of the contemporary operations are conducted within specific environment such as urban and build-up areas (Siberia, Ukraine), etc., which significantly limit ordinary way of reconnaissance and surveillance. Such environment requires new approaches to gather all necessary information and process them in order to support Military decision process (in case of battalion level and above) or Troops leading procedure (in case of company level and below).

One of the crucial steps of the commanders' decision making is surveillance. It is possible to state, that surveillance is a continuous process which begins during the planning and decision making procedure. It provides critical information for commander's

J. Mazal et al. (Eds.): MESAS 2020, LNCS 12619, pp. 231–249, 2021.
https://doi.org/10.1007/978-3-030-70740-8_15

decision making. Ordinarily, it is conducted by special teams deployed in the depth of the enemy area. Apparently, deployment of such a team or teams is demanding on their training and preparation. Moreover, information flow from such teams are delayed and does not have to be precise, which could have a tremendous impact on the mission. In contemporary operations, new technologies such as unmanned aerial vehicles (UAVs) are used in order to gather almost online information and support commanders' decision making. The use of UAVs has tremendous impact on speed and quality of decision making. In addition to that, this information gathering save human resources. More information to this issue can be found for example in [1–7].

This article proposes the model of autonomous surveillance using a swarm of small UAVs (sUAV). The goal is to cover as large area of interest as possible by sensors of UAVs in the swarm. Each UAV is deployed in the exact location (waypoint) in the area of operations monitoring a portion of the area of interest. The model also allows the situation when more than one sensor is needed for the detection of some object of interest (further on referred to as the simultaneous detection). Furthermore, the surveillance is assumed to be conducted in complex area of operations (including urban environments, build-up areas, or mountain environments with very uneven terrain) where occlusions caused by obstacles or terrain may occur often.

Countless scientific works are focused on the use of swarms of UAVs for many purposes. There are several topics important to solve such complex problems like reconnaissance or surveillance conducted with an UAV swarm. Path planning for such a task is one of the crucial issue. Yao et al. [8] proposed a hybrid approach based on the Lyapunov Guidance Vector Field (LGVF) and the Improved Interfered Fluid Dynamical System (IIFDS), to solve the problems of target tracking and obstacle avoidance in three-dimensional cooperative path planning for multiple UAVs. Lamont et al. [9] designed and implemented a comprehensive mission planning system for swarms of UAVs. This system integrates several problem domains including path planning, vehicle routing, and swarm behavior as based upon a hierarchical architecture. Shanmugavel et al. [10] examined the problem of path planning for simultaneous arrival on a target.

Another crucial topic connected with UAVs is their reliability and failure protection. Military commanders must be ready to fulfil the mission in any unexpected situation. The use of a swarm of UAVs for surveillance tasks is a very important issue regarding precise critical information gathering. There are no specific scientific works focusing on this topic, however, there are several interesting papers which should be taken into the consideration. Triharminto et al. [11] developed an algorithm for moving target intercept with obstacle avoidance in 3D. The algorithm which is called L+Dumo Algorithm integrates a modified Dubins Algorithm and Linear Algorithm. Such approach could be modified in order to diminish impact of a UAV failure to finish the surveillance mission. Sampedro et al. [12] focused on scalable and flexible architecture for real-time mission planning and dynamic agent-to-task assignment for a swarm of UAVs. The proposed mission planning architecture consists of a Global Mission Planner (GMP) which is responsible of assigning and monitoring different high-level missions through an Agent Mission Planner (AMP), which is in charge of providing and monitoring each task of the mission to each UAV in the swarm. Sujit et al. [13] addressed the problem of generating feasible paths from a given start location to a goal configuration for multiple

UAVs operating in an obstacle rich environment that consist of static, pop-up and moving obstacles. The path planning system in the environment with pop-up and moving obstacles provides an inspiration to solve UAV swarm failure during the surveillance mission in a complex environment including build-up areas or mountain terrain.

2 Problem Formulation

In this section, the problem examined in this article is defined and formulated. In the first part, the model of the aerial surveillance is revised; this model is transformed from the reconnaissance model proposed during the previous research of the authors [14]. Then, the model is extended by the simultaneous detection principle.

2.1 Model of Aerial Surveillance

Let AoI be the area of interest to be monitored during the surveillance operation. AoI is represented by a polygon deployed in the area of operations ($AoI \subseteq AoO$). The goal of the surveillance operation is to monitor as much portion of AoI as possible using a swarm of UAVs.

Let $U = \{U_1, U_2, \ldots, U_N\}$ be a finite set of UAVs in the swarm where $N \geq 1$ is their number. The UAVs in the swarm are deployed in their base position in the area of operations and they are prepared to start the surveillance at the beginning of the operation. Each UAV is equipped with a sensor which is able to monitor a portion of the area of interest directly below. All UAVs (and their sensors) in the swarm are assumed to be identical. Each sensor is characterized by two parameters:

- Angular field of view α_{fov}.
- Maximum distance from an object of interest d_{max}.

These parameters specify detection range of the sensor – see the green area in Fig. 1. The ground objects which are located somewhere in this range are detected by an UAV.

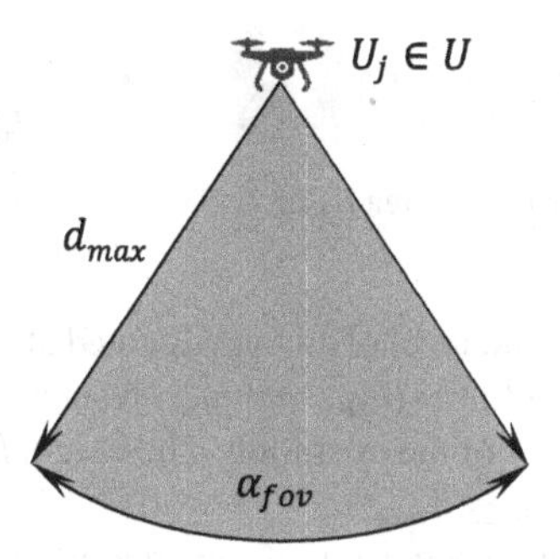

Fig. 1. Detection range of a sensor of an UAV.

Let $W = \{W_1, W_2, \ldots, W_N\}$ be a finite set of waypoints deployed in the area of operations. Each waypoint $W_i \in W$ is defined by:

- Coordinate x_i (x-coordinate).
- Coordinate y_i in the plane (y-coordinate).
- Height above the ground level h_i (z-coordinate).

Each UAV is assigned to its own waypoint from which it participates in the surveillance of the area of interest. Prior to the operation, each UAV flies from their base position to its waypoint. The operation starts when all the UAVs are located at their waypoints and it continues for a specified duration; during this time, all the UAVs are hovering at their waypoints and monitoring the area. At the end of the operation, the UAVs return back to their base position.

Every UAV in the swarm must be a vertical take-off and landing (VTOL) aircraft, i.e., one that can hover, take off, and land vertically. The duration of the operation must comply with the flight parameters of the UAVs, i.e., it must not take longer that the maximum flight time (including the time necessary to fly to their waypoints prior to the operation, and back to their base positions after the operation).

An UAV $U_i \in U$ located at its waypoint $W_i \in W$ during the operation observes some portion of the area of interest. This means that objects of interest are detected if located in the observed area at any time of the operation. However, the terrain and obstacles can cause occlusions. The principle is illustrated in Fig. 2. The green line represents the observed area (further on referred to as visible), the red line shows the area outside the detection range or the area occluded by obstacles.

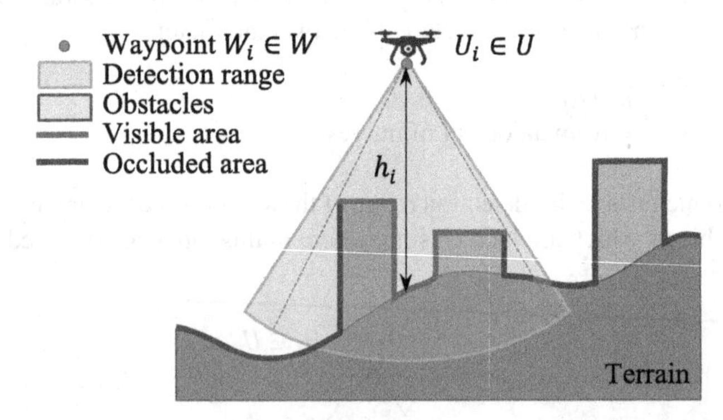

Fig. 2. Principle of monitoring the area by an UAV from a waypoint. (Color figure online)

The height above the ground level of an UAV located at any waypoint can be limited. The minimum and maximum limits (h_{min} and h_{max} respectively) can be set as a tactical requirement of the commander of the operation. The height h_i at waypoint $W_i \in W$ must be within the limits: $h_{min} \leq h_i \leq h_{max}$.

Each UAV $U_i \in U$ observes a portion of the area of interest $V_i \subseteq AoI$. Every point of the visible area V_i must satisfied the conditions as follows:

- Points in V_i must lie in the area of interest ($V_i \subseteq AoI$).
- Points in V_i must be within the detection range of the sensor of UAV U_i (see Fig. 1).

- There must be a visual line of sight (VLOS) between the sensor of UAV U_i and all points in V_i (see Fig. 2).

The principle is illustrated in Fig. 3. It shows the real situation from the top view. The area of interest is enclosed by a blue polygon; grey objects inside are obstacles (buildings in this case). The green area represents the visible area V_i observed from waypoint W_i. The occlusions caused by obstacles can be seen inside the green area.

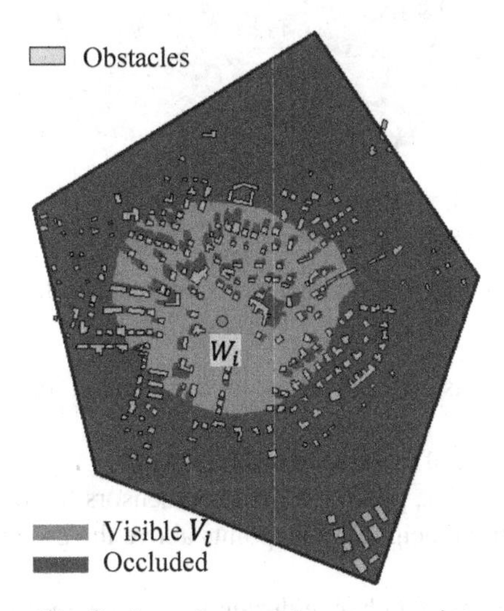

Fig. 3. Area observed from a waypoint.

The entire visible area V observed from all the waypoints is calculated according to formula (1) by unification of visible areas observed from individual waypoints. The coverage of the area of interest is computed according to formula (2) as the proportion of the size of the visible area $|V|$ to the size of the area of interest $|AoI|$.

$$V = \bigcup_{i=1}^{N} V_i \tag{1}$$

$$C = \frac{|V|}{|AoI|} \tag{2}$$

The principle is shown in Fig. 4. It is the same situation as in Fig. 3; this time, however, AoI is observed from five waypoints. The coverage (in percent) in case of using one waypoint is $C_{\%} = 20.07\%$ whereas it is $C_{\%} = 89.38\%$ in case of using five waypoints.

Let X be a particular solution in the state space. This solution is composed of N waypoints: $X = \{W_1, W_2, \ldots, W_N\}$; each waypoint has its exact position in the area of operations: $W_i = \{x_i, y_i, h_i\}$. Thus, a solution is composed of $3N$ independent continuous optimization variables: $X = \{x_1, y_1, h_1, x_2, y_2, h_2, \ldots, x_N, y_N, h_N\}$.

Beside the optimization variables, a surveillance operation is characterized by a number of constant parameters. These are as follows:

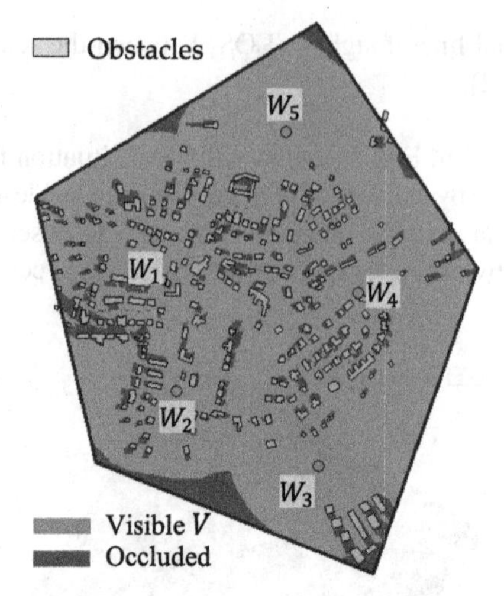

Fig. 4. Coverage of the area of interest.

- The terrain and database of obstacles in the area of operations determined by geographical data.
- Size, shape and position of the area of interest (AoI).
- Number of available UAVs, parameters of their sensors (N, α_{fov}, d_{max}).
- Minimum and maximum height of waypoints above the ground level (h_{min}, h_{max}).

The constant parameters do not change during the whole operation. Thus, the value of coverage C in a particular operation is computed for values of variables in solution X: $C = f(X)$.

The goal of the surveillance problem is to find such a solution X (that means to find values for $3N$ optimization variables) so that the coverage C is as large as possible, i.e., to maximize C – see the optimization criterion in formula (3).

$$\text{maximize}(C) \tag{3}$$

2.2 Model of Aerial Surveillance with Simultaneous Detection

In this section, the model of aerial surveillance is extended by the simultaneous detection principle. This principle ensures planning the surveillance operation in case that more than one sensors are required for the detection of the objects located in the area of interest.

In order to incorporate the simultaneous detection into the model, two new parameters are defined:

- Minimum number of sensors F for detection.
- Minimum permitted distance between waypoints D.

Both parameters are new constant parameters of the particular operation set according to tactical and technical requirements of the commander. Parameter F determines the number of sensors necessary to simultaneously observe all the points inside the AoI ($1 \le F \le N$). Parameter D ensures that the minimum space between every pair of sensors is kept – see constraint (4). This parameter is included because the model have a tendency to deploy waypoints close to one another for $F > 1$; this is not always desirable from the tactical point of view.

$$\left|W_i - W_j\right| \ge D \text{ for all } W_i \in W \text{and} W_j \in W;\; W_i \ne W_j \tag{4}$$

The calculation of the visible area V as defined in Sect. 2.1 needs to be reformulated since formula (1) is no longer applicable. Let V^F be a visible area observed simultaneously from at least F waypoints by F sensors. Let S be a set of all visible areas observed from individual waypoints: $S = \{V_1, V_2, \ldots, V_N\}$; set S has N distinct elements. Let S^i be an F-combination of set S (i.e., a subset of F distinct elements of S); $i = 1, 2, \ldots, C(N, F)$. Let S^i_j be an element of set S^i; $j = 1, 2, \ldots, F$. Then, the calculation of V^F is given in formula (5). Let C^F be the resulting coverage of the area of interest computed according to formula (6).

$$V^F = \bigcup_{i=1}^{C(N,F)} \left(\bigcap_{j=1}^{F} S^i_j \right) \tag{5}$$

$$C^F = \frac{|V^F|}{|AoI|} \tag{6}$$

The principle is shown on an example with $N = 4$ waypoints, i.e., $S = \{V_1, V_2, V_3, V_4\}$. Then, formulae (7) to (10) show the calculations of V^F for $F = 1, 2, 3, 4$.

$$V^1 = V_1 \cup V_2 \cup V_3 \cup V_4 \tag{7}$$

$$V^2 = (V_1 \cap V_2) \cup (V_1 \cap V_3) \cup (V_1 \cap V_4) \cup (V_2 \cap V_3) \cup (V_2 \cap V_4) \cup (V_3 \cap V_4) \tag{8}$$

$$V^3 = (V_1 \cap V_2 \cap V_3) \cup (V_1 \cap V_2 \cap V_4) \cup (V_1 \cap V_3 \cap V_4) \cup (V_2 \cap V_3 \cap V_4) \tag{9}$$

$$V^4 = (V_1 \cap V_2 \cap V_3 \cap V_4) \tag{10}$$

The example in Fig. 5 shows the same situation as in Fig. 4; there are five waypoints deployed in the area of operations for five UAVs in the swarm. Figure 5(a) presents the coverage C^F for $F = 1$ (i.e., one sensor is sufficient for detection); the intensity of the green color distinguishes between the number of sensors observing the area. Figure 5(b) shows the coverage for the same deployment of waypoints for $F = 2$ (i.e., at least two sensors are needed for detection); the orange color represents the area which is observed by too low number of sensors (less than F). In the former case, the coverage is $C^1_{\%} = 89.38\%$ whereas in the latter it is only $C^2_{\%} = 11.50\%$. The reason for this high reduction in the quality of the surveillance operation is that the positions of waypoints were optimized for $F = 1$.

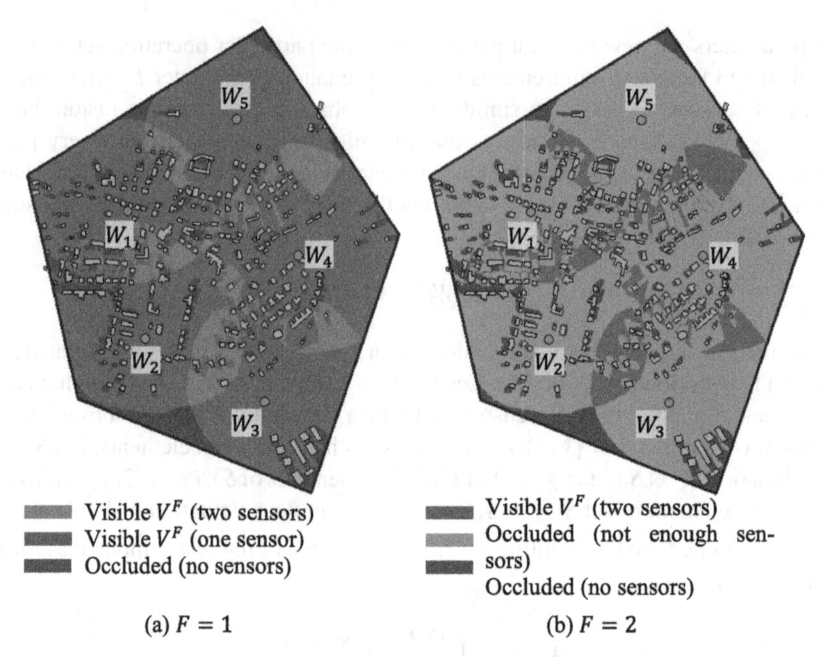

(a) $F = 1$ (b) $F = 2$

Fig. 5. Coverage for the deployment of waypoints optimized for $F = 1$.

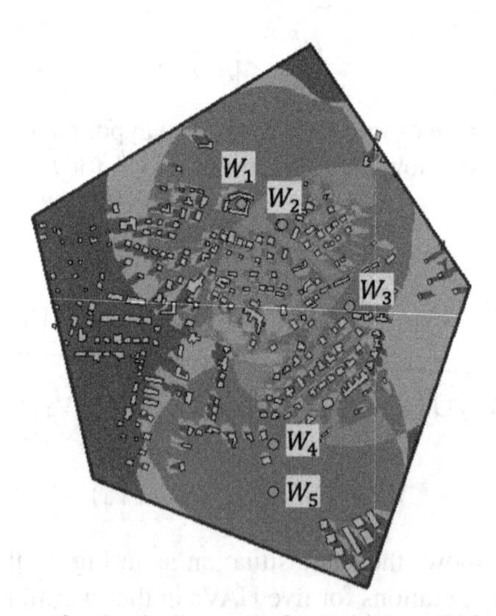

Fig. 6. Coverage for the deployment of waypoints optimized for $F = 2$.

Figure 6 presents the same surveillance operation with the same number of available UAVs in the swarm $N = 5$, but this time the positions of waypoints are optimized for higher number of sensors for detection: $F = 2$ and $D = 100$m. The new deployment of waypoints provides the increase in the coverage from $C^2_{\%} = 11.50\%$ (see Fig. 5(a)) to

$C^2_{\%} = 57.21\%$ (see Fig. 6). If bigger coverage is required, more sensors (i.e., UAVs in the swarm) are necessary.

3 Solution Algorithms

For the deployment of sensors, a metaheuristic algorithm based on the simulated annealing principle was proposed. The simulated annealing is a probabilistic technique inspired by the process of controlled cooling of a material in metallurgy to increase the size of its crystals and reduce their defects.

The simulated annealing is one of many possible stochastic approaches to solve complex optimization problems. The metaheuristic methods are often problem dependent; a method could be very successful in solving a certain type of a problem, nevertheless it does not cope well with others. The reason why the authors chose the simulated annealing is that it proved to be very effective in deployment problems, see for example [15, 16].

3.1 Evaluation of a Solution

The most critical part of the optimization from the time consumption point of view is the evaluation of a particular solution $X = \{x_1, y_1, h_1, x_2, y_2, h_2, \ldots, x_N, y_N, h_N\}$, i.e., calculation of $C^F = f(x_1, y_1, h_1, x_2, y_2, h_2, \ldots, x_N, y_N, h_N)$. It includes testing all the points laying inside the area of interest: they must be within the detection range of any sensor (or a number of sensors for $F > 1$) and the VLOS must exist between the point and this sensor (sensors).

There are an infinite number of points laying inside the area of interest; such a number of points cannot be evaluated from the practical point of view. Therefore, a rasterization using the Sukharev grid was chosen as an approach to estimate the value of C^F. Only points laying in the middle of rasterization squares are evaluated. Then, the value of C^F is calculated as a proportion of the number of visible points in rasterization squares to the total number of these squares. The precision of the estimated value depends on the size of the rasterization step d_{rast}.

A lot of effort was put into the effectiveness of the solution evaluation. Several approaches were used to make the algorithm as efficient as possible. The implemented algorithm manages to evaluate tens of millions of points per second on computers with common configurations. These main approaches were implemented:

- The VLOS between a sensor and a point is tested using the floating horizon algorithm. This algorithm is able to evaluate a number of points laying on the straight line between the point and the sensor with the linear computational complexity.
- Operations with integers only are employed in the most critical parts of the algorithm.
- The evaluation of points can be parallelized. Thus, it is distributed on cores of a multi-core processor.

3.2 Deployment of Waypoints

The algorithm for the waypoints deployment works in iterations. The idea behind the simulated annealing is in gradual cooling of the temperature in successive iterations. The temperature influences the probability of accepting newly generated solutions; even worse solutions can be accepted. This principle extends the search space and prevents to be stuck in some local optimum.

The algorithm employs a set of parameters controlling its behavior – see Table 1. The algorithm is presented in Fig. 7 in pseudocode. The algorithm calls several standalone functions as follows:

Table 1. Parameters of the algorithm for the deployment of waypoints.

Parameter	Description
T_{max}	Initial temperature (used in the first iteration)
T_{min}	Lower limit of temperature (termination condition)
γ	Cooling coefficient
n_{max}	Maximum number of transformations per iteration
m_{max}	Maximum number of replacements per iteration

- **Generate_Random_Solution**: generation of an initial (first) solution.
- **Evaluate_Solution**: calculation of $C^F = f(X)$ (see Sect. 3.1).
- **Transform_Solution**: transformation of the current solution into the new solution.
- **Calculate_Probability**: calculation of the probability to replace the current solution by the transformed solution.

The algorithm starts by generating an initial solution X_{cur} and its evaluation (lines 1 and 2). Each variable of the solution vector is set by the pseudo-random number generator with uniform distribution in the permitted range (given by the width and height of the area of interest for variables x_i and y_i, and between h_{min} and h_{max} for variables h_i). Then, the current temperature is set to its higher limit (line 3).

The algorithm works in iterations (lines 4 to 18). The temperature does not change within an iteration. The algorithm is terminated when the current temperature drops below the lower limit (line 4). In an iteration, a number of transformations and replacements are performed (lines 6 to 17). This number is controlled by parameters n_{max} (maximum number of transformations per iteration) and m_{max} respectively (maximum number of replacements per iteration) – see the termination condition of an iteration (line 6).

The transformation of the current solution X_{cur} into the new solution X_{new} (line 7) is carried out as follows. The solution vector $X = \{x_1, y_1, h_1, x_2, y_2, h_2, \ldots, x_N, y_N, h_N\}$ is composed of $3N$ independent variables. All the variables remain the same except one which is randomly selected. The selected variable changes its value in its permitted

Deploy_Waypoints()

Output: C^F, X

Constant parameters: $N, F, D, AoI, \alpha_{fov}, d_{min}, h_{min}, h_{max}, d_{rast}$

Algorithm settings: $T_{max}, T_{min}, \gamma, n_{max}, m_{max}$

1. $X = X_{cur}$ = Generate_Random_Solution()
2. $C^F = C^F_{cur}$ = Evaluate_Solution(X_{cur})
3. $T_{cur} = T_{max}$
4. **while** $T_{cur} \geq T_{min}$ **do** // Iterations
5. $n = m = 1$
6. **while** $n \leq n_{max}$ **and** $m \leq m_{max}$ **do** // Transformations
7. X_{new} = Transform_Solution(T_{cur}, X_{cur})
8. C^F_{new} = Evaluate_Solution(X_{new})
9. $p(X_{new} \to X_{cur})$ = Calculate_Probability($T_{cur}, C^F_{cur}, C^F_{new}$)
10. **with** probability $p(X_{new} \to X_{cur})$ **do** // Replacements
11. $X_{cur} = X_{new}$
12. $C^F_{cur} = C^F_{new}$
13. $m = m + 1$
14. **if** $C^F_{cur} > C^F$ **then do** // Save the best solution
15. $X = X_{cur}$
16. $C^F = C^F_{cur}$
17. $n = n + 1$
18. $T_{cur} = \gamma \cdot T_{cur}$
19. **return** C^F, X

Fig. 7. Algorithm for the deployment of waypoints

range by adding a random value obtained using the pseudo-random number generator with the normal distribution with the zero mean and the standard deviation σ calculated according to formula (11). *Range* is the difference between the maximum and minimum limits for the selected variable (for x and y it is determined by the size of the area of interest, for h it is the difference between the maximum and minimum permitted heights h_{max} and h_{min}). The current temperature influences the size of the change; the bigger the temperature, the bigger the change. This results in the extensive search in the state space at the beginning of the optimization, and tuning the solution towards its end.

$$\sigma = \frac{(T_{cur} - T_{min}) \cdot \left(\frac{range}{3}\right)}{T_{max} - T_{min}} \tag{11}$$

The transformation may lead to the violation of the constraint in formula (4), i.e., the requirement on the minimum distance D between all combinations of pairs of sensors (if set). In this case, sensors which do not satisfy the distance condition are disabled, i.e., they do not participate on the surveillance.

The new transformed solution replaces the current solution (lines 10 to 13) with probability $p(X_{new} \to X_{cur})$ computed according to formula (12) using the Metropolis criterion (line 9). If the transformed solution is better than the original, it is always replaced. Otherwise, the probability depends on the difference between the two solutions (the lower the difference, the bigger the probability) and the current temperature (the bigger the temperature, the bigger the probability).

$$p(X_{new} \rightarrow X_{cur}) = \begin{cases} 1 & \text{for } C^F_{new} \geq C^F_{cur} \\ e^{-\frac{C^F_{cur} - C^F_{new}}{T_{cur}}} & \text{otherwise} \end{cases} \quad (12)$$

When an iteration ends, the current temperature is decreased by the cooling coefficient $0 < \gamma < 1$ (line 18). The best solution found during the optimization is stored (lines 14 to 16) and returned at the end of the optimization (line 19).

4 Experiments and Results

This section presents the results from experiments carried out using simulations. First, it describes the scenarios used as benchmark surveillance operations in experiments. Then, the results of the algorithm achieved on the benchmark operations are presented. Finally, a discussion about the performance and behavior of the algorithm concludes this section.

4.1 Benchmark Surveillance Operations

Five benchmark operations (labelled e01 to e05) for experiments were created based on the typical scenarios of the military surveillance operations. In simulations, real geographic data were provided by the Military Geographic and Hydrometeorologic Office of the Ministry of Defense of the Czech Republic. Two models were used as follows:

- Digital Elevation Model (DEM): a representation of the terrain surface in the form of a heightmap. In its last version, the distance between neighboring elevations is 2.5 m and the elevation precision is 0.3 m.
- Topographic Digital Data Model (TDDM): a database of topographic and other objects represented by a polygon and parameters (e.g., object height). The layer of buildings in this model was used as a database of obstacles which may cause possible occlusions.

Table 2 shows the parameters (terrain and obstacles) of the physical environment for individual operations. The environment varies for individual operations both in parameters describing the terrain (from flat to very uneven surface) and obstacles (from low density to very high density). The variety of operations can be seen on these examples: scenario e04 is a city center located in a relatively flat terrain with high density of tall buildings and narrow streets; opposite to this, scenario e05 is a mountain area with very uneven terrain and low density of obstacles.

Table 2. Parameters of the physical environment of the area of operations.

Operation	Terrain			Obstacles	
	Minimum elevation	Maximum elevation	Elevation difference	Number of obstacles	Avg. height of obstacles
e01	201 m	245 m	44 m	14	14.6 m
e02	412 m	491 m	79 m	266	6.4 m
e03	236 m	365 m	129 m	533	5.1 m
e04	197 m	244 m	47 m	936	10.6 m
e05	950 m	1,604 m	654 m	8	9.4 m

Table 3 presents the parameters of the surveillance operations: number of UAVs in the swarm at the disposal of the commander of the mission, parameters of their sensors, minimum required distance between deployed sensors, minimum and maximum limits of the flight height, and size of the area of interest to be surveilled. The operations suppose that a large swarm of small UAVs are available (120 in case of scenario e04).

Table 3. Parameters of the surveillance operations.

Operation	Number of UAVs N	Sensors		Sensor space D	Flight height		Area of interest
		α_{fov}	d_{max}		h_{min}	h_{max}	
e01	20	90°	100 m	40 m	50 m	150 m	0.1 km^2
e02	40	90°	160 m	20 m	50 m	200 m	0.7 km^2
e03	60	120°	220 m	50 m	50 m	300 m	2.8 km^2
e04	120	120°	220 m	50 m	50 m	300 m	5.3 km^2
e05	40	120°	600 m	100 m	100 m	600 m	6.8 km^2

For the solution evaluation (see Sect. 3.1), the rasterization step d_{rast} was set for individual benchmark operations as a compromise between the optimization speed and the precision of the coverage estimation. The values of d_{rast} are shown in Table 4 together with the resulting number of points inside the area of interest which needs to be processed when computing the coverage C^F for a particular solution X.

4.2 Simulation Results

The optimizations of the deployment of waypoints were conducted via the proposed algorithm using the computer with the configuration: CPU Intel i7-7700 3.5 GHz (4 cores). The parameters of the algorithm as defined in Table 1 were set empirically as follows: $T_{max} = 10^{-2}$, $T_{min} = 10^{-6}$, $\gamma = 0.9$, $n_{max} = 200$ (for e01, e02, e05)/$n_{max} = 500$ (for e03, e04), $m_{max} = 20$ (for e01, e02, e05)/$m_{max} = 50$ (for e03, e04).

Table 4. Size of the rasterization step selected for the surveillance operations.

Operation	Rasterization step d_{rast}	Number of points
e01	2 m	19,375
e02	3 m	75,442
e03	5 m	110,266
e04	5 m	213,891
e05	10 m	68,050

The optimizations of the individual benchmark operations were conducted independently for different number of sensors necessary for the simultaneous detection: $F = 1$ (Table 5), $F = 2$ (Table 6), and $F = 3$ (Table 7). In total, 50 trials per experiment were performed and the best solution, the mean and the standard deviation are recorded. The last columns in Tables 5 to 7 show the average execution time of the algorithm per optimization.

Table 5. Results achieved by the proposed algorithm for $F = 1$.

Operation	Best solution $C_{\%}^{F}$	Mean $C_{\%}^{F}$	Standard deviation	Execution time
e01	100%	100%	0%	28 s
e02	99.99%	99.98%	0.01%	57 s
e03	100%	99.99%	0.01%	341 s
e04	99.45%	99.38%	0.04%	628 s
e05	100%	100%	0%	92 s

Table 6. Results achieved by the proposed algorithm for $F = 2$.

Operation	Best solution $C_{\%}^{F}$	Mean $C_{\%}^{F}$	Standard deviation	Execution time
e01	91.85%	89.82%	1.36%	32 s
e02	86.55%	84.15%	1.05%	117 s
e03	84.99%	81.94%	1.83%	764 s
e04	87.54%	86.00%	0.87%	1210 s
e05	100%	99.93%	0.08%	120 s

Table 7. Results achieved by the proposed algorithm for $F = 3$.

Operation	Best solution $C_{\%}^{F}$	Mean $C_{\%}^{F}$	Standard deviation	Execution time
e01	69.45%	66.39%	2.53%	32 s
e02	53.66%	51.43%	1.29%	109 s
e03	55.94%	49.58%	4.20%	427 s
e04	56.52%	54.43%	1.02%	929 s
e05	89.42%	86.01%	1.35%	147 s

From the experiments, it is clear that there are enough UAVs to cover the whole *AoI* when $F = 1$. The coverage is over 99% in all cases. Moreover, 100% coverage was achieved in all optimization trials in case of experiments e01 and e05. The situation when $F = 2$ shows the reduction in the coverage (expect for operation e05); the coverage exceeds 80% in all cases Another reduction is obvious when $F = 3$; e.g., the number of UAVs are sufficient to cover only about 50% in case of experiments e02, e03, e04.

Figure 8 illustrates the results of the surveillance operation e01. Figure 8(a) shows the layout of the obstacles and the area of interest which is small and laying in the relatively flat terrain; Fig. 8(b) shows the image of the real physical environment; Fig. 8(c), 8(d) and 8(e) present the best solution for $F = 1, 2, 3$.

4.3 Analysis and Discussion

This section analyses various features and behavior of the proposed algorithm. First, the convergence of the algorithm is examined. Figure 9 shows the example of the progress of an average optimization on task e01 in dependence on the value of coefficient F. The convergence curves show the fast improvement of the solution at the initial phases of the optimization and the tuning of the solution in the final phases. Also, the curves differ depending on the value of F; the reason is the influence of the number of sensors available for the task (there are more than enough sensors to cover the entire *AoI* when $F = 1$, enough to cover about 90% when $F = 2$, and less than 70% when $F = 3$). Another aspect causing the worse solutions at the beginning of the optimizations and slower convergence to some local optimum for bigger values of F is the constraint not allowing waypoints to be too close to one another (see formula (4)). This constraint has the effect especially when $F > 1$; the bigger the value of F, the bigger the effect.

The impact of the constraint forcing the minimum space between waypoints is also apparent when comparing the execution times of the algorithm for different values of F. This comparison is shown in Fig. 10. In some tasks (e02, e03), it took more than twice as long for $F = 2$ compared to $F = 1$. This is caused by the different number of transformations resulting in total number of solution evaluations.

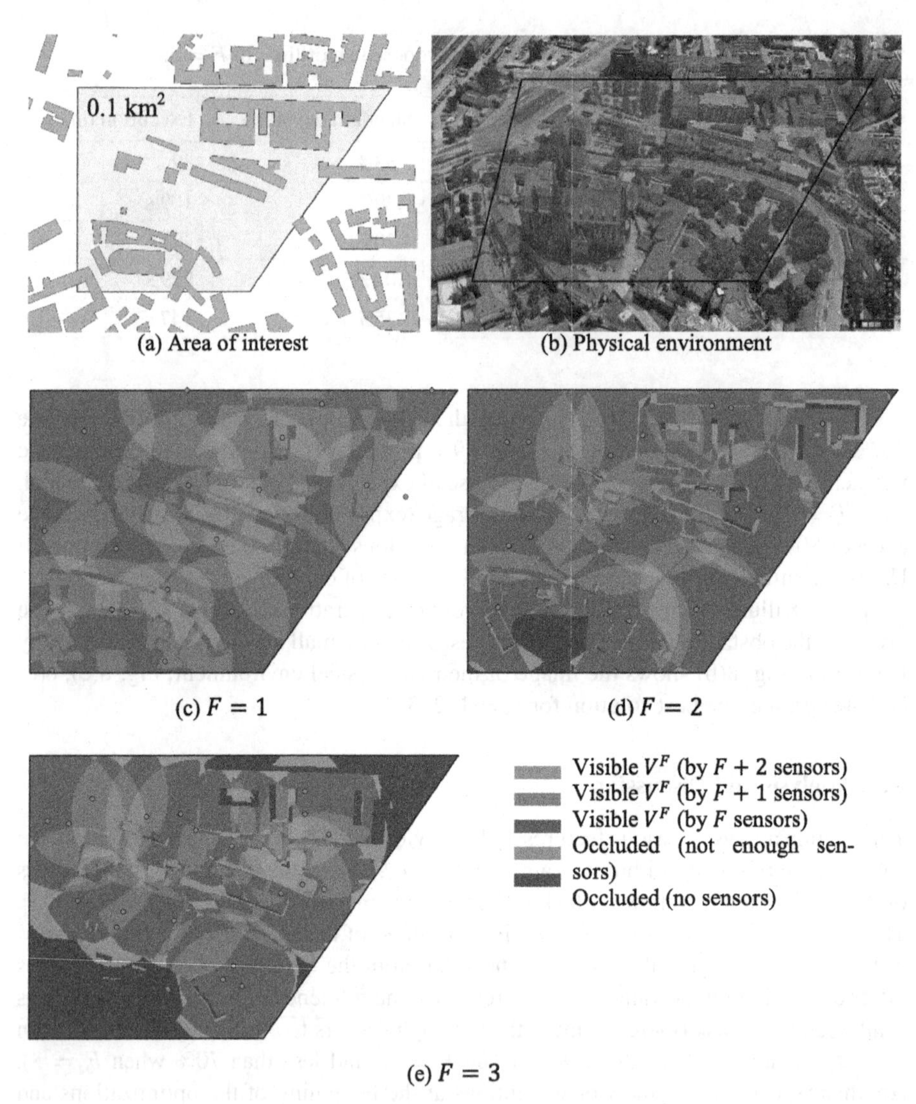

Fig. 8. Experiment e01.

The accumulation of evaluations with the increasing number of iterations is shown on the example of experiment e02 in Fig. 11. In average, the total number of solution evaluations is 6,233 when $F = 1$, 13,140 when $F = 2$, and 11,912 when $F = 3$. The number of evaluations during the later phases of the optimization for $F > 1$ is corresponding to the value of n_{1max} (there are not enough replacements so that an iteration is terminated by exceeding the permitted number of transformations). This is not true for $F = 1$ where it corresponds approximately to half of n_{1max} (an iteration is terminated by exceeding the permitted number of replacements). The reasons for this is the logical tendency to place waypoints close to one another for $F > 1$ (the constraint limiting the

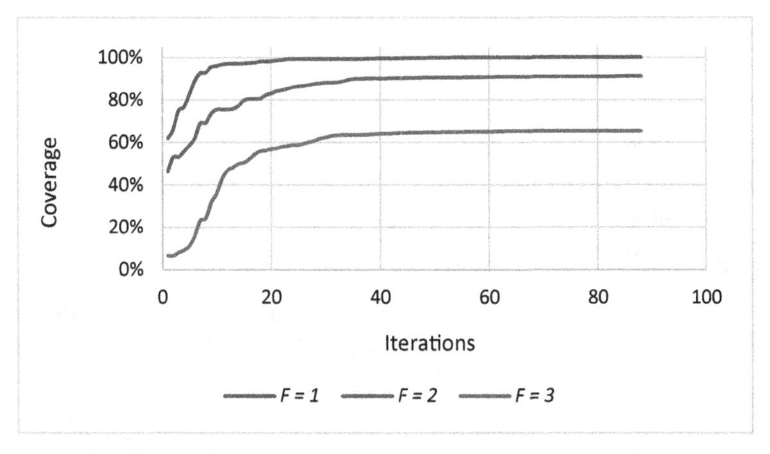

Fig. 9. Optimization progress in an example of experiment e01.

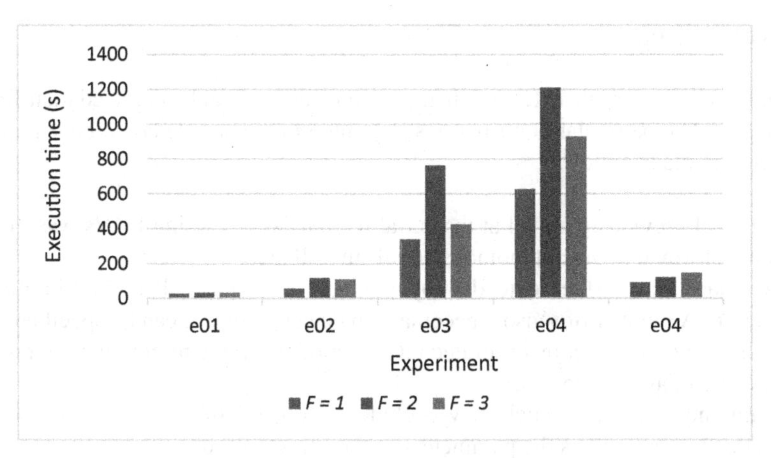

Fig. 10. Execution times of the algorithm.

space between waypoints can be easily violated which leads to disabling those sensors and results in the lower probability to accept the worse solutions); however, this is not the case for $F = 1$. The average time necessary to evaluate a single solution can be calculated; it is about 9 ms in all cases of experiment e02. There are 75,442 points to be evaluated in case of experiment e02 (see Table 4); i.e., more than 8 million points are evaluated per second.

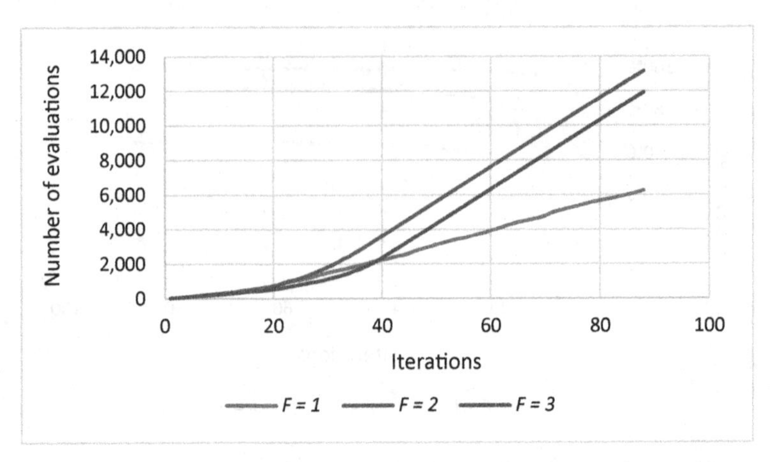

Fig. 11. Accumulation of the number of evaluations in experiment e02.

5 Conclusions

The authors of this article extended their previous research and completed their findings in this area. The particular contributions presented in this article could be summarized in several points as follows:

- The model previously aimed at the aerial reconnaissance using UAVs was modified and transformed to the autonomous aerial surveillance.
- The model was further extended by the new principle called the simultaneous detection. A number of sensors necessary to detect an object can be specified.
- Another new parameter incorporated into the model was the minimum required space between all pairs of sensors.
- Terrain and obstacles, which may occlude the objects of interest, were taken into consideration as well as the parameters of the sensors used.
- The metaheuristic algorithm based on the simulated annealing principles was proposed for deploying waypoints so that as large of the area of interest as possible is monitored by available UAVs in the swarm. New parameters connected with the simultaneous detection were incorporated into the algorithm.
- A set of scenarios was designed for verification. They use the real geographic data and are based on the parameters and features of the real typical surveillance operations.
- The simulation results of the experiments are discussed and some of the parameters of the algorithm are analyzed.

The choice of the metaheuristic method based on the simulated annealing has been supported by the experience of the authors obtained in their previous research. In general, metaheuristic methods are problem dependent; this means that some family of algorithms may be very successful in solving some type of a problem; however, they do not provide good results for another different type of a problem.

The future work of the authors will be focused on implementation of different metaheuristic methods for the solution of the problem examined in this article. Some of the basic principles which will be considered for implementation are as follows: Particle Swarm Optimization (PSO), genetic algorithms, tabu search, and bio-inspired algorithms (Ant Colony Optimization, Artificial Bee Colony, Bat algorithm). Also, the possibility of their hybridization will be investigated.

References

1. Bruzzone, A.G., Massei, M., Di Matteo, R., Kutej, L.: Introducing intelligence and autonomy into industrial robots to address operations into dangerous area. In: Mazal, J. (ed.) MESAS 2018. LNCS, vol. 11472, pp. 433–444. Springer, Cham (2019). https://doi.org/10.1007/978-3-030-14984-0_32
2. Hodický, J., Prochazka, D., Prochazka, J.: Training with and of autonomous system – modelling and simulation approach. In: Mazal, J. (ed.) MESAS 2017. LNCS, vol. 10756, pp. 383–391. Springer, Cham (2018). https://doi.org/10.1007/978-3-319-76072-8_27
3. Nohel, J., Flasar, Z.: Maneuver control system CZ. In: Mazal, J., Fagiolini, A., Vasik, P. (eds.) MESAS 2019. LNCS, vol. 11995, pp. 379–388. Springer, Cham (2020). https://doi.org/10.1007/978-3-030-43890-6_31
4. Otřísal, P., et al.: Testing methods of assessment for the chemical resistance of insulating materials against the effect of selected acids. Materiale Plastice **55**(4), 545–551 (2018)
5. Šilinger, K., Blaha, M.: Conversions of METB3 meteorological messages into the METEO11 format. In: International Conference on Military Technologies, Brno, pp. 278–284 (2017)
6. Stodola, P., Mazal, J.: Tactical decision support system to aid commanders in their decision-making. In: Hodicky, J. (ed.) MESAS 2016. LNCS, vol. 9991, pp. 396–406. Springer, Cham (2016). https://doi.org/10.1007/978-3-319-47605-6_32
7. Stodola, P., Nohel, J., Mazal, J.: Model of optimal maneuver used in tactical decision support system. In: International Conference on Methods and Models in Automation and Robotics, Miedzyzdroje, pp. 1240–1245(2016)
8. Yao, P., Wang, H., Zikang, S.: Cooperative path planning with applications to target tracking and obstacle avoidance for multi-UAVs. Aerosp. Sci. Technol. **54**, 10–22 (2016)
9. Lamont, G.B., Slear, J.N., Melendez, K.: UAV swarm mission planning and routing using multi-objective evolutionary algorithms. In: IEEE Symposium on Computational Intelligence in Multi-Criteria Decision-Making, Honolulu, pp. 10–20 (2007)
10. Shanmugavel, M., Tsourdos, A., Zbikowski, R., White, B.A.: 3D path planning for multiple UAVs using pythagorean hodograph curves. In: AIAA Guidance, Navigation, and Control Conference, Hilton Head, USA, pp. 1576–1589 (2007)
11. Triharminto, H.H., Adji, T.B., Setiawan, N.A.: Dynamic UAV path planning for moving target intercept in 3D. In: International Conference on Instrumentation Control and Automation, Bandung, pp. 157–161 (2011)
12. Sampedro, C., et al.: A flexible and dynamic mission planning architecture for UAV swarm coordination. In: International Conference on Unmanned Aircraft Systems, Arlington, USA, pp. 355–363 (2016)
13. Sujit, P.B., Beard, R.: Multiple UAV path planning using anytime algorithms. In: American Control Conference, St. Louis, pp. 2978–2983 (2009)
14. Stodola, P., Drozd, J., Šilinger, K., Hodický, J., Procházka, D.: Collective perception using UAVs: autonomous aerial reconnaissance in a complex urban environment. Sensors **20**(10), 2926 (2020)
15. Stodola, P., Drozd, J., Nohel, J., Hodický, J., Procházka, D.: Trajectory optimization in a cooperative aerial reconnaissance model. Sensors **19**(12), 2823 (2019)
16. Stodola, P., Mazal, J.: Model of optimal cooperative reconnaissance and its solution using metaheuristic methods. Defence Sci. J. **67**(5), 529–535 (2017)

Initial Solution Constructors for Capacitated Green Vehicle Routing Problem

Viktor Kozák[✉], David Woller, Václav Vávra, and Miroslav Kulich

Czech Institute of Informatics, Robotics, and Cybernetics, Czech Technical University in Prague, Jugoslávských partyzánů 1580/3, 160 00 Praha 6, Czech Republic

viktor.kozak@cvut.cz

http://imr.ciirc.cvut.cz

Abstract. This paper presents an analysis of the initial feasible solution constructions for the Capacitated Green Vehicle Routing Problem (CGVRP). CGVRP is a more challenging variant of the conventional Vehicle Routing Problem (VRP), where Alternative Fuel Vehicles (AFVs) are limited by their fuel capacity and the scarce availability of Alternative Fueling Stations (AFSs). The problem also imposes a constraint on the maximum carrying capacity of the vehicle. Presented methods can be used as a starting point for more advanced metaheuristics. Some of the methods can be seen as a generalization of the traveling salesmen problem, where one can draw on the numerous techniques described and known in this domain, other methods belong to the class of divide and conquer techniques, where the subproblems are some variants of the VRP. We use a two-phase approach with several different methods for both the initial VRP route construction and the following CGVRP tour validation and repair procedure. The methods are compared on various instances and their advantages are discussed in relation to common variants of the VRP, together with a possibility of their improvements or extensions.

Keywords: Capacitated green vehicle routing problem · Combinatorial optimization · Traveling salesmen problem

1 Introduction

Route optimization techniques are a powerful tool for reducing costs and pollution caused by transport and mobility operations. The Vehicle Routing Problem (VRP) is an NP-hard combinatorial optimization programming problem designed to minimize the total traveled distance of a fleet of vehicles/agents while visiting a set of customers. It generalizes the well-known Travelling Salesman Problem (TSP) and helps to reduce fuel usage and increase the efficiency of delivery routes. Because the problem is NP-hard, the size of a problem that can be solved optimally by an exhaustive search is limited and various heuristics have to be used to reach a solution that is close to optimal.

J. Mazal et al. (Eds.): MESAS 2020, LNCS 12619, pp. 250–268, 2021.
https://doi.org/10.1007/978-3-030-70740-8_16

Real word applications impose additional constraints on the problem, which resulted over the years in numerous variations of the VRP. Since VRP is often used for distribution companies, one of the most popular variants is the Capacitated Vehicle Routing Problem (CVRP), in which each customer has a specified demand and the maximum carrying capacity of the vehicle is considered. The classical CVRP considers a singe depot, where vehicles can reload their cargo. Other popular variants include a multi-depot variant [2] or a VRP with Pickups and Deliveries (VRPPD), where customers have both positive and negative demands [13].

Another variant reacts to the recent effort to reduce environmental impact by utilizing Alternatively Fueled Vehicles (AFVs), which are, however, limited by their fuel capacity and the scarce availability of Alternative Fueling Stations (AFSs). The Green Vehicle Routing Problem (GVRP) introduced in [8] considers the need for refueling and incorporates intermediate stops at AFSs to extend the possible length of a route traveled by a single vehicle. The same problem also applies to Electric Vehicles (EVs), which are limited by their battery capacity. Recently, the GVRP is especially relevant, since a large number of transportation companies are converting their gasoline-powered vehicle fleets to AFV fleets to reduce environmental impact or to meet new environmental regulations.

The Capacitated Green Vehicle Routing Problem (CGVRP) combines the CVRP and GVRP problems by considering both the maximum carrying capacity and the need for AFS visits. A specific variant of this problem, using EVs is referred to as the Electric Vehicle Routing Problem (EVRP), which often imposes additional constraints specific for EVs [7]. The CGVRP is especially suited for delivery companies operating fleets of AFVs.

This paper provides an evaluation of initial CGVRP route constructors using a two-phase heuristic, in which we combine initial VRP route constructions and subsequent CGVRP repair procedures. Several different methods were developed for each phase and tested on a benchmark data set provided in [14]. Both the intermediate VRP routes and the resulting CCGVRP routes are compared in regards to their performance, complexity and applicability to different VRP variants. The developed methods were then discussed as a base for advanced optimization methods.

2 Related Work

The VRP problem in its numerous variants has already been a subject of research for several decades. The CVRP is one of its more common variants, which offers a large number of methods for its solution developed over the years, as well as numerous test instances for the problem. For example, the CVRP library [21], created by Ivan Xavier and the authors of [18], provides a collection of the most popular test instances in a common framework.

Bard et al. [2] formulated the VRP with Satellite Facilities (VRPSF), an extension of CVRP in which vehicles constrained by their maximum carrying capacity have the option to stop at satellite facilities to reload their cargo, this problem is alternatively referred to as the multi-depot VRP [3,11,17]. Techniques developed for the multi-depot VRP were of great relevance for the more recent GVRP, which was first introduced in [8]. While numerous works addressed the classical VRP with capacity and distance constraints, GVRP added the capability to extend a vehicle's distance limitation by visiting an AFS en route.

We draw on methods presented in [19,20,22], developed specifically for the CGVRP, which combines the cargo capacity constraints with the possibility of extending the vehicle's driving range by an AFS visit. In this paper, we work with the basic CGVRP definition, with the intent to provide a general baseline for initial construction methods in this field. Developed methods are evaluated on testing instances provided in [14].

A specific case of this problem is the EVRP, which considers electric vehicles (EVs) instead of general AFSs. While the exact definitions differ, EVRP can be considered an extension to the classical CGVRP problem, and many developed methods apply to both problems. A survey on different EVRP variants was presented in [7]. Those variants extend the original CGVRP by considering the use of hybrid vehicles, heterogenous vehicle fleets, dynamic traffic conditions, different charging technologies, non-linearity of the charging function, et cetera. Out of the EVRP oriented works, [15] should be mentioned. Although this work uses an additional criterion (non-linear fuel consumption based on the current load) the problem is also evaluated on the extended version of the dataset [14].

3 Problem Definition

The CGVRP is a challenging NP-hard combinatorial optimization problem extending the original VRP with the addition of several other constraints. It can be seen as a mixed-integer programming problem for which the mathematical model with its parameters has already been formulated in [22].

The problem can be expressed using a fully connected weighted graph $G = (V, E)$, where V is a set of nodes and E is a set of weighted edges connecting these nodes. The node set V contains a depot D, a set of customers I and a set of AFSs F. A non-negative distance value d_{ij}, representing the distance between nodes i and j, is associated with each edge. Each customer $i \in I$ is assigned a positive demand. We assume a homogeneous fleet of vehicles, where each vehicle is constrained by its maximal carrying capacity C, maximal fuel capacity Q and a set fuel consumption rate h. It is assumed that the fuel consumption rate is constant and each traveled edge consumes the amount hd_{ij} of the remaining fuel.

The objective of CGVRP is to find a set of routes, where each vehicle starts and ends at the depot, and that minimizes the total distance traveled. Every customer can be visited only once and the corresponding demand needs to be satisfied after this visit. For every route, the total demand of visited customers can not exceed the maximal carrying capacity of the vehicle.

After a vehicle leaves a refueling station, the fuel consumption cannot exceed the maximal fuel capacity of the vehicle until it reaches another refueling station or ends in the depot. It is assumed, that the depot is capable of both refueling the vehicles and reloading their cargo. When a vehicle visits either an AFS or the depot, it is refueled to the full tank capacity. It is also assumed that AFSs can be visited multiple times.

4 Selected Methods

As the problem is NP-hard, it is impossible to solve larger instances to optimality in a reasonable time. Therefore, it is suitable to deploy a heuristic approach with polynomial time complexity. The CGVRP is a constrained variant of the VRP, for which many effective construction methods were proposed over the years. Some of the methods presented in this paper are modified variants of constructions proposed for different variants of the VRP, while others are newly designed specifically for CGVRP or EVRP.

Most of the CGVRP constructions presented in this paper are split into two phases. The first phase creates a feasible solution to a TSP problem, either totally or selectively disregarding the AFS nodes and the load and energy constraints. The output of the first phase is typically an invalid CGVRP tour, therefore, the second phase can be seen as a repair procedure. The two-phase approach is convenient for problem modularization which in turn led to robust and comparative results, as it allowed to freely test different combinations of individual methods.

Individual TSP and VRP generating methods were later augmented using the Density-Based Clustering (DBC) metaheuristic. DBC exploits the spatial properties of VRP and considers the distribution of nodes over space. This proved to further improve the construction methods by providing global spatial awareness about local point density which is in common unavailable to basic construction methods.

The description of individual methods follows, together with example solutions on a selected instance from [14], presented in Fig. 2. The example presents valid TSP and CVRP solutions, and two different CGVRP solutions.

4.1 Initial TSP Constructions

Nearest Neighbour Algorithm (NN). The NN algorithm was one of the first algorithms used to approximately solve the TSP and is commonly used for the initial TSP construction phase for vehicle routing problems [10]. A set of all the customers I is used as an input for the algorithm. The algorithm starts at a random node and moves on to the nearest unvisited node, this way the remaining nodes are greedily added to the tour. The output of the algorithm is a sequence of nodes $T = \{n_1, n_2, ..., n_n\}$, which starts at a random customer and visits every customer exactly once. The AFSs and the depot are not considered in this phase. A TSP tour created using the NN algorithm can be seen in Fig. 2b.

Minimum Spanning Tree Algorithm (MST). The MST-based approximation algorithm [12] starts with a construction of the MST over the set of all customers I. Then a root node is chosen arbitrarily and all nodes are then visited by depth-first search, generating a sequence of visited nodes. Only the first visit of each node is taken from this sequence, and the root node is added at its end, thus generating a feasible TSP solution.

The MSP can be used to assign weights to selected edges, providing a useful tool for algorithms based on ant colony optimization [12]. Another advantage of MST-based methods is, that they can be used to provide both the upper and the lower bound for the problem. When the cost function satisfies the triangular inequality, an approximate algorithm can be designed to generate a tour, whose cost is never more than twice the cost of an optimal tour, while the cost of the best possible TSP tour is never less than the cost of the MST.

Modified Clarke-Wright Savings Algorithm (CWS). This method was originally developed as a heuristic for solving the VRP with a central repository and an unfixed number of vehicles [6]. The heuristic works over the set V' composed of all the customers I and the depot D ($V' = \{I \cup D\}$). An individual route from the depot and back is created for each customer, disregarding the energy constraints and AFSs. The algorithm then starts with a node that is furthest from the depot, and individual routes are then merged while utilizing the saving distance $S_{i,j}$ defined in Eq. 1 as a difference between the original and resulting routes. Only the outmost customer nodes of the route are considered during the merge process. In the equation, $d_{i,j}$ represents a distance of an edge from node i to node j.

$$S_{i,j} = d_{i,j} - d_{depot,j} - d_{depot,i} \tag{1}$$

The original CWS algorithm intended for VRP was modified for CVRP by stopping the route merging procedure when there are no more customers that could be connected to the current route without exceeding the maximum cargo capacity, a new route is then created from the remaining customers. The consideration of the depot position in Eq. 1 proved to reduce the overall cost of node-to-depot edges created by this approach. Since this is the main advantage of this approach, the capacitated version of the CWS algorithm (C-CWS) will be used in this paper. The output of this algorithm is a valid CVRP tour, however, the satisfaction of the energy constraints is not guaranteed, therefore AFSs generally have to be inserted afterward by one of the CGVRP tour repair procedures. A valid CVRP route, without the consideration of energy constraints, created by the C-CWS algorithm is shown in Fig. 2c.

One Route Each (ORE). We have implemented a simple CGVRP route construction method as a simple baseline, intended primarily for comparison. It is also the only stand-alone CGVRP tour construction method presented in this paper. The method takes a list of all customers as an input, and a separate route from the depot and back is planned for each customer. If the customer cannot be reached by traveling directly from the depot and back, an AFS closest to the customer is added to the route before and after visiting the customer. It is assumed here that all AFSs are directly reachable from the depot and the demand of individual customers is lower than the maximum carrying capacity of the vehicle, which holds for all testing instances used in this paper. This method can also be used to verify if a specific CGVRP instance is solvable. An example of a valid CGVRP tour constructed by this method is shown in Fig. 2a.

Density-Based Clustering Algorithm (DBC). This metaheuristic exploits the spatial properties of the VRP and considers the distribution of nodes over space. This allows us to divide the original set of nodes V into several sub-sets, that can be solved separately for each subproblem, and thus downsize the original planning problem. The implemented DBC algorithm builds on concepts from the Density-Based Spatial Clustering of Applications with Noise (DBSCAN) proposed in [9]. The use of this method in combination with initial TSP constructions was inspired by [8].

The DBC algorithm separates nodes based on a given neighborhood radius ϵ and a density threshold δ. We define the ϵ-neighborhood $N_\epsilon(n_i)$ as a set of nodes within a radius of ϵ from n_i and introduce the condition $|N_\epsilon(n_i)| \geq \delta$ for potential cluster core candidates, wherein all nodes in the ϵ-neighborhood of a core candidate node n_i are considered as directly density-reachable from n_i. Initial clusters are formed by identifying sets of density-reachable nodes based on the core candidates. Nodes outside of these initial clusters are typically considered as outliers and can be treated as separate entities or assigned to the nearest cluster. An illustration of node clustering can be seen in Fig. 1, where core candidate nodes are depicted in red, the border nodes in yellow and noise nodes in gray. The δ parameter for the illustration was set to 4 and the ϵ parameter can be seen from the ϵ-neighborhood around the C node. We can see that while the border nodes are directly reachable from the core candidates, the same doesn't hold in reverse.

The algorithm generates a number of cluster sets corresponding to each pair of (ϵ, δ) parameter combinations. Fractions of the maximal distance that can be traveled by a vehicle on a full tank will be used as potential values for the ϵ parameter, since a suitable value for this parameter is largely dependent on individual problem instances. Potential values of the δ parameter are set as integers in an interval $[\delta_{min}, \delta_{max}]$.

The generated cluster sets V_i are subsequently augmented with a depot and AFSs as $V_i' = \{V_i \cup D \cup F\}$ and used as an input for the developed methods for initial TSP tour construction. The Nearest Neighbour and C-CWS initial construction methods have been modified to support cluster sets generated from the DBC algorithm as inputs.

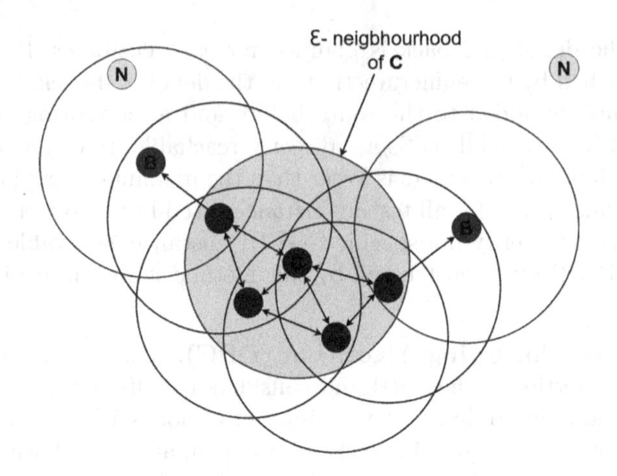

Fig. 1. Illustration of a cluster created by the DBC algorithm

4.2 CGVRP Tour Repair Procedures

Separate Sequential Fixing (SSF). SSF is a novel EVRP tour repair procedure proposed in [20], the procedure is also applicable to the CGVRP. This procedure is split into two subphases. It starts with an initial feasible TSP solution generated by one of the initial TSP construction methods described in Sect. 4.1, on this tour the load constraint is sequentially checked, and whenever the next customer cannot be satisfied, a depot is inserted, thus, the constraints on the load are fixed.

The energy constraints based on fuel consumption are fixed next. The current tour is again sequentially checked, and if the next customer is reachable from the current node and the vehicle will not get stuck in it (meaning that it can still reach the closest AFS), the customer is added to the final valid tour. Otherwise, SSF adds the AFS closest to the current customer, the AFS closest to the next customer, and any intermediate AFSs in between, if necessary. This AFS sequence is obtained as the shortest path on a graph of all AFSs. After that, the next customer can be safely added.

Relaxed Two-Phase Heuristic for CGVRP. Another repair procedure is the two-phase heuristic for CGVRP proposed by Zhang, Gajpal and Appadoo in [22] (further referred to as ZGA). The first phase is the TSP route construction solved using the NN algorithm. This phase is already covered by several construction methods in Sect. 4.1, therefore, our focus will be on the second phase, the CGVRP repair procedure.

The ZGA repair procedure starts from the depot, iterates through the TSP route, and inserts the depot or AFS nodes when needed. The depot is inserted when the demand of the next customer can't be satisfied. An AFS node is inserted if the depot can't be reached via the next customer. If not, the nearest AFS is inserted into the route at the current position. If even the nearest AFS

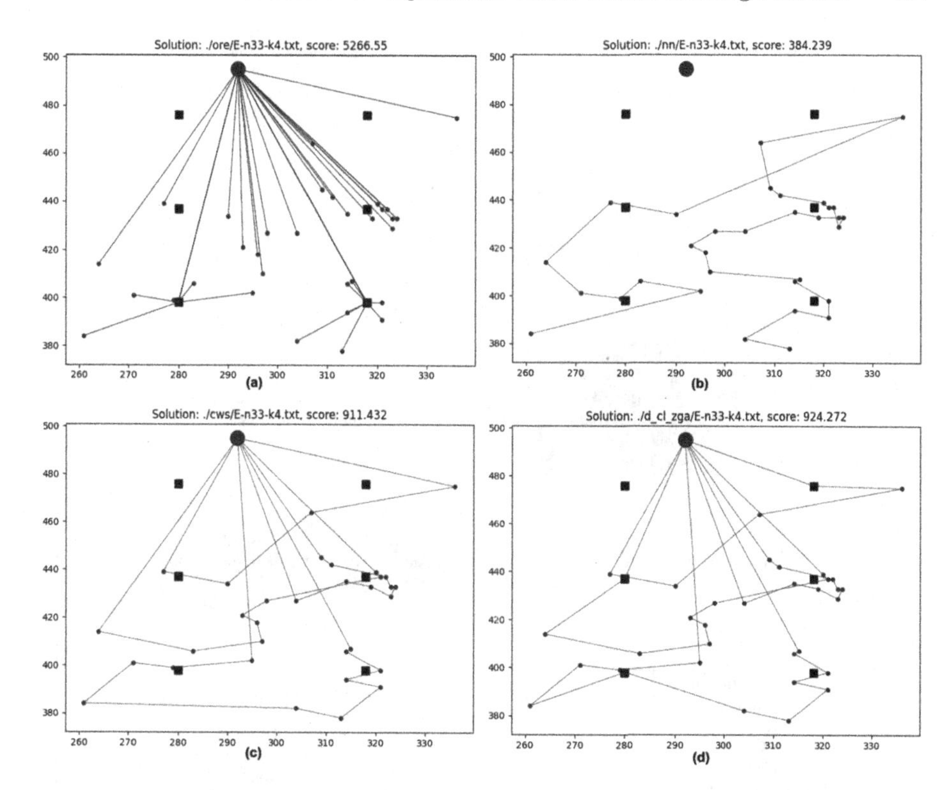

Fig. 2. Example constructions on the E-n33-k4 instance. The depot is represented by the red dot, AFSs are represented by black squares and customers are depicted as the blue dots. Individual constructions were generated by a) the ORE method, b) the NN algorithm, c) the C-CWS algorithm, and d) the C-CWS algorithm originating from cluster sets generated by the DBC metaheuristic, followed by the Relaxed ZGA repair procedure. (Color figure online)

cannot be reached, the algorithm backtracks as it changes the current node to the previous one.

While the method as described in [22] works well in general, there are situations when it can get stuck in an endless cycle. It happens when the demand is satisfied, the depot can't be reached via the next customer and the closest AFS can be reached. After adding the closest AFS this sequence repeats for the same customer. This can happen if there exists a node in the instance such that a route from the closest AFS to the depot via this node is too long (demands more energy than the maximum fuel level allows).

We modified the original algorithm by checking whether the closest AFS from the next customer can be reached via the next customer, instead of the original condition which checked only the reachability of the depot. In reaction to this change, an additional option for an AFS visit was added in situations where an AFV is returning to the depot. The modified ZGA algorithm for the second

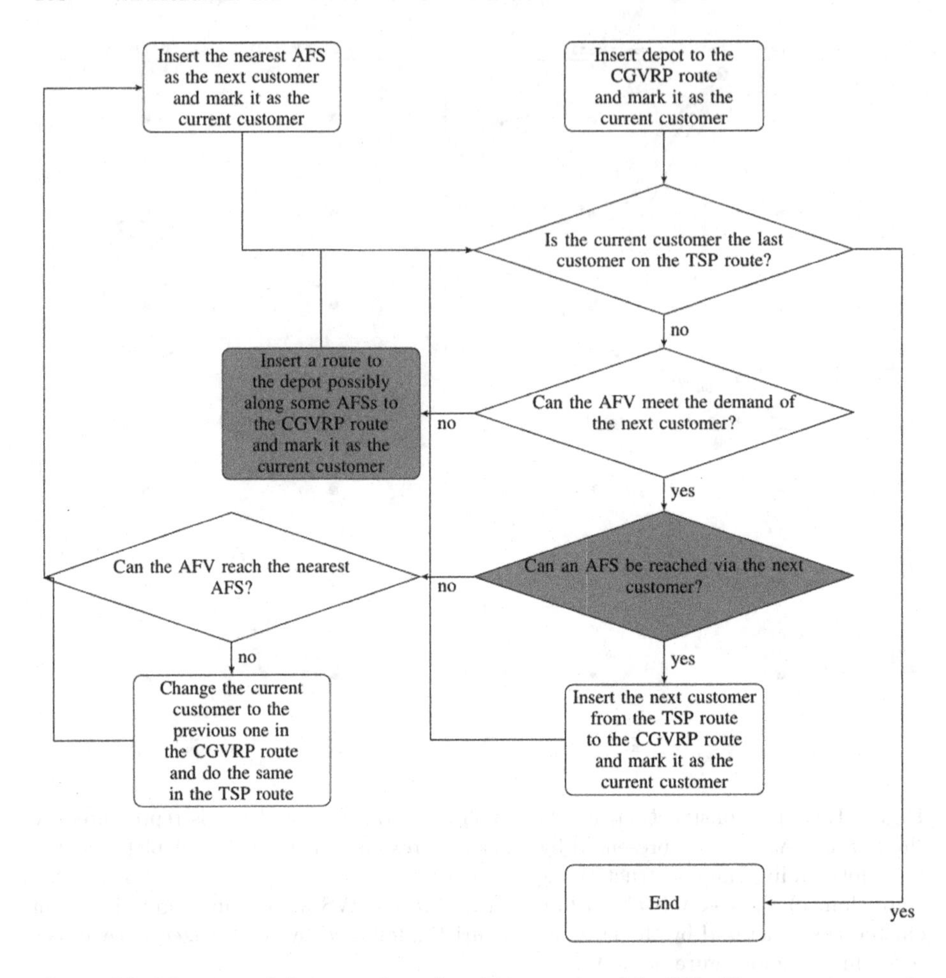

Fig. 3. Modification of the two-phase heuristic proposed in [22]. Changes were made to the orange nodes. (Color figure online)

phase is visualized in Fig. 3. Since the first TSP construction phase was omitted, we will further be referring to the modified algorithm as *Relaxed ZGA*. A valid CGVRP route, generated using the C-CWS algorithm originating from cluster sets generated by the DBC method used in combination with the Relaxed ZGA, can be seen in Fig. 2d.

WL Decoding Method. The last repair procedure for an initial feasible solution construction was proposed by Wang and Lu (further referred to as WL) [19]. This procedure works on a similar principle as the ZGA method, iterating through the TSP route and inserting AFSs and the depot when needed. However, once again a situation may arise when the method gets stuck in an infinite cycle. This situation arises on occasions when there is a path $P = \{depot, n_1, n_2, ..\}$ on which the first customer n_1 can not be reached directly from the depot, therefore an AFS has to be inserted directly after the depot, and the total demand of the following customers becomes greater than the carrying capacity of the vehicle before visiting another AFS.

In such a situation, the algorithm goes from the depot directly to an AFS before continuing on the TSP route. Customers are served until the sum of the total demand of visited customers and the demand of the next customer exceeds the maximum carrying capacity, in which case the algorithm tries to go to the depot. However, the depot can't be reached directly from the current node, therefore the algorithm backtracks through the previous nodes back to the AFS node. The algorithm then proceeds to the depot, leaving all the customer nodes unvisited. This situation then results in an endless cycle.

We can speculate that the authors considered all customers as directly reachable from the depot, which would eliminate the possibility of the aforementioned situation. Creating the return route to the depot via an AFS when needed could fix this particular issue, however, it would affect the basic principles of the WL algorithm. Therefore, we use this method in its original variant and evaluate its result only on instances where it successfully generated a valid solution.

5 Results and Discussion

The VRP route generating methods described in Sect. 4 were applied to various test instances and the results are presented in this section. The main evaluation criteria used in this paper is the total distance traveled on the generated route, further referred to as the tour fitness value. Section 5.2 provides an overall evaluation of developed methods, together with their time complexity, the complexity of individual components w.r.t. the number of customers and AFSs, and the number of node-to-node distance evaluations.

5.1 Experiment Setup and Parameters

This paper was inspired by [14], which provides a benchmark set of 17 instances for the EVRP, however, the problem definition for the provided instances is identical to CGVRP. The instances are split into two sets, small (E) and large (X), where the number of customers ranges from 21 to 100 in the E-instances and from 142 to 1000 in the X-instances. All of the instances contain exactly one depot, while the number of AFSs varies. A more detailed description of the instances can be found in [16], wherein the optimal upper-bound values for the E-instances are provided. It should be noted that the E-instances were generated by extending the well-known instances from [5] and the X-instances are an extension of the more recent [18], both used for the conventional CVRP problem.

The NN and MST based algorithms used for initial TSP constructions were implemented in their non-deterministic variants. Both algorithms select a starting node at random, which can result in different TSP routes each time. Therefore, 20 independent runs with random seeds were performed for each instance and the final score was then evaluated based on both the minimal achieved fitness value and the average fitness value.

Values for the ϵ and δ parameters used for the DBC algorithm are shown in Table 1. The ϵ parameter is problem specific and set using fractions of the maximal distance that can be traveled by the vehicle on a full tank, which we define here as *reach*. Values for the δ parameter were set based on our evaluation of the initial results on the test instances. There's a total of 8 ϵ and 4 δ parameter values, which can result in up to 32 different cluster sets. We perform the full initial construction procedure with all unique cluster sets and evaluate the results based on the construction with the best final tour fitness value.

Table 1. Parameter values for the DBC method.

Parameter	Values
ϵ	$[\frac{1}{2}, \frac{1}{3}, \frac{1}{4}, \frac{1}{6}, \frac{1}{8}, \frac{1}{10}, \frac{1}{15}, \frac{1}{20}] \times reach$
δ	2, 3, 4, 5

The experiments were performed on a computer with an Intel Xeon(R) E3-1240 v5 (8M Cache, 3.50 GHz) processor and 32 GB RAM.

5.2 Initial Construction Comparison

The presented methods were able to generate valid solutions for all test instances, with the exception of the WL tour repairing method, which failed at specific instances for reasons closely described in Sect. 4.2. We rank the construction methods based on their average and minimal scores and provide a more in-depth evaluation based on the VRP variant, computing complexity, and the overall results.

Table 2 presents the best achieved tour fitness values for each specific VRP variant over all instances together with the CGVRP benchmark values provided for the E-instances. The difference in VRP variants can be seen in Table 2, wherein, due to the influence of additional constraints imposed on the problem, the CVRP and CGVRP fitness values are much higher than for the classical TSP route. As some of the developed methods are fully deterministic, the minimal score value is omitted when it's identical to the average value, or completely omitted, when it's irrelevant to the problem.

The initial method evaluation is presented in Table 3, which presents the number of times each method generated a tour with the best score achieved for the instance. The results are further differentiated based on the VRP variant applicable to the generated route. It is not uncommon that more methods generated a route with the same fitness value, especially on smaller instances. The MST-based methods and the ORE method did not achieve the best score on any of these instances.

Table 2. Best achieved scores for individual VRP variants.

Instance	Best TSP		Best CVRP	Best CGVRP		Provided CGVRP benchmark values
	Minimal	Average		Minimal	Average	
E-n22-k4	235.832	256.875	446.539	441.641	452.443	384.955
E-n23-k3	409.893	453.598	630.508	629.25	656.747	571.947
E-n30-k3	311.731	366.52	574.632	–	576.976	509.47
E-n33-k4	384.239	408.869	911.432	–	924.272	840.146
E-n51-k5	467.596	495.138	624.531	–	659.213	532.225
E-n76-k7	574.858	612.257	816.158	845.395	879.895	697.438
E-n101-k8	713.977	758.734	982.484	–	1040.34	836.847
X-n143-k7	9959.79	10656.4	19275.7	–	21339.4	–
X-n214-k11	7152.51	7659.63	13575.3	13896.3	14142.1	–
X-n351-k40	10910.5	11253.7	29298.7	–	30266.1	–
X-n459-k26	12783.7	13118.4	30638.1	30593.5	31624.2	–
X-n573-k30	11093.6	11714.9	56064.0	–	57863.0	–
X-n685-k75	21507.8	22147.1	78800.2	–	81786.1	–
X-n749-k9	18412.7	18744.8	85979.5	–	89326.5	–
X-n819-k171	19656.4	19994.5	167862.0	–	170603.0	–
X-n916-k207	24843.1	25382.8	345951.0	–	350725.0	–
X-n1001-k43	27817.0	28443.4	83036.2	–	85730.7	–

Table 3. The number of best minimal and average scores achieved by individual construction methods (out of 17 testing instances).

Construction method	Best TSP		Best CVRP	Repair procedure	Best CGVRP	
	Minimal	Average			Minimal	Average
Random-seed NN	**12**	9	–	SSF	1	0
				Relaxed ZGA	3	0
				WL	1	0
NN from DBC	9	**11**	–	SSF	0	0
				Relaxed ZGA	2	0
				WL	0	0
C-CWS	0	0	7	SSF	0	0
				Relaxed ZGA	5	6
				WL	1	1
C-CWS from DBC	0	0	**17**	SSF	1	2
				Relaxed ZGA	**10**	**14**
				WL	2	4

Fitness values from the best TSP and CGVRP methods were used as a reference and qualitative results provided in Table 4 are given as a ratio of the computed fitness in relation to these values. Minimal TSP scores generated by the Random-seed NN are used as a baseline for TSP routes, while CGVRP

Table 4. Initial TSP and CGVRP constructions comparison - relative average score over all instances. Methods using the WL repair procedure did not generate valid solutions for all the instances, and should, therefore, be taken with reservations.

Construction method	TSP constructions avg. score	Repair procedure	CGVRP routes avg. score
ORE	–	–	6.8440
Random-seed NN	1.0580	SSF	1.1644
		Relaxed ZGA	1.1157
		WL	1.1629
Random-seed MST	1.1620	SSF	1.2065
		Relaxed ZGA	1.1610
		WL	1.1958
C-CWS	3.5396	SSF	1.0850
		Relaxed ZGA	1.0385
		WL	1.1707
NN from DBC	**1.0520**	SSF	1.1235
		Relaxed ZGA	1.0839
		WL	1.1135
C-CWS from DBC	3.4880	SSF	1.0349
		Relaxed ZGA	**1.0**
		WL	1.0465

routes are evaluated in relation to scores generated using the C-CWS algorithm with DBC generated cluster sets, in combination with the Relaxed ZGA repair procedure. The table presents a relative comparison of all VRP and CGVRP route scores for individual construction methods. The scores are averaged over all testing instances using Eq. 2, in which F_i^j represents the average fitness of method j achieved on instance i and F_i^{ref} is the reference value for this instance.

$$F^j_{rel\_avg} = \frac{1}{n} \sum_{i=1}^{n} \frac{F_i^j}{F_i^{ref}} \tag{2}$$

Methods using the WL repair procedure are an exception, since these methods generated only an average of 6 valid solutions out of the 17 testing instances, and should, therefore, be taken with reservations. Further comparison can be seen in Figs. 4, 5 and 6, where relative minimal and average scores are presented in relation to the problem size on individual instances.

Table 4 shows that solutions generated by the ORE method are far from optimal. While this method is clearly inferior in the initial construction phase, it is the only stand-alone method and has the lowest complexity from the presented methods, it can also be used to ascertain if the problem has a valid solution. The ORE method generates a separate route for each customer, while it may be impractical as a final VRP solution, it may prove beneficial to use the generated route as an input for certain optimization methods, since methods based on local search operators, or evolution algorithms often benefit from more freedom in initial conditions.

We can separate the remaining construction methods by two main aspects. First are the methods used to generate the TSP tour (NN, C-CWS, and MST), second are the tour repair procedures that create a valid CGVRP tour from the original TSP (SSF, Relaxed ZGA and WL). The influence of the DBC metaheuristic will be discussed separately.

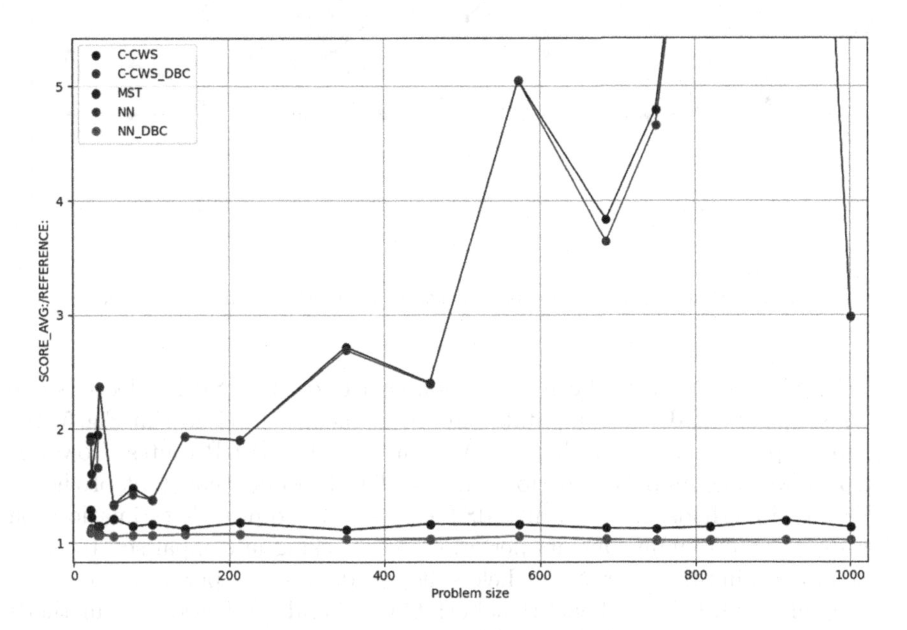

Fig. 4. Comparison of TSP generating methods on all instances.

Methods using the TSP tour generated by C-CWS generally tend to produce better results for the CGVRP, while their performance is largely inferior for the classical TSP. The reason for this is that C-CWS is the only method that directly incorporates the carrying capacity during the TSP tour creation, generating a valid CVRP route. The difference between the TSP and CVRP methods can be seen in Fig. 4, were the lengths of C-CWS generated routes are several times higher than for the TSP generated ones, this only increases with the number of customers. The main advantage of the C-CWS method is that it considers the depot position during the initial VRP route creation phase, generating a route more optimal for frequents visits to this specific node.

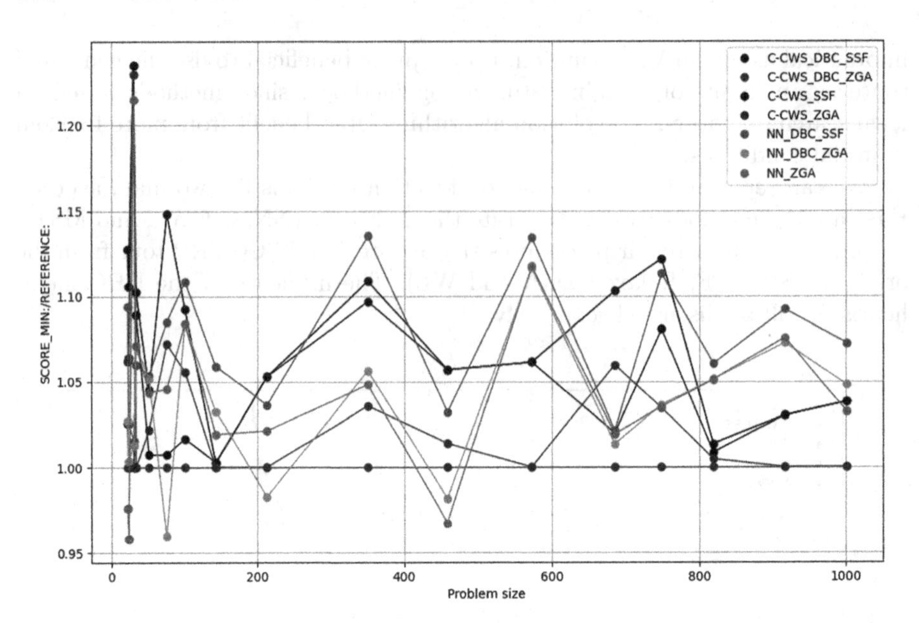

Fig. 5. Minimal score comparison of the top 7 methods on all instances.

The NN algorithm has been a common choice for the TSP for decades due to its simplicity and low computational requirements. As shown in Fig. 5, the algorithm performed better than C-CWS on several CGVRP routes. However, Fig. 6 shows that its overall performance is slightly worse than that of the C-CWS based methods. Lastly, while MST can be used to provide estimations on the cost of an optimum tour, its performance is lacking in comparison to other methods used in this paper. Nevertheless, it provides a stable performance as can be seen in Fig. 4. It is our belief that both C-CWS and NN based TSP methods are generally more suitable for the VRP.

From the repair procedures, the Relaxed ZGA method proved to generate the best results. One of its advantages over the SSF is its approach regarding the carrying capacity. While SSF divides the tour according to the carrying capacity in its first step, Relaxed ZGA updates the current capacity dynamically during the whole tour repairing process. This means that Relaxed ZGA sometimes picks the depot as the nearest AFS for refueling and, at the same time, loads the vehicle, this case is not accounted for in the SSF method. There is also a slight difference in the AFS selection during the process. The WL method presented in [19] failed to generate a valid CGVRP tour on most of the instances. Despite that, the method shows great promises, since it reached the best resulting scores in several instances (see Table 3).

Lastly, we can see that initial constructions originating from clusters generated by the DBC method have a significant chance for improvement over the original methods. Although the improvement is clear, this comes at the cost of higher computational complexity. It would be interesting to compare this to

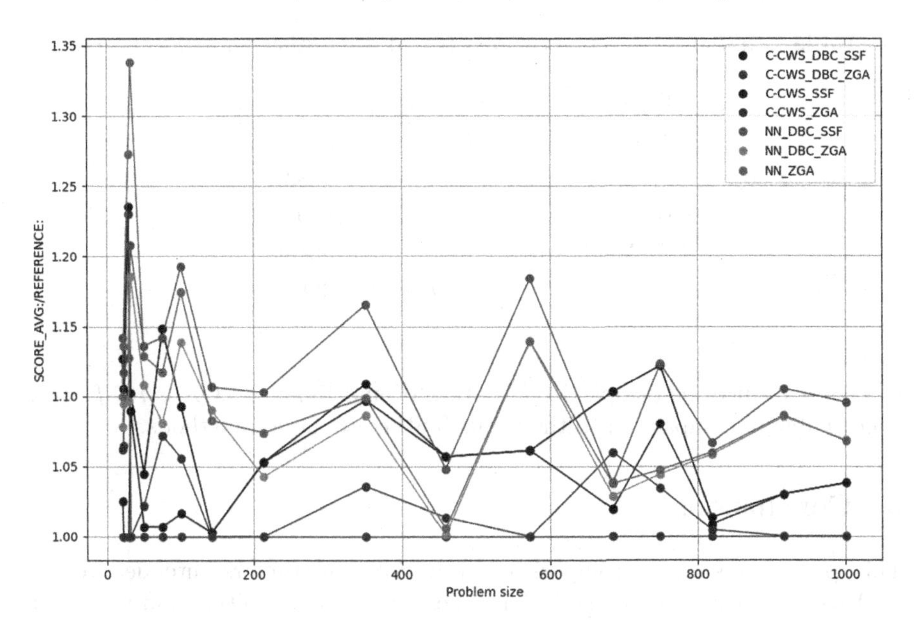

Fig. 6. Average score comparison of the top 7 methods on all instances.

other clustering methods. For example, the sweep algorithm, which generates feasible clusters by rotating a ray centered at the depot, proved to have good results for the CVRP [1,4]. However, we work with the problem defined by a distance matrix, which is unsuitable for this approach. Therefore, we have chosen the DBC algorithm, which is more general and applicable to a wider range of problems.

The complexity of individual initial construction methods can be seen in Table 5. The complexity for benchmark instances provided in [14] is evaluated as the number of requests for a distance between nodes a and b. Defining the problem as $V = \{D \cup I \cup F\}$ where D is the depot, I denotes the set of customers and F denotes the set of AFSs, we can define $n = \|V\|$, $i = \|I\|$ and $f = \|F\|$ as the size of the problem, number of customers and number of AFSs respectively. In our complexity estimates, we work with the assumption that i is significantly higher than f.

Table 5 also contains the maximal computational time T_{max} for each type of the initial TSP construction methods. The time value presented in the table is the highest computational time out of 20 independent runs over all the test instances, needed to achieve a valid CGVRP solution. The time necessary for a valid CGVRP construction for the X-instances was generally in single or double digits (ms), while for the smaller E-instances most methods constructed a valid tour in under one millisecond.

The complexity for constructions initiating from the DBC depends on the spatial distribution of nodes in the problem and the underlying construction method, generally $k \in (10, 17)$. We do not include the comparison for repair

Table 5. Initial construction complexity and maximal computational times.

Type	Complexity	$T_{max}[ms]$
ORE	$i \times f$	1.524
NN-based	i^2	15.83
MST-based	i^2	40.94
C-CWS-based	i^2	19.86
DBC-based	$k \times complexity$	129.9

procedures, since the complexity of both SSF and Relaxed ZGA is $i \times f$, which is negligible in comparison to the initial TSP construction methods.

6 Conclusions

This paper presents a comparison of initial constructors intended for the CGVRP. A variety of methods for generating a valid initial route construction over a set of customers was developed with the intent to minimize the total traveled distance while respecting the maximum load capacity of the vehicle and incorporating AFSs en route when necessary.

A two-phase heuristic for the CGVRP was presented. This heuristic consists of the initial TSP route construction and the subsequent CGVRP repair procedure. Various methods ware implemented for each phase and their combinations were tested on selected benchmark instances. The results were evaluated based on their performance and complexity, and discussed in regard to different applications and VRP variants.

The developed methods provide a good, fast and reliable solution, making them suitable for a stand-alone TSP or CGVRP route constructions. The repair procedures provide a reliable tool for fast reconstruction of a VRP route. The developed methods can also serve as an initial solution for advanced optimization methods and heuristics. Presented results and discussion should provide insight into these techniques and should serve as a good starting point for the design and development of advanced planning methods.

The two-phase heuristic is applicable to any alternatively fueled or electric vehicle and while the developed methods are focused on CGVRP, the model could be extended to other variants of the VRP. The outcomes motivate future research of initial construction methods in combination with more advanced metaheuristics.

Acknowledgements. This work has been supported by the European Union's Horizon 2020 research and innovation program under grant agreement No. 688117. The research leading to these results has received funding from the Czech Science Foundation (GAČR) under grant agreement no. 19-26143X. The work has been also supported by the Grant Agency of the Czech Technical University in Prague, grant No. SGS18/206/OHK3/3T/37.

References

1. Akhand, M.A.H., Jannat, Z., Murase, K.: Capacitated vehicle routing problem solving using adaptive sweep and velocity tentative PSO. Int. J. Adv. Comput. Sci. Appl. **8**, 288–295 (2017). https://doi.org/10.14569/IJACSA.2017.081237
2. Bard, J., Huang, L., Dror, M., Jaillet, P.: A branch and cut algorithm for the VRP with satellite facilities. IIE Trans. **30**, 821–834 (1998). https://doi.org/10.1023/A:1007500200749
3. Chan, Y., Baker, S.: The multiple depot, multiple traveling salesmen facility-location problem: vehicle range, service frequency, and heuristic implementations. Math. Comput. Model. **41**, 1035–1053 (2005). https://doi.org/10.1016/j.mcm.2003.08.011
4. Chen, M., Chang, P., Chiu, C., Annadurai, S.P.: A hybrid two-stage sweep algorithm for capacitated vehicle routing problem. In: 2015 International Conference on Control, Automation and Robotics, pp. 195–199 (2015). https://doi.org/10.1109/ICCAR.2015.7166030
5. Christofides, N., Eilon, S.: An algorithm for the vehicle-dispatching problem. J. Oper. Res. Soc. **20**(3), 309–318 (1969). https://doi.org/10.1057/jors.1969.75
6. Clarke, G., Wright, J.W.: Scheduling of vehicles from a central depot to a number of delivery points. Oper. Res. **12**(4), 568–581 (1964). https://doi.org/10.1287/opre.12.4.568
7. Erdelic, T., Carić, T., Lalla-Ruiz, E.: A survey on the electric vehicle routing problem: variants and solution approaches. J. Adv. Transp. (2019). https://doi.org/10.1155/2019/5075671
8. Erdogan, S., Miller-Hooks, E.: A green vehicle routing problem. Transp. Res. Part E Logist. Transp. Rev. **48**(1), 100–114 (2012)
9. Ester, M., Kriegel, H.P., Sander, J., Xu, X.: A density-based algorithm for discovering clusters in large spatial databases with noise. In: Proceedings of the Second International Conference on Knowledge Discovery and Data Mining, KDD 1996, pp. 226–231. AAAI Press (1996). https://doi.org/10.5120/739-1038
10. Gutin, G., Yeo, A., Zverovich, A.: Traveling salesman should not be greedy: domination analysis of greedy-type heuristics for the tsp. Discrete Appl. Math. **117**, 81–86 (2002). https://doi.org/10.1016/S0166-218X(01)00195-0
11. Kek, A., Cheu, R., Meng, Q.: Distance-constrained capacitated vehicle routing problems with flexible assignment of start and end depots. Math. Comput. Model. **47**, 140–152 (2008). https://doi.org/10.1016/j.mcm.2007.02.007
12. Kheirkhahzadeh, M., Barforoush, A.: A hybrid algorithm for the vehicle routing problem, pp. 1791–1798 (2009). https://doi.org/10.1109/CEC.2009.4983158
13. Martinovic, G., Aleksi, I., Baumgartner, A.: Single-commodity vehicle routing problem with pickup and delivery service. Math. Prob. Eng. **2008** (2009). https://doi.org/10.1155/2008/697981
14. Mavrovouniotis, M.: CEC-12 competition on electric vehicle routing problem. https://mavrovouniotis.github.io/EVRPcompetition2020/ (2020). Accessed 23 Nov 2020
15. Mavrovouniotis, M., Menelaou, C., Timotheou, S., Ellinas, G.: A benchmark test suite for the electric capacitated vehicle routing problem. In: 2020 IEEE Congress on Evolutionary Computation (CEC), pp. 1–8 (2020). https://doi.org/10.1109/CEC48606.2020.9185753

16. Mavrovouniotis, M., Menelaou, C., Timotheou, S., Panayiotou, C., Ellinas, G., Polycarpou, M.: Benchmark Set for the IEEE WCCI-2020 competition on evolutionary computation for the electric vehicle routing problem. Technical report, KIOS Research and Innovation Center of Excellence, Department of Electrical and Computer Engineering, University of Cyprus, Nicosia, Cyprus (2020). https://mavrovouniotis.github.io/EVRPcompetition2020/TR-EVRP-Competition.pdf
17. Tarantilis, C., Zachariadis, E., Kiranoudis, C.: A hybrid guided local search for the vehicle-routing problem with intermediate replenishment facilities. INFORMS J. Comput. **20**, 154–168 (2008). https://doi.org/10.1287/ijoc.1070.0230
18. Uchoa, E., Pecin, D., Pessoa, A., Poggi, M., Vidal, T., Subramanian, A.: New benchmark instances for the capacitated vehicle routing problem. Euro. J. Oper. Res. **257** (2016). https://doi.org/10.1016/j.ejor.2016.08.012
19. Wang, L., Lu, J.: A memetic algorithm with competition for the capacitated green vehicle routing problem. IEEE/CAA J. Automatica Sinica **6**(2), 516–526 (2019). https://doi.org/10.1109/JAS.2019.1911405
20. Woller, D., Kozák, V., Kulich, M.: The grasp metaheuristic for the electric vehicle routing problem (2020)
21. Xavier, I.: Cvrplib - capacitated vehicle routing problem librar. http://vrp.atd-lab.inf.puc-rio.br/ (2020). Accessed 23 Nov 2020
22. Zhang, S., Gajpal, Y., Appadoo, S.S.: A meta-heuristic for capacitated green vehicle routing problem. Ann. Oper. Res. **269**, (1–2) 753–771 (2018). https://doi.org/10.1007/s10479-017-2567-3

Artificial Intelligence in the Defence Sector

Antonio Carlo(✉)

Tallinn University of Technology, Tallinn, Estonia
antonio.carlo@pec.it

Abstract. In the last decade, Artificial Intelligence (AI) has been at the centre of attention for its civilian and military applications. Even though nowadays it is referred to as "weak AI", due to its narrow range of functions, in the long-term AI has the potential to play a critical role in every aspect of National Security. Recognising this potential, NATO Member States have already started to invest in this technology and have incorporated it in their defence strategy.

Its strategic importance leads several nations to develop AI for an array of military functions such as Command, Control, Communications, Computers, Intelligence, Surveillance, and Reconnaissance (C4ISR) and cyber operations as well as a variety of autonomous and semi-autonomous machines. This new technology has the potential to play a vital role in the acquisition and processing of Open-source Intelligence (OSINT) and the information available on the internet as the first filter for data that will be later evaluated by personnel. Thus, AI can mitigate the biggest hindrance relating to these fields of intelligence, which is the sheer amount of information that is available. The incorporation of AI in the decision-making process at every level of warfare (tactical, operational and strategic) can provide leadership with calculated input that is distanced from emotion and other judgement-affecting factors to which humans are susceptible. Furthermore, AI can be used to create simulations and models that allow for different strategies to be tested and evaluated. Taking into consideration the aforementioned as well as the momentum that AI is gaining globally, a multitude of ethical dilemmas emerge relating to its application.

The objective of this paper is to analyse existing AI technologies as well as future technological developments, states' financial investments and the potential impact on the defence sector with regards to opportunities and vulnerabilities.

Keywords: AI · Defence · NATO · OSINT

1 Excursus

In the last decade, the development and use of Artificial Intelligence (AI) has become a hype. As an emerging technology that has the ability to transform social fabric, it has captured the attention of decision-makers, engineers, and philosophers, to name but a few. What was once strictly reserved for the domain of science fiction, has become a part of our reality. Today, the use of AI is being explored in diverse sectors ranging from telecommunications and transportation, to security and healthcare. The military and defence industry are no exception. While lethal autonomous weapons systems (LAWS)

J. Mazal et al. (Eds.): MESAS 2020, LNCS 12619, pp. 269–278, 2021.
https://doi.org/10.1007/978-3-030-70740-8_17

are perhaps the most cited application of military AI, it is noteworthy that AI can perform an array of other functions including conflict prediction and prevention.

Before delving deeper into the development and implementation of AI in the military and defence context, it is important to provide an overview of what AI actually entails.

Defining AI

AI is often used as an umbrella term for a variety of disciplines encompassing machine learning, computer vision, natural language processing (NLP), deep learning and cognitive computing, all of which can be distinguished from one another. Despite the fact that there is no commonly accepted definition of AI, fundamentally, it refers to algorithms that can be "leveraged to collect, compile, structure, process, analyse, transmit and act upon increasingly large data sets" [1]. While this definition places big data at the core, other interpretations compare AI with human intelligence. Along these lines, AI has also been defined as a "simulation of human intelligence on a machine, so as to make the machine efficient to identify and use the right piece of 'knowledge' at a given step of solving a problem" [2]. Whereas both definitions are inherently different, parallels can nonetheless be drawn. One commonality is the autonomous action taken on received information.

Further differentiation should be made on the type of AI – narrow (weak), general (strong) and super. The distinction is given by the range of functions and capabilities of each of the three supports. In the present-day context, the most progress has been made in the field of weak AI, specialised on a narrow range of functions, such as voice-activated and pre-programming assistance. Put differently, weak AI repeats similar codes that were predefined by their makers and classifies them accordingly. This kind of AI has entered private homes and is now widely used through smart devices such as smart-homes, phones and cars. Strong AI aims to duplicate human intellectual abilities while even more advanced, super AI seeks to outperform human intelligence.

2 Strategic Approach to Military and Defence AI

As mentioned in the introduction of this paper, AI is already a reality in the defence and national security sector. Recognising its potential, some NATO Member States have already started investing in this technology and incorporating it in their defence and military strategies. AI potentials in addressing complex digital and physical security challenges are limitless, however these are connected with the Research and development (R&D) in this field. "In 2018, global defence spending reached USD 1.67 trillion, a 3.3% year-on-year increase" [3]. "Two trends make AI systems increasingly relevant for security: the growing number of digital security attacks and the skills shortage in the digital security industry" [4]. As a consequence of these trends, AI systems are necessary in the detection and response of these threats.

US

The US has already incorporated AI technology into military operations (Iraq and Syria) and has established the Joint Artificial Intelligence Center (JAIC) to accelerate the delivery of AI-enabled capabilities [5]. In 2018, the US increased the funding devoted to the

AI programmes "to include the JAIC's $1.75 billion six-year budget and the Defense Advanced Research Projects Agency's (DARPA's) $2 billion multiyear investment in over 20 AI programs". In more detail, the US expects that AI will be particularly useful in the intelligence sector due to the large data sets available for analysis. GEOINT and SIGINT are fed with a large number of data that allows them to create patterns to analyse [6].

The Department of Defense, "given warnings from the Defense Innovation Advisory Board and the Defense Science Board, appears interested in more systematically determining how to integrate AI" [7].

The US Government started a close collaboration between military and civilian personnel, providing the essential know-how in order to contrast and enhance resilience in case of threats. Moreover, the US stated in their 2018 AI strategy that "strong partnerships are essential at every stage in the AI technology pipeline, from research to deployment and sustainment". In order to achieve this, the US will be partnering with leading private sector technology companies, academia, and global allies and partners [6].

UK

In 2016, the UK Ministry of Defence (MoD) launched the Defence Innovation Initiative, whereby it pledged to spend £800 million over a ten-year period to increase the pace at which AI innovations are developed and brought into service. Two years later, in a paper entitled 'Mobilising, Modernising and Transforming Defence', the UK MoD reasserted its ambition with regard to AI by emphasising that it is "pursuing modernisation in areas like artificial intelligence, machine-learning, man-machine teaming and automation to deliver the disruptive effects we need in this regard" [8]. Whereas such initiatives are commendable, the UK has acknowledged that aside from a brief overview of LAWS and its normative implications, it has not addressed military and defence AI "with the thoroughness and depth that only a full inquiry into the subject could provide" [9].

At the 2018 World Economic Forum Annual Meeting, UK Prime Minister Theresa May, announced "we are establishing the UK as a world leader in Artificial Intelligence...We are absolutely determined to make our country the place to come and set up to seize the opportunities of Artificial Intelligence for the future" [10].

France

France has committed EUR 1.5 billion over five years to AI R&D and announced the creation of Interdisciplinary Institutes of Artificial Intelligence, connecting public and private researchers. France designed an AI national strategy in order to avoid defining the path for national security, and not to be "dependent on technologies over which it has no control" [11].

EU

In December 2018, Member States of the European Union (EU) joined forces with the European Commission on a Coordinated Plan on AI for increased cooperation that will boost AI in Europe. The EU has committed EUR 1.5 billion in the period between 2014 and 2019 with the aim to reach EUR 20 billion investments in the next 10 years. Following the revision and publishing of the Guidelines, the Commission launched the

Communication on Building Trust in Human-Centric Artificial Intelligence, ensuring European values were at the heart of creating the right environment of trust for the successful development and use of AI.

As a result of the Coordinated Plan, European policies are endorsing the stimulation, promotion and usage of AI technologies. Moreover, EU governments have discussed their national approach to maximise the benefits of AI and its positive impacts on the country system and, on the other hand, to mitigate the inevitable risks associated with the discipline [12].

China

In July 2017, China's State Council released the country's strategy for developing AI, entitled "New Generation Artificial Intelligence Development Plan". China has planned for major national R&D projects including AI projects. This strategy is articulated in three timely steps which aim to make China the world leader in AI by 2030. "By 2020 China aims to maintain competitiveness with other major powers and optimise its AI development environment…By 2025, China aims to have achieved a 'major breakthrough' (as stated in the document) in basic AI theory and to be world-leading in some applications…By 2030, China seeks to become the world's innovation centre for AI" [13]. This strategy predicts that China will have a $147 billion growth in AI technology. Moreover, China is investing in a series of measures for technology R&D, application and industrial development. This has allowed China to rank "second in the number of published technology papers and patents and making important breakthroughs in the core technology of some areas" [14].

Chinese ICT companies are investing in different areas of the world such as Europe, Africa and South America. This has allowed those regions to have an exchange with the Chinese know-how in these States connecting them directly to Chinese technology.

In addition, Chinese efforts regarding investments in US companies that focus on developing "militarily relevant AI applications" cause speculation in the U.S. AI market [6]. In 2016, Chinese investment in the US market reached $45.6 billion. Overall, since 2000 these investments in the US have exceeded $100 billion [15]. According to one estimate, "the availability of massive amounts of data is also considered a strategic edge, since China is on track to have 20% of the world's data by 2020 and 30% by 2030" [16].

Russia

On 1 November 2017, in an official statement, Viktor Bondarev, Chairman of the Federation Council's Defense and Security Committee, stated that "artificial intelligence will be able to replace a soldier on the battlefield and a pilot in an aircraft cockpit" and later noted that "the day is nearing when vehicles will get artificial intelligence" [6]. The Russian military aims to acquire more unmanned military systems for its air, land and naval forces [17].

In 2018, Russia ranked 20th in the World [18] by number of AI start-ups. Moreover, Russia made extensive use of AI in surveillance and propaganda within its borders. Compared to the US and China, Russia has published fewer research papers and documents. On the other hand, Russia has made important investments in the cyber ecosystem reinforcing its resilience and offensive capabilities. However, regardless of its ambitions, "it may be difficult for Russia to make significant progress in AI development given fluctuations in Russian military spending" [6] which, between 2017 and 2018, dropped by 20%. Although in 2019 Russian spending slightly increased, it is expected to once again to drop between 2020 to 2022 [19].

3 Strategic Importance of AI

Economic Impact and Partnerships

The strategic importance of AI can be observed by its global economic impact, the worth of which is estimated to be between USD 1.5 and 3 trillion in the decade of 2016 to 2026 [20]. AI is a discipline that deals with the development of software systems which are able to act in the physical and digital dimension, in order to acquire, interpret and learn data. AI has the capability to revolutionise entire industrial sectors including defence, justice, healthcare and civil government. Its potential is such that many governments of the industrialised countries have adopted an AI strategy in recent years.

Governments will use AI to reduce backlogs, redistribute resources and optimise workforce. It will be pivotal in the supply of human services, law enforcement and healthcare. This is possible thanks to the Private Public Partnerships that allow elaboration of data at large scale. AI is an essential step towards the creation of Smart Cities that apply smart infrastructure to streetlights, water use, traffic management etc., exploiting connected devices (IoT) and 5G. Thanks to new data generated by sensing infrastructure, the city can make "smarter decisions".

Apart from the extraordinary and undeniable opportunities that it entails, if misused, AI can have harmful consequences for the economy, the environment and society. It is important to evaluate the risk of unintentional injury to individuals or to society as a whole. In particular, it has now been shown that the misuse of AI can amplify forms of bias and discrimination on a social level.

Big Data

Big Data is a huge data set of different kinds of information. The concept of Big Data cannot be considered new. Before the era of computers and social media, Big Data was already collected and stored in physical archives, libraries and museums. Since then, the collection and storage of data did not only change but evolved in a more structured and defined way, the so-called "digital dimension". Big Data can be considered a continuation of that tendency that began decades ago. Data sets can provide new insights that offer opportunities for fine grained detail that was not accessible beforehand. As defined, Big Data could capture the whole of a domain and provide full resolution of it. Moreover, human bias can be prevented thanks to the use of Big Data. Information is available to a large number of users and can be used for economic and social research. Big Data is being elaborated and is growing thanks to the millions of users that implement it daily. It can be recorded and used in order to analyse past events and predict future ones. On the other hand, Big Data carries a number of challenges. The complexity and the large number of information needs to be elaborated and analysed by an expert. Moreover, focal problems are the security of data (sensitive information), privacy issues and ethics, data sovereignty, and biased information. Not all the resources of Big Data contain updated or reliable information. This may lead to an imprecise outcome, and it is important to take into consideration that data cannot be fully objective. This means that there should be a logical explanation to the empirical observations. Data needs to be carefully managed due to ethical and legal concerns (privacy). The analysis of the information is key to prevent the use of manipulated data. The use of data sets can provide new insights that are frequently overestimated due to a lack of understanding. Mass control and surveillance also result in a lack of data protection.

Use Of Data (Data Protection)

The General Data Protection Regulation (GDPR) is a regulation in EU law on data protection and privacy in the European Union and the European Economic Area implemented on 25 May 2018 that has led to the affirmation of fundamental principles, such as that of data minimisation and that of data control by the user, which the world of AI must now take into account. On the other hand, awareness has emerged that Open Data and its policies are aimed at achieving the free flow of data. This allows in many cases for the appropriation of data by large IT companies, which already have significant quantities of data, and is often difficult for governments themselves to control and prevent.

Space and Airspace Sectors

Space, a domain that is getting increasingly more funds and international attention, would gain a great advantage from the use of AI in the monitoring of space debris. Due to its importance, the Office of Debris Monitoring is continuously operating, even though the large number of debris (natural or artificial) prevents accurate monitoring. With the increase of international players in the space sector the possibilities of Anti-Satellite Weapons (ASAT) are increasing, which is why close monitoring is necessary to neutralise the threat before causing damages and raising legal actions.

With regards to airspace defence, a strong AI implemented in the monitoring of the airspace border would increase efficiency and reduce human error to a minimum. On 13 January 2018 in the state of Hawaii (USA), a false missile alert was issued via the Emergency Alert System and Commercial Mobile Alert System. This alert was issued due to a human error. Such a notable example of miscommunication could be prevented in the future with the help of AI, working in parallel with the assign personnel.

Analysis and Fake News

In military terms, AI can considerably boost the speed of analysis and action of both humans and machines. AI would improve the quality of the decision to support the chain of command in the field.

AI has great potential for the analysis of a large amount of information in order to be classified into two groups: “true” or “fake”. In their research on fake news, Anne Cybenko and George Cybenko (2018) consider two groups of people. The first group considers fake news to be true while the second regards it as fake. In this case, AI is taught to divide the news according to the person’s preferences and settled beliefs [21].

Some AI or other machine learning solutions can be used to help at least partially classify the trustworthiness of the information. For example, algorithms can be used to add tags that the posted article has no cited sources or that it was posted by an unverified author. The same systems can also be used to adjust the way in which recommendation algorithms work. For example, critical information concerning health risks such as those resulting from the COVID-19 pandemic can be prioritised via recommendation algorithms. Prioritisation should allow trustworthy sources to be shown regardless of users' preferences.

4 Future Developments

AI in the Military field "has the potential to be a transformative national security technology, on par with nuclear weapons, aircraft, computers, and biotech" [22].

Cyberspace Operation

The cyberspace is a dynamic environment with a wide range of threat actors. It is vital that defence is agile and resilient in the cyber domain in order to enable Mission Assurance for operational commanders. To this end security monitoring, incident response and integration of advanced technologies with legacies should be established.

In 2016, U.S. Cyber Command Admiral Michael Rogers stated before the Senate Armed Services Committee for which he gave testimony on encryption and cyber matters, that relying on human intelligence alone in cyberspace is "a losing strategy" [22]. The cyber ecosystem is in constant transformation and the human capital does not have the power and capabilities to keep up with its development. Conventional cyber defence analyses patterns and events that are already transcribed. On the other hand, AI can be designed to prevent any kind of anomaly and patch the problem before the system would be subject to a cyber-event.

Lethal Autonomous Weapons

Lethal Autonomous Weapon Systems (LAWS) are systems that can independently search and engage without human intervention. With the auxiliary of AI, these systems would reach unprecedented precision to identify their target without any human intervention. The Russian Military Industrial Committee has already approved an aggressive plan whereby 30% of Russian combat power will consist of entirely remote controlled and autonomous robotic platforms by 2030 [12].

Command and Control

AI can be used in the Multi-Domain Command and Control which aims to centralise planning and execution of air-, space-, cyberspace-, sea-, and land-based operations. This would allow for more effective communication and cooperation between the different domains. This has been carried out through cooperation between the US Airforce and Lockheed Martin.

The Defence sector in the future should use data-driven techniques in order to exploit AI to the maximum. The data will need to be stored in predefined establishments in order to prevent any kind of manipulation and will be the key element for the evolution and use of AI.

In the defence sector, AI will be directly used by the MoD and commanders in order to evaluate the best possible decision to take in the field. This could be for geographical and Supply Chain purposes. Supply Chain is a cross-functional approach that involves a company and its suppliers in the production and distribution of a specific product to the final user. Due to the high number of suppliers in the defence sector, the trend is considered of high criticality.

Training and Simulations

Training and Simulations are a key point for the future development. The quick and continuous change of this field does not provide any security. In order to be always up to date, reports on best practices, emerging technology development and the training needs to be elaborated by internal and external entities. This distinction allows to have a double vision of the state of the art and the way forward, that may change day to day. Such programmes entail awareness to the new emerging threats.

Ethics and Risk

AI is an essential technology for the development of security and defence however it has a fragile ecosystem that can collapse on itself due to our society being characterised by bias and inequality.

Governments that will develop the use of AI will face different issues such as the immense number of resources (memory, data and computer powering) needed. Moreover, a constant monitoring of the systems by Computer Emergency Response Teams (CERTs), Computer Security Incident Response Teams (CSIRTs) and Security Operations Centres (SOCs) should be put in place to prevent bad test/training data and data poisoning.

For the future development of AI, ethics and risk should be taken into consideration in order to prevent any misconception and treated:

- *Ethics:* Stimulating reflection on the need to protect individuals and groups at the most basic level; new innovation and technology that seeks to foster ethical values and improve individual flourishing and collective wellbeing by generating prosperity, value creation and wealth maximisation.
- *Risk:* While AI offers many positive benefits, it can lead to significant unintended (or maliciously intended) consequences for individuals, organisations and society.

AI is already used in the military context, however what would be the ethical issues arising from an even greater use? An ethical dilemma would arise if human judgement is completely removed. Where would this lead us to? Could this constitute a violation of International Humanitarian Law?

There should be risk mitigation in the use of data that will be implemented in the AI algorithm. This needs to be done in order to prevent any kind of bias in the system that would produce an escalation of events provoking a mass conception of an AI unable to be used due to its faulty creator. In June 2019, AI and autonomous systems were at the heart of the agenda of the Sub-Committee on Technology Trends and Security (STCTTS) visit to the UK.

5 Conclusion

In recent years, scientific and technological progress has led to the blossoming of new technologies that have deeply impacted the defence sector. This development brings not only strategic advantages, but also triggers a series of drawbacks in terms of security challenges.

In critical situations and due to the growing availability of data, the role of AI cannot be underrated. Humans will need more and more the support from machines capable of providing intelligence through the analysis of data from multiple sources and the gathering of actionable decision-making information in an efficient manner.

AI is a developing technology at the embryonic stage in the defence sector. Due to its proven high efficiency, it has the potential to enable new types of low-cost and high-impact military solutions. However, these activities need to be monitored by highly qualified personnel working in SOCs, CERTs and CSIRTs. Cooperation between future CERTs and CSIRT are essential to monitor and support a more resilient and secure AI and cyberspace.

In the wake of sophisticated and unpredictable kinetic and cyber threats, governments and organisations need to adapt their military and defence mechanisms to the modern disruptive, geo-political landscape.

AI in the defence sector will see increasing autonomy in the future that will change the character of the battlefield and combat operations such as C4ISR.

Raising awareness and fostering large investments in AI technology are essential for governments and international organisations because of AI's continuous evolution. It is critical that organisations such as NATO maintain their superiority and react to the challenges posed by this new technology.

Whether AI will generate a new industrial revolution or enable technology development, what is clear is that AI technology remains essential. The importance of this technology has led to a worldwide AI race due to extensive investments in national and foreign R&D. The former to establish technological supremacy, the latter to gain control over the adversary's AI related technology in the long run.

As with any technological advancements, R&D is not the only field that requires further development. Policy recommendations should be drafted nationally and internationally in order to define the perimeter of AI in the defence sector.

References

1. International Institute for Strategic Studies: Big data, artificial intelligence and defence. In: The Military Balance 2018, pp. 10–13. International Institute for Strategic Studies, London (2018)
2. Shukla, A., Tiwari, R., Kala, R.: Real Life Applications of Soft Computing, pp. 3–40. CRC Press, Boca Raton (2010)
3. Janes: Global defence spending to hit post-Cold War high in 2018, IHS Mar-kit, 18 December 2017. https://ihsmarkit.com/research-analysis/global-defence-spending-to-hit-post-cold-war-high-in-2018.html
4. OECD: Artificial Intelligence in Society. OECD, Paris (2019). https://doi.org/10.1787/eedfee77-en
5. Deputy Secretary of Defence: Establishment of the Joint Artificial Intelligence Centre. Memorandum for Chief Management Officer of the Department of Defence (2018). https://admin.govexec.com/media/establishment_of_the_joint_artificial_intelligence_center_osd008412-18_r....pdf

6. Sayler, K.M.: Artificial Intelligence and National Security. US Congressional Research Service (2020). https://fas.org/sgp/crs/natsec/R45178.pdf
7. Horowitz, M., Kania, E.B., Allen, G.C., Scharre, P.: Strategic Competition in an Era of Artificial Intelligence. Center for a New American Security (2018). https://www.cnas.org/publications/reports/strategic-competition-in-an-era-of-artificial-intelligence
8. UK Ministry of Defence: A report on the modernising defence programme, p. 16. Mobilising, Modernising & Transforming Defence (2018). https://assets.publishing.service.gov.uk/government/uploads/system/uploads/attachment_data/file/931705/ModernisingDefenceProgramme_report_2018_FINAL.pdf
9. House of Lords: AI in the UK: ready, willing and able? Select Committee on Artificial Intelligence: Report of Session 2017-19, p. 101. Authority of the House of Lords (2017). https://publications.parliament.uk/pa/ld201719/ldselect/ldai/100/100.pdf
10. May, T.: PM's speech at Davos 2018. World Economic Forum, Davos, 25 January 2018. https://www.gov.uk/government/speeches/pms-speech-at-davos-2018-25-january
11. Ministere des Armees: Artificial Intelligence in Support of Defence. Report of the AI Task Force (2019). https://www.defense.gouv.fr/content/download/573877/9834690/Stratégie%20de%20l%27IA-UK_9%201%202020.pdf
12. European Parliament: The Ethics of Artificial Intelligence: Issues and Initiatives. Scientific Foresight Unit (STOA), Brussels (2020). https://www.europarl.europa.eu/RegData/etudes/STUD/2020/634452/EPRS_STU(2020)634452_EN.pdf
13. Roberts, H., Cowls, J., Morley, J., Taddeo, M., Wang, V., Floridi, L.: The Chinese approach to intelligence: an analysis of policy, ethics, and regulation, 2 (2020) https://doi.org/10.1007/s00146-020-00992-2
14. Department of International Cooperation Ministry of Science and Technology: China Science and Technology Newsletter No 17. China Association for International Science and Technology Cooperation, Beijing (2017). https://fi.china-embassy.org/eng/kxjs/P020171025789108009001.pdf
15. Brown, M., Singh, P.: China's Technology Transfer Strategy: How Chinese In-vestments in Emerging Technology Enable a Strategic Competitor to Access the Crown Jewels of U.S. Innovation. Defence Innovation Unit Experimental, p. 4 (2018). https://admin.govexec.com/media/diux_chinatechnologytransferstudy_jan_2018_(1).pdf
16. Kania, B.E.: Battlefield Singularity: Artificial Intelligence, Military Revolution, and China's Future Military Power. Center for a New American Security, Washington DC (2017). https://s3.us-east-1.amazonaws.com/files.cnas.org/documents/Battlefield-Singularity-November-2017.pdf?mtime=20171129235805&focal=none
17. Bendett, S.: Should the U.S. Army Fear Russia's Killer Robots? The National Interest (2017)
18. ASGARD: Artificial Intelligence: A Strategy for European Start-Ups. Roland Berger GMBH, Munich (2018)
19. Stockholm International Peace Research Institute: Military expenditure by country. SIPRI, Stockholm (2020). https://www.sipri.org/sites/default/files/Data%20for%20all%20countries%20from%201988–2019%20in%20constant%20%282018%29%20USD.pdf
20. Allen, G., Chan, T.: Artificial Intelligence and National Security, p. 1. Belfer Center for Sceince and Internatioanl Affairs, Cambridge (2017). https://www.belfercenter.org/sites/default/files/files/publication/AI%20NatSec%20-%20final.pdf
21. Cybenko, A.K., Cybenko, G.: AI and fake news. IEEE Intell. Syst. 1–5 (2018). https://doi.org/10.1109/MIS.2018.2877280
22. United States Senate: Hearing to Receive Testimony on Encryption and Cyber Matters, p. 39. Alderson Reporting, Washington (2016). https://www.armed-services.senate.gov/imo/media/doc/16-68_09-13-16.pdf

Artificial Intelligence and Robotics Addressing COVID-19 Pandemic's Challenges

Walter David[1(✉)] and Michelle King-Okoye[2]

[1] Italian Army Training Specialization and Doctrine Command, 00143 Rome, Italy
walter.david@esercito.difesa.it
[2] Ronin Institute, Montclair, NJ 07043, USA
michelle.king-okoye@ronininstitute.org

Abstract. There is a growing awareness that the unfolding Covid-19 pandemic will deeply change people's lives, while in the humanitarian system the gap between available resources and need is widening. Authors aim to investigate the ways new technologies can be effective in addressing global challenges. A session has been conducted at the United Nations conference HNPW 2020 where humanitarian experts have recognized the potential for Artificial intelligence (AI) and robotics to support response, decision-making, logistics and health services. In effect, one of the differences between Covid-19 and previous epidemics, consists in the massive deployment of technologies' applications for monitoring, surveillance, detection, prevention, and mitigation. Areas of concern have been identified in bias, accuracy, protection and use of data, citizens' privacy and legal gaps. Provided that such issues are addressed in every new project, authors propose to link AI and robotics with the triple nexus concept of the Humanitarian-Development-Peace (HDP) aiming to bridge the divide between humanitarian assistance, development agenda and peacebuilding.

Keywords: Robotics · Artificial intelligence · Autonomous systems · COVID-19 pandemic · Data protection · Privacy · Humanitarian-Development-Peace (HDP) Nexus

1 Introduction

1.1 Background and Identification of the Problem

Given the severe acute respiratory syndrome coronavirus 2 (SARS-CoV-2) infection lingering for an undetermined period of time, it is hard to believe how drastically our lives have changed during this difficult year 2020. The unusual severity of the novel corona virus disease (Covid-19) pandemic has been determined by key drivers such as the exponential pace of transmission, fast global transportation systems and interconnections, urbanization (since 2008 more than one half of the world population live in towns) [1, 2] and the capacity of national health systems [3].

We are trained to plan for and respond to events in one or more countries, but a pandemic affects and threatens many nations, despite not causing damage to the infrastructures [4]. Countermeasures have included the lockdown of many businesses and

J. Mazal et al. (Eds.): MESAS 2020, LNCS 12619, pp. 279–293, 2021.
https://doi.org/10.1007/978-3-030-70740-8_18

pushed employees to take part in the biggest experiment of remote working in history. Even the military had to reduce training and operations in order to protect their most valuable assets, their service women and men, and prioritize the support to the government authorities and health services.

Covid-19 devastating effects hit hard on the economy, in particular in less developed countries, due to suppression measures [3]. Rising food prices and shortages, international migration and internal displacement could threaten social and political stability. Due to the duration of previous (protracted) crises, the humanitarian system was already stretched beyond capacity before the pandemic, with an average length of a humanitarian appeal at seven years.

In effect, even before the pandemic, the gap between the available resources and the needs was widening, driving a renewed interest on nexus concept thinking approaches [5], aiming to operate effectively across the humanitarian assistance, development agenda and peacebuilding (HDP) triple nexus [6].

1.2 Aim, Purpose and Methodology

Authors aim to investigate and research on the role played and the issues identified from the development and the deployment of innovative technologies applications in support of the fight against Covid-19.

The paper draws from the productive lead of a session at United Nations HNPW Conference 2020 in Geneva, by Ronin Institute and United Drone Community in partnership with the United Nations Office for the Coordination of Humanitarian Affairs (OCHA) in one of the last live conferences that took place in 2020.

The session aimed to focus and discuss the impact of the increased capabilities of new technologies, to explore opportunities for supporting the humanitarian action, the development agenda and the peacebuilding.

UN humanitarian, emergency and Red Cross officers were asked to describe the main opportunities and challenges from their perspectives. The session was oriented towards collecting insights and suggestions on practical actions.

Later, attendees have been invited to convey their priority programmes and partnering needs, and to suggest the appropriate channels of communication for those who may wish to collaborate.

Interesting questions involve the possibility to leverage artificial intelligence (AI), robotics and autonomous systems to effectively support needs assessment, preparedness and response in humanitarian settings, including the challenges posed by the Covid-19 pandemic.

Authors examined press and social media reports from China, Asian countries, European Union and United States about how AI angettingd robotics are being used during the crisis. Interactions and digital discussions with humanitarian experts and scientists helped in gaining better insight in their perceptions about the role of these technologies.

An author is involved in the UN established Global Information Management, Assessment and Analysis Cell (GIMAC) (https://www.gimac.info/) multi-stakeholder initiative that is co-lead by OCHA, UNHCR, IOM, WHO & Global Health Cluster. This group is providing technical support to prioritized countries' (that is the 25 countries included in the WHO Global Humanitarian Response Plan) [7] on needs assessment, analysis and response planning decisions related to the humanitarian impact of Covid-19.

The group's focus goes beyond the disease, including the effects of pandemic, economic, socio-economic, and political issues as well as supply chain, food market, labor market. The aim is to realign existing primary data collection with the Covid-19 realities, while secondary data analysis will fill information gaps.

Finally, in a novel approach, this paper examines the role of artificial intelligence (AI) and robotics technologies in support of the fight against Covid-19, links such technological implementations with the humanitarian-development-peace (HDP) triple nexus concept, and considers the shortcomings of ethics, privacy and bias towards safeguarding confidential information.

2 New Technologies in Support of the Fight Against COVID-19

If we compare the response to Covid-19 with those to previous pandemics, we realize that many applications of innovative converging technologies have been deployed [8]. There are many ways new technologies are being exploited to manage the crisis and fight Covid-19 [9].

AI, robotic and unmanned systems are deployed, facilitated by the internet of things' (IoT) sensors, available data, affordable hardware prices and online open resources for developers [10, 11], cloud computing, modelling & simulation (M&S), virtual reality, 5G networks, 3D printing and blockchain [8].

In this global challenge, many projects are under development and, rapidly, technical solutions exploiting digital technologies have already been implemented in just few months by governments, academia, industry, and startups for supporting *monitoring, surveillance, detection* and *prevention* (Table 1).

Table 1. Technologies' for monitoring, surveillance, detection and prevention of Covid-19

Digital tech/IoT	Big Data	Artificial intelligence	Robots & drones
Real-time tracking; Online databases updates	Modelling disease, potential growth and areas of spread	Covid-19 detection (chest imaging X-ray)	Quarantine Enforcement; Identification of infected
At-risk vicinity Live tracking	Modelling countries' preparedness and vulnerability	Disease progression prognosis	Public services announcements Disinfection of public spaces; Monitoring traffic flow; Critical infrastructures protection

Novel applications based on Internet of Things (IoT) sensors' networks have been tested and quickly implemented, taking advantage of unmanned aerial vehicles (UAV) to enable accurate traffic control, contact tracing and mapping the movements of citizens, in particular during enforced lockdown and quarantine.

Data and analytics have proved to be useful in combating the spread of disease, in prediction, detection, response and recovery due to possibility for national governments and health authorities to access ample data on population's health and travel history [4].

In fact, artificial intelligence solutions have been deployed to identify, track and forecast outbreaks, to identify non-compliance with social distancing rules and identify infected individuals [12].

Robotics is playing a crucial role in the coronavirus crisis and is already providing lessons for future applications [13], for example for the disinfection of large public areas, hospitals and infrastructures, and for the management of medical waste [2].

Data provide opportunities for performing modeling studies of viral activity and for supporting national decision makers to enhance preparation.

Data can be exploited by deep learning systems for forecasting healthcare trends, modeling risk associations and predicting outcomes of implemented countermeasures [8]; big data help to identify and assess the risk of individuals based on their travel history, the time they have spent in virus hotspots, and their potential exposure to the virus [12].

Such technological applications have been deployed also to mitigate the impact of the rapid spread of infection on the functioning of the national health systems, including those aspects that are not directly related to Covid-19 (Table 2).

While companies and public administrations have pushed many of their staff to work from home, promoting agile working and online meetings, smartphones, IoT devices and sensors support public health agencies access to data for monitoring the pandemic in hospitals and clinics, thus facilitating the establishment of a highly interconnected digital ecosystem and enabling real-time collection of primary data.

Artificial intelligence can help scientists and medical doctors gain deeper knowledge about diseases and enable national health services to make better and faster decisions [4].

Front line radiologists are also using AI deep learning systems that learn from experience with large data sets to make better treatment decisions based on CT scan imaging [4].

Machine learning and deep learning ability to process very large amounts of patients' data are key drivers to improve the detection and diagnosis of Covid-19 in addition to testing.

Digital technologies enhance public-health education and communication [8], support medical doctors to provide remote assistance to patients during lockdown of clinics and private studios, while e-commerce platforms provide food and essential goods to quarantined citizens [2].

Table 2. Technologies in support to Health services.

Digital tech/IoT	Big data	Artificial intelligence	Robots & drones
Virtual clinics	Business modelling on medicine, food and essential supplies	Automatic diagnose (non-Covid-19)	Clinical care: Disinfection of facilities; Health worker tele-presence; Prescription/meal dispensing; Patient intake, visitors, socializing
Dissemination of public information	Modelling operating theatres and clinics with workforce projections	Medical chatbots	Non hospital care: Delivery to quarantined; Socialising for quarantine/nursing homes, Off –site testing; Testing & care - nursing homes
Prioritising health information and advices (e.g. symptoms/home treatment)	Identifying those most at risk using publicly held data and making sure they have access to services during isolation	Tackling misinformation online	Lab & supply chain: Delivery; Handling infectious materials; Manifacture or Decon PPE Lab automation
	Tracking longer-term impact of Covid-19 on other health conditions (e.g. diabetes, heart conditions)	Predict potential responses of the virus to drugs. Analysing Covid-19 patient information to predict risks of patients progressing to severe disease	Delivery; Socializing; Tele-commerce; Robot assistants;

3 Artificial Intelligence, Drones and Robots

The session on artificial intelligence and robotics in the humanitarian space conducted at UN HNPW 2020 aimed to raise awareness of the potential benefits and the risks connected with autonomous systems, on the ethical and legal levels.

Brainstorming and a *Strengths Weaknesses Opportunities Threats* (SWOT) analysis (Table 3) helped in recognizing the great potential for such technologies to support response and decision-making, in particular by providing better situational awareness.

Table 3. SWOT analysis of AI and robotics in the humanitarian space.

STRENGTHS	WEAKNESSES
Large amounts of data to be stored and used Verifying data Efficient surveillance Early warning Mapping Tracking	Security Collection, storage and use of data
OPPORTUNITIES	**THREATS**
Humanitarian logistics Medicine supply Cargo delivery Health support Local communities support Jobs creation Training Situational awareness	Abusing information Legal gaps Bias and discrimination Lack of framework for integration of AI and drones

We found that aerial drones and ground robots are already playing an important role in almost every aspect of managing a crisis. In effect, they support logistics, decision-making, disaster response, humanitarian action. They are transforming the way we think from early warning to evacuation models and they are behind a number of promising innovations for risk reduction, search and rescue and health.

Unmanned systems offer new opportunities to disaster responders, humanitarian actors and local communities. Digital humanitarians are already saving lives exploiting drones for health services and delivery of medicines to small communities worldwide [14].

AI and drones can also improve decision–making. In fact, success in operation depends from the ability to execute the *Observe-Orient-Decide-Act* (OODA) loop (Fig. 1) as fast as possible, to decide at a higher pace. AI will support time–critical decision-making, by improving intelligence (due to the possibility to exploit multiple sources from sensors to social media, web and repositories' contents), prediction, classification and analysis capabilities and finally, the provision of course of actions' choices.

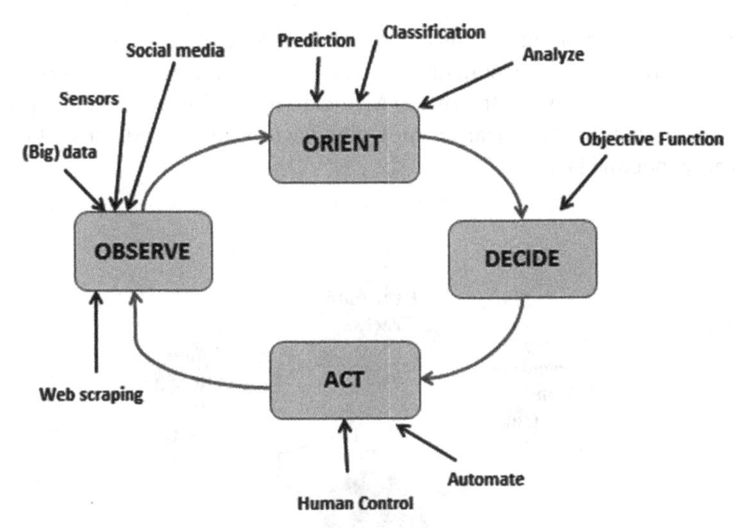

Fig. 1. The OODA loop supported by artificial intelligence.

4 Lessons from the Pandemic for the Post-pandemic World

The experience from the last months provide an opportunity to learn. Robots are already playing crucial roles in the Covid-19 crisis and they can help even more in the future [13]. Every disaster is different, but the experience of using robots in this pandemic, monitoring the situation, assess the needs and the achievements of the implemented countermeasures is indispensable to rapidly adjust the response and to prepare for the post-pandemic new normal.

Covid-19 will accelerate the use of available robots and their adaptation to new niches, but it will also lead to the deployment of new robots. For instance, new opportunities are emerging in the automation of laboratories and supply chain, public spaces disinfection; more robots will be deployed to work side by side with the humans in health care and relief front lines [15].

One important lesson is that during a disaster, robots empower and support humans, either performing tasks that are dangerous for human workers or reduce the workload and the time required for repetitive jobs.

Men and women are still involved in the OODA loops of unmanned and semi-autonomous systems; the crucial meaningful human control is still retained [16]. The second lesson is that preferred robots used during an emergency are commercial off-the-shelf already in common use before the disaster. The availability of flexible and user-friendly payloads is a must because customizing available robots for new tasks is generally more cost effective and faster than building specialized prototypes [17].

5 Ethics, Bias and Privacy

With the current mortality rates and increasing susceptibility to Covid-19 infections, due to the nature of transmission and potential spread among populations, artificial intelligence algorithms, machine learning and robotics have been instrumental towards governmental policy-making, healthcare decision-making, social distancing measures, quarantine control and treatment updates for suspected and diagnosed cases [8, 18].

For example, the Cognitive Internet of Medical Things (CIoMT) highlights the areas (Fig. 2) where emerging medical technologies have been utilised for addressing the coronavirus pandemic [18].

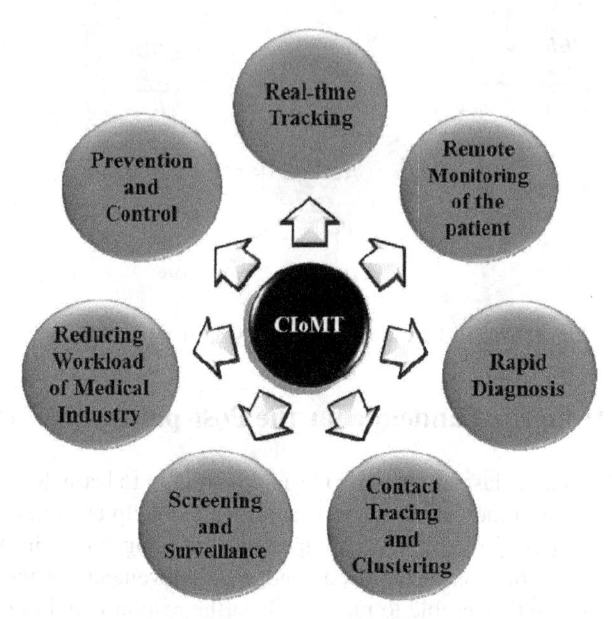

Fig. 2. Cognitive Internet of Medical Things (CIoMT) for tackling coronavirus disease 2019 (COVID-19) cited from Swayamsiddha and Mohanty, 2020.

Screening, thermal imaging facial recognition and surveillance software [8, 19], coronavirus genetic testing kits, deep learning models for imaging [20], tracking and tracing tools and remote monitoring of patients towards the rapid diagnosis [21] and treatment of suspected cases of coronavirus have proved very promising for patients.

While some existing studies have noted potential benefits of using these technologies, critics have highlighted significant issues in accuracy, particularly with facial recognition and surveillance software [22–24].

Technical and societal risks, including (but not limited to) accuracy, discrimination and privacy breaches, need to be carefully assessed and weighed against the potential opportunities AI technologies can offer. As the Ada Lovelace Institute [25] summarised, “To create systems that work for the public, these challenges and concerns need to be acknowledged and explored, rather than discounted and silenced, and complexities must be designed in.” AI has enabled wide population outreach and reduce the human workload, thus controlling the spread of the coronavirus and saving lives [18].

Whilst artificial intelligence, machine learning and robotics have been revolutionary in the diagnosis, control of spread and treatment measures for Covid-19 disease, there is great potential for bias.

Firstly, this is related to data quality. There is lack of historical data [26] related to Covid-19 due to limited information that exists both for new and emerging pathophysiological trends of the virus. Algorithms will be helpful for patients that fit the prescribed data sets. However, standard algorithms may prove detrimental for emerging clinical presentations that do not fit the standard codes [26]. As such, there is an urgent need for audit trails, risk assessment procedures and contingency plans to guide the reliability and trustworthiness and ongoing updates of data sets to minimize risk of bias and pitfalls within healthcare. It is also clear that fairness needs to be a key consideration when auditing for algorithmic accountability [27, 28].

Secondly, one would question the ethical principles that guide AI machine ethics [29], including artificial and robotic procedures relating to human testing and treatment since algorithm data sets are unable to utilise moral judgement, empathy and respect for human participants [30] as well as address cultural, social and ethnicity factors that are critical components to diagnosis and treatment. There is a high risk that stereotypical codes may be implemented by human error if these data sets fail to undergo rigorous procedures to maintain the principles of beneficence and non-maleficence and impartiality [31].

Whilst human involvement is fundamental to promote explainability and transparency, coordinated international ethics review committees reflective of diverse health experts from multiple ethnic and racial backgrounds that can govern procedures are significant to ensure neutrality, eliminate racial bias and discrimination, minimise risks and maximise benefits for consumers of healthcare [32]. In assessing ethics and potential impact of AI technology in a crisis, it is vital to acknowledge that, "The (Covid-19) technologies under discussion are not viewed as neutral. They must be conceived and designed to account for their social and political nature" [25].

Thirdly, a great challenge is the issue of privacy, data protection and anonymity [33]. The AI Barometer report of the UK Government's Centre for Data Ethics and Innovation [34] identified a series of opportunities and barriers to AI's use in five key sectors, including healthcare. Across all five sectors it examined, privacy of personal data was found to be both the one of the risks most likely to occur and that which carried the highest danger of harmful impact to the public.

Privacy risks and regulations concerns associated with AI are not only in the awareness of some organisations, but have also been voiced by some citizens around the world [25, 35, 36].

From existing literature and experiments, it is clear that privacy and discrimination risks are linked to algorithmic datasets which reveal more about citizens' personal lives than many realise [37, 38].

Algorithmic bias and inference can be a "veiled type of discrimination that deepens existing social inequalities" [37]. This is a particular concern in a crisis, where decisions can determine life or death.

In addition, it is important to note the legal aspects of capturing, storing and processing personal data. While this is a global concern, it is particularly regulated in some regions. For example, biometric data and medical data are classed as 'sensitive data,' under the European Union's General Data Protection Regulation (GDPR) and the California Consumer Privacy Act (CCPA). This means that any public or private organisation capturing, storing or processing health data of individuals must have a specific and defined legal purpose for doing so [39].

With these concerns around existing AI technology, further questions remain unanswered about the use of new or existing technology during a pandemic, such as: What procedures are in place to safeguard confidentiality of information and protect vulnerable populations from infringement of human rights?

Who controls and access private health information across populations? How will this data be used post pandemic?

With the widespread usage of technology authorised by governmental and private entities globally in a bid to control the spread of coronavirus, there remains the dilemma of collecting personal health data with or without consent across populations [40].

There is the issue of trust regarding the manipulation of data for Governmental and private entities' purposes that may not be beneficial to the general population [41]. European Digital Rights (EDRi) [42] issued a call on member states and institutions of the European Union (EU) to implement measures towards anonymity and data privacy during the Covid-19 pandemic.

The Global Privacy Assembly [43] initiated the GPA Covid-19 Response Repository to promote transparency and measures to control data protection and privacy. Whilst these are steps in the right direction, utilising AI in healthcare for the effective management of patients with suspected or confirmed Covid-19 has provided benefits but its yet in its early stages and is still premature towards utilising highest standards to control and mitigate issues and avoid exploitation of data privacy, control and access [44].

6 Linking AI and Robotics with the HDP Nexus

National governments and international organizations have been largely caught off-guard by the scale of impact of the unfolding Covid-19 [45]. The limited basic healthcare coping capacity of national governments will cause an exacerbation of already existing crises [45].

Grounded airplanes and businesses' lockdown create unprecedented obstacles to international staff to deploy while local and national leaders have a central role in leading the humanitarian response and the development cooperation (https://werobotics.org/covid/) [46].

Covid-19 will have a devastating impact on the world's most fragile economies. With an increasing number of crisis that are more and more protracted (the average length of a humanitarian appeal now at seven years), and even though more funding might be provided globally, the gap between scarce donors' resources and increasing needs is growing.

This require to take a long-term and holistic approach to coordinating response to present and future crises [45]. Simplified innovative pre-arranged flexible financing mechanisms and behavior are proving to be critical [5, 47] for delivering more and better by stronger planning.

Humanitarian, development and peacebuilding actors have to come together to jointly assess the effects of Covid-19 crisis on social, economic and conflict dynamics [45]. Working across the humanitarian-development-peace (HDP) nexus is a must in high-risk environments for responding effectively to this crisis and protecting the most vulnerable [48, 49].

In fact, operating effectively across the HDP nexus can help save precious lives and avert economic losses [6] by promoting collaboration and exploiting synergies, reuse of capabilities, involvement of local communities and businesses.

The Ebola experience in the Democratic Republic of Congo (DRC) shows how this approach can lead to real operational win-wins for the actors involved in the containment of public health crises [6].

To manage security risks that could escalate at any moment, intelligence data on violence trend could be shared with peacekeepers; within the United Nations Organization Stabilization Mission in the DR Congo (MONUSCO) this approach across the *triple nexus* helped adjust the supervision of development projects and implementation strategies [6].

The deployment of robotic systems would address old infrastructures gaps that are challenging medical and humanitarian logistics operations, support development by providing temporary or even permanent jobs in drone industry, manufacturing, maintenance and operations at local level, while at the same time strengthen the resilience of those communities.

Decision-making requires a combination of software tools and human judgment [50] but currently, humans cannot go through the *Observe-Orient-Decide-Act* (OODA) decision making model steps instantly [51].

Intelligence from drones can be supplemented with near/real-time information from satellites, IoT sensors, social media, crowd-mapping platforms for fast, real-time information about unfolding disasters and the accurate location of civilians in need [10, 50], incidents and armed fighting situations, thus supporting armistice agreements and even helping in preventing new armed conflicts (Fig. 3).

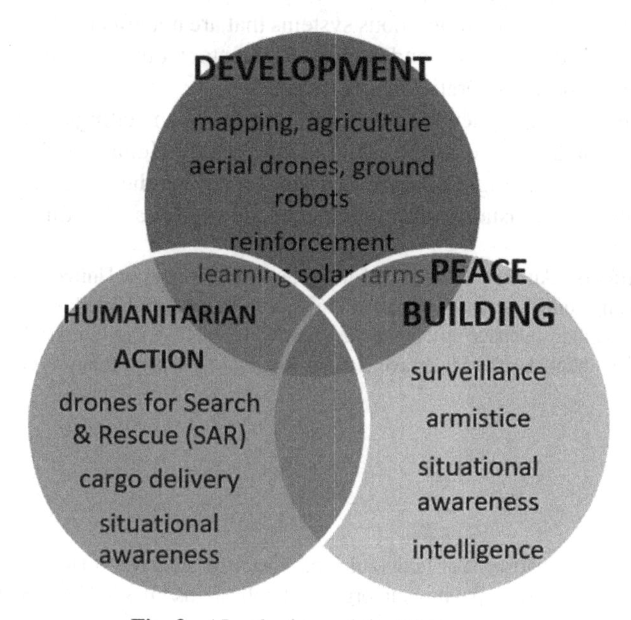

Fig. 3. AI, robotics and the HDP nexus.

7 Conclusions

The post-Covid-19 world will be significantly different from before; our experience will help in accepting automation as a part of our lives [15, 52]. For example, the delivery of faster results will improve the acceptance of medical outcomes generated with the support of artificial intelligence (AI) [6].

AI and robotics play a major positive role in controlling the pandemic. It is envisaged that their role in humanitarian response will become increasingly fundamental towards controlling the spread of Covid-19 infection, disease surveillance and crisis management.

In particular, there are more and more drone-based initiatives but critical thinking is necessary. Some solutions can even cause harm if they do not follow the guidelines or if they create a false sense of safety [53]. In fact, there remain ethical issues, including privacy and bias, thus contributing to questionable management and control of data gathered by governments and private industries.

Responsible development and deployment of AI is especially crucial in a crisis, when existing inequalities may already be amplified, and new ones may develop [54].

In order to reduce the risk of harm, especially to citizens, it is vital to acknowledge that, "The (Covid-19) technologies under discussion are not viewed as neutral. They must be conceived and designed to account for their social and political nature" [25].

In conclusion, this paper presents the outcomes from the brainstorming, analysis workshop performed at the United Nations conference HNPW 2020. AI and drones are already behind a number of promising innovations for risk reduction, emergency logistics, search and rescue, health, supporting disaster responders, humanitarian actors and local communities.

Provided that data protection and citizens' privacy must be safeguarded, investing in and deploying robotics and autonomous systems that are not affected by human frailties like the vulnerability to viruses and biological threats, would reduce the exposure of humans while improving operational effectiveness.

Finally, authors recommend that towards humanitarian settings, user-friendly and affordable technologies have the potential to support the Humanitarian-Development-Peace (HDP) nexus paradigm concept that aims to bridge the divide in programming among humanitarian assistance, development agenda and peacebuilding.

Acknowledgements. Authors wish to thank Ms. Alexandra Stefanova, United Drone Community, for her outstanding contribution as co-lead of the session on Artificial intelligence and robotics in military and humanitarian space at United Nations/OCHA HNPW Conference 2020, Geneva.

Authors also thank Laura Musgrave, Ronin Institute, for her literature contributions and suggestions.

References

1. United Nations Department of Economic and Social Affairs, World Urbanization Prospects Revision 2018. https://population.un.org/wup/Publications/Files/WUP2018-Highlights.pdf. Accessed 21 Jul 2020

2. Paradox Engineering, When smart technologies combat Covid-19 and contribute to urban. https://www.pdxeng.ch/2020/03/31/smart-technologies-Covid-19-urban-resilience/. Accessed 18 Jun 2020
3. Collins, A., Marie-Valentine Florin, M.-V., Renn, O.: COVID-19 risk governance: drivers, responses and lessons to be learned. J. Risk Res. (2020). https://doi.org/10.1080/13669877.2020.1760332
4. Barnett, D.J., Rosenblum, A.J., Strauss-Riggs, K., Kirsch, T.D.: Readying for a post–COVID-19 world. the case for concurrent pandemic disaster response and recovery efforts in public health. J. Public Health Manage. Pract. **26**(4), 310–313, July/August 2020. https://doi.org/10.1097/phh.0000000000001199
5. Tarpey, F.: Why the timing is right to address the humanitarian–development nexus, 16 March 2020. https://devpolicy.org/why-the-timing-is-right-to-address-the-humanitarian-development-nexus-20200316/. Accessed 18 Jun 2020
6. Grumelard, S., Paul, M., Bisca, P.M.: Can humanitarians, peacekeepers, and development agencies work together to fight epidemics? (2020). https://blogs.worldbank.org/dev4peace/can-humanitarians-peacekeepers-and-development-agencies-work-together-fight-epidemics. Accessed 15 Jun 2020
7. United Nations Office for the Coordination of Humanitarian Affairs-OCHA, COVID-19 Global Humanitarian Response Plan 2020. https://www.unocha.org/sites/unocha/files/Global-Humanitarian-Response-Plan-COVID-19.pdf. Accessed 15 Aug 2020
8. Ting, D.S.W., Carin, L., Dzau, V., et al.: Digital technology and COVID-19. Nat. Med. **26**, 459–461 (2020). https://doi.org/10.1038/s41591-020-0824-5
9. Kritikos, M.: Ten technologies to fight coronavirusEPRS | European Parliamentary Research Service https://doi.org/10.2861/632553. https://www.europarl.europa.eu/RegData/etudes/IDAN/2020/641543/EPRS_IDA(2020)641543_EN.pdf. Accessed 15 Jun 2020
10. David, W., Pappalepore, P., Stefanova, A., Sarbu, B.A.: AI-powered lethal autonomous weapon systems in defence transformation. impact and challenges. In: Mazal, J., Fagiolini, A., Vasik, P. (eds.) MESAS 2019. LNCS, vol. 11995, pp. 337–350. Springer, Cham (2020). https://doi.org/10.1007/978-3-030-43890-6_27
11. David, W., Pappalepore, P., Rozalinova, E., Sarbu, B.A.: The rise of the robotic weapon systems in armed conflicts, 7th CMDR INTERAGENCY, CMDR COE, Sofia (2019)
12. Marr, B.: Coronavirus: how artificial intelligence, data science and technology is used to fight the pandemic, FORBES (2020). https://www.forbes.com/sites/bernardmarr/2020/03/13/coronavirus-how-artificial-intelligence-data-science-and-technology-is-used-to-fight-the-pandemic/. Accessed 15Jun 2020
13. Murphy, R.R., Adams, J., Gandudi V.B.M.: Robots are playing many roles in the coronavirus crisis and offering lessons for future disasters (2020). https://theconversation.com/robots-are-playing-many-roles-in-the-coronavirus-crisis-and-offering-lessons-for-future-disasters-135527. Accessed 15 Jun 2020
14. WeRobotics. https://werobotics.org/covid/. Accessed 07 Aug 2020
15. Howard, A., Borenstein, J.: AI, robots, and ethics in the age of COVID-19, 12 May 2020. https://sloanreview.mit.edu/article/ai-robots-and-ethics-in-the-age-of-Covid-19/. Accessed 15 Jun 2020
16. International Committee of the Red Cross - ICRC: Autonomy, artificial intelligence and Robotics: Technical aspects of Human Control, ICRC. Geneva (2019)
17. Financial Times. https://www.ft.com/content/291f3066-9b53-11ea-adb1-529f96d8a00b, The role of robots in a post-pandemic world The Editorial Board, 21 May 2020. Accessed 15 Jun 2020
18. Swayamsiddha, S., Mohanty, C.: Application of cognitive Internet of Medical Things for COVID-19 pandemic. Diab. Metab. Synd. Clin. Res. Rev. **14**(5), 911–915 (2020). https://doi.org/10.1016/j.dsx.2020.06.014

19. Vaishya, R., Haleem, A., Vaish, A., Javaid, M.: Emerging technologies to combat COVID-19 pandemic. J. Clin. Exp. Hepatol. 2020 (2020). https://doi.org/10.1016/j.jceh.2020.04.019
20. Toğaçar, M., Ergen, B., Cömert, Z.: COVID-19 detection using deep learning models to exploit social mimic optimization and structured chest X-ray images using fuzzy color and stacking approaches. Comput. Biol. Med. **121**, 103805 (2020). https://doi.org/10.1016/j.compbiomed.2020.103805
21. Shi, F., et al.: Review of artificial intelligence techniques in imaging data acquisition, segmentation and diagnosis for COVID-19. IEEE Rev. Biomed. Eng. **2**(1), 220e35 (2020)
22. Buolamwini, J., Gebru, T.: Gender shades: intersectional accuracy disparities in commercial gender classification. In: Proceedings of Machine Learning Research, vol. 81, pp. 1–15. Conference on Fairness, Accountability, and Transparency (2018). http://proceedings.mlr.press/v81/buolamwini18a/buolamwini18a.pdf. Accessed 07 Aug 2020
23. Fussey, P., Murray, D.: Independent Report on the London Metropolitan Police Service's Trial of Live Facial Recognition Technology. University of Essex Human Rights Centre (2019). http://repository.essex.ac.uk/24946/. Accessed 12 Jul 2020
24. Waltz, E.: Entering a building may soon involve a thermal scan and facial recognition. IEEE Spectrum (2020). https://spectrum.ieee.org/the-human-os/biomedical/devices/entering-a-building-may-soon-involve-a-thermal-scan-and-facial-recognition. Accessed 12 July 2020
25. Ada Lovelace Institute. No green lights, no red lines: Public perspectives on COVID-19 technologies (2020). https://www.adalovelaceinstitute.org/our-work/covid-19/covid-19-report-no-green-lights-no-red-lines. Accessed 15 Jun 2020
26. Naudé, W.: Artificial intelligence vs COVID-19: limitations, constraints and pitfalls [published online ahead of print, 2020 Apr 28]. AI Soc. 1–5 (2020). https://doi.org/10.1007/s00146-020-00978-0
27. O'Neil, C.: Audit the algorithms that are ruling our lives. Financial Times (2018). https://www.ft.com/content/879d96d6-93db-11e8-95f8-8640db9060a7. Accessed 18 Jul 2020
28. Raji, I.D., et al.: Closing the AI accountability gap: Defining an end-to-end framework for internal algorithmic auditing. In: FAT* 2020 - Proceedings of the 2020 Conference on Fairness, Accountability, and Transparency, pp. 33–44 (2020). https://doi.org/10.1145/3351095.3372873
29. Kramer, M.F., Borg, J.S., Conitzer, V., Sinnott-Armstrong, W.: When do people want AI to make decisions? In: Proceedings of the AAAI/ACM Conference on AI, Ethics, and Society (AIES) (New Orleans, LA) (2018). https://doi.org/10.1145/3278721.3278752
30. Schramowski, P., Turan, C., Sophie Jentzsch, S., Rothkopf, C., Kersting, K.: The moral choice machine. Front. Artif. Intell. (2020). https://doi.org/10.3389/frai.2020.00036
31. Caliskan, A., Bryson, J.J., Narayanan, A.: Semantics derived automatically from language corpora contain human-like biases. Science **356**, 183–186 (2017). https://doi.org/10.1126/science.aal4230
32. West, D.: The role of corporations in addressing AI's ethical dilemmas (2018). https://www.brookings.edu/research/how-to-address-ai-ethical-dilemmas/. Accessed 25 Jun 2020
33. Kerry, C.F., Chin, C.: Hitting refresh on privacy policies: Recommendations for notice and transparency, The Brookings Institution (2020). https://www.brookings.edu/blog/techtank/2020/01/06/hitting-refresh-on-privacy-policies-recommendations-for-notice-and-transparency/. Accessed 27 Jun 2020
34. Centre for Data Ethics and Innovation. AI Barometer (2020). https://www.gov.uk/government/publications/cdei-ai-barometer. Accessed 12 Jul 2020
35. Pew Research Center (2020). https://www.pewresearch.org/internet/2020/02/21/hopeful-themes-and-suggested-solutions/. Accessed 22 Aug 2020
36. Boyon, N.: Widespread concerns about artificial intelligence. IPSOS (2019). https://www.ipsos.com/en/widespread-concern-about-artificial-intelligence. Accessed 21 Aug 2020

37. Wachter, S., Mittelstadt, B., Russell, C.: Why fairness cannot be automated: bridging the gap between EU Non-Discrimination Law and AI. SSRN Electron. J. 1–72 (2020). https://doi.org/10.2139/ssrn.3547922
38. Mattu, S., Hill, K.: The house that spied on me. Gizmodo (2018). https://gizmodo.com/the-house-that-spied-on-me-1822429852. Accessed 12 Jul 2020
39. Information Commissioners Officer (ICO). Special Category Data. Retrieved (2020). https://ico.org.uk/for-organisations/guide-to-data-protection/guide-to-the-general-data-protection-regulation-gdpr/lawful-basis-for-processing/special-category-data/
40. Harari, Y.: The world after coronavirus. Finan. Times (2020). https://www.ft.com/content/19d90308-6858-11ea-a3c9-1fe6fedcca75. Accessed 27 Jul 2020
41. Ienca, M., Vayena, E.: On the responsible use of digital data to tackle the COVID-19 pandemic. Nat. Med. (2020). https://doi.org/10.1038/s41591-020-0832-5
42. EDRi. EDRi calls for fundamental rights-based responses to COVID-19 (2020). https://edri.org/covid19-edri-coronavirus-fundamentalrights/. Accessed 27 Jun 2020
43. Global Privacy Assembly – GPA. COVID-19 Response Repository (2020). http://globalprivacyassembly.org/covid19/. Accessed 27 Jun 2020
44. Bullock, J., Luccioni, A., Pham, K.H., Lam, C.S.N., Luengo-Oroz, M.: Mapping the landscape of artificial intelligence applications against COVID-19. arxiv, https://arxiv.org/abs/2003.11336v1 (2020)
45. Inter-Agency Standing Committe-IASC. Looking at the coronavirus crisis through the nexus lens – what needs to be done, 29 Apr 2020. https://reliefweb.int/report/world/looking-coronavirus-crisis-through-nexus-lens-what-needs-be-done. Accessed 27 Jun 2020
46. WeRobotics. https://werobotics.org/covid/. Accessed 13 Aug 2020
47. European Commission, Humanitarian-Development Nexus: Strengthening preparedness and response of the health system addressing the COVID-19 Pandemic Humanitarian-Development Nexus: https://ec.europa.eu/trustfundforafrica/region/horn-africa/sudan/humanitarian-development-nexus-strengthening-preparedness-and-response_en. Accessed 18 Jun 2020
48. World Bank. https://blogs.worldbank.org/dev4peace/can-humanitarians-peacekeepers-and-development-agencies-work-together-fight-epidemics. Accessed 15 Jun 2020
49. Howe, P.: The triple nexus: a potential approach to supporting the achievement of the sustainable development goals? World Dev. **124**, 104629 (2019). https://doi.org/10.1016/j.worlddev.2019.104629
50. David, W., et al.: Giving life to the map can save more lives. wildfire scenario with interoperable simulations. Adv. Cartogr. GIScience Int. Cartogr. Assoc. **1**, 4 (2019). https://doi.org/10.5194/ica-adv-1-4-2019
51. Anderson, W.R., Husain, A., Rosner, M.: The OODA Loop: Why timing is everything, Cogn. Times (2017). https://www.europarl.europa.eu/cmsdata/155280/WendyRAnderson_CognitiveTimes_OODA%20LoopArticle.pdf
52. Hooda, D-S.: In General's Jottings Lessons from pandemic: Robotics and readiness for info warfare. Times of India (2020). https://timesofindia.indiatimes.com/blogs/generals-jottings/lessons-from-pandemic-robotics-and-readiness-for-info-warfare/. Accessed 21 Jul 2020
53. WeRobotics. https://blog.werobotics.org/2020/03/30/thinking-of-using-drones-covid-19-why/. Accessed 18 Jul 2020
54. The Lancet. Editorial: Redefining vulnerability in the era of COVID-19, 395(10230), P1089, 04 April 2020. https://www.thelancet.com/journals/lancet/article/PIIS0140-6736(20)30757-1/fulltext. Accessed 21 Aug 2020

ORB-SLAM2 Based Teach-and-Repeat System

Tomáš Pivoňka(✉) and Libor Přeučil

Czech Institute of Informatics, Robotics and Cybernetics, Czech Technical University in Prague, Jugoslávských partyzánů 1580/3, 160 00 Praha, Czech Republic
{Tomas.Pivonka,Libor.Preucil}@cvut.cz

Abstract. The teach-and-repeat task is autonomous navigation along the trajectory taught in advance. A robot is manually driven along the desired trajectory during a teaching phase, and it stores information about it. In this paper, there is proposed a simple teach-and-repeat system based on visual simultaneous localization and mapping system ORB-SLAM2. The map is created during the teaching phase, and it is saved together with a taught trajectory. Afterward, the robot follows the trajectory based on SLAM localization. In the experimental part, the system is compared with SSM-Nav appearance-based system with bearing-only navigation. Besides the presented system, the direct comparison of the SSM-Nav with a SLAM-based system is the main contribution of the paper.

Keywords: Visual SLAM · Teach-and-repeat · Navigation

1 Introduction

Teach-and-repeat task is divided into a teaching and a navigating phase. The trajectory is taught manually by controlling a robot along the desired path. During this phase, a robot creates a map of an environment or stores information about the trajectory. In the second step, a robot follows the taught path autonomously. The task can be performed with a variety of sensors as simple wheel encoders, cameras, or satellite navigation systems. The system presented in this work is based on a monocular camera only.

The cameras can work in indoor, outdoor, or underwater environments, whereas satellite systems are restricted to outdoor environments only. Another cameras' advantage is availability. In comparison with wheel odometry, camera systems are resistant to drifts. The intricacy of cameras is gaining spatial information from camera images and its representation for matching the same places between images. Common environments' appearance is changing due to variable lighting, weather and season changes, or moving objects. The robustness to all mentioned cases is necessary for a long life autonomy of systems.

There is presented a new teach-and-repeat system in this paper. It is based on the state-of-the-art SLAM (simultaneous localization and mapping) system

J. Mazal et al. (Eds.): MESAS 2020, LNCS 12619, pp. 294–307, 2021.
https://doi.org/10.1007/978-3-030-70740-8_19

ORB-SLAM2 [9], which is used for creating a map of robot surroundings and its localization in it. The map is created only during the teaching phase, and in the navigation phase, the robot is localized in the saved map. Based on the returned position, the teach-and-repeat system computes an action command to follow the stored trajectory.

In the experimental part, the precision and robustness of the proposed system were tested and compared with a novel teach-and-repeat system SSM-Nav presented in [3]. SSM-Nav system is built on the visual place recognition [2], and it uses bearing only navigation. The comparison of this approach with a SLAM-based one used in the proposed system is one of the paper's main contributions, and it extends experiments presented in [3].

Summary of ORB-SLAM2 principles is presented in Sect. 2.1. Basic teach-and-repeat approaches are described in Sect. 2.2 and the SSM-Nav teach-and-repeat system is introduced in Sect. 2.3. In Sect. 3, the proposed system is presented. Experiments comparing the system with SSM-Nav are in Sect. 4. Finally, the results are discussed in Sect. 5.

2 State of the Art

2.1 ORB-SLAM2

The presented system is built on ORB-SLAM2 [9], which is an advanced open-source SLAM system. It supports monocular, stereo, and RGB-D cameras. It follows on from previous system ORB-SLAM [8] designed for a monocular camera only. It is a feature-based system, which uses ORB features [12] to create a map and determine a camera position. The system runs in three parallel computing threads used for localization, mapping, and loop closure detection.

The localization thread returns the current position of the camera. It is isolated from mapping and loop closure to work at a high rate with low delay. ORB features extracted in a current frame are matched to the features stored in the map. At the beginning, the features from previous frames are matched to the features detected in the current frame. The position is estimated from the previous velocity and refined by a motion-only bundle adjustment. Further, the position is optimized again based on other visible points in a map, and a motion-only bundle adjustment is performed again to minimize a reprojection error of 3D points from the map. If the previous position is unknown, a place recognition based on a bag-of-words approach is used to find a corresponding keyframe from the map. Besides, new keyframes for the map are selected in the thread.

The selected keyframe is incorporated into the map in the mapping thread. At first, a co-visibility graph between particular keyframes is updated, and the bag-of-words description of the frame is computed. New points are added to the map if they are not matched to an existing locality in the map, and they are corresponding to an unmatched point in another keyframe. The point's position is triangulated, and it is reprojected to other keyframes to find correspondences. Finally, the map is locally optimized by local bundle adjustment. Besides, the points and keyframes are filtered to remove outliers from the map and to maintain the map representation efficient.

Every new keyframe is checked for a loop closing (reaching a previously visited place). Loops are detected by a place-recognition method based on a bag-of-words approach. If the closure is recognized, the key-frames positions are refined, and map points representing the same point in space are merged.

The ORB-SLAM2 offers a localization mode, which disables mapping and loop closure threads. The camera is localized based on a previously created map. The approach can be used only in previously mapped areas, but it ensures stable localization and prevents a changing of a position due to a map refinement.

2.2 Teach-and-Repeat

Teach-and-repeat systems are divided into position-based (quantitative) or appearance-based (qualitative) ones [14]. Position-based systems directly localize a robot in a space based on a stored map. The appearance-based approach does not require creating a map, and stored information about trajectory is used directly to compute an action. The robot displacement can be determined by a comparison of current and stored images of the same place.

The system presented in [5] demonstrates that a position-based system can be used to follow multi-kilometer trajectories. It uses a stereo camera for mapping, and the map is created during the teaching phase by a SLAM algorithm. Further, the method was updated about the teaching of new features in a navigation phase [11] and a selection of features from a map captured in similar weather and lighting conditions [7] to localize a robot. The similarity is based on a bag-of-words [13] comparison of the current image with keyframes.

The appearance-based systems do not need a map for localization, and it controls only robots heading reactively. In addition, the approach does not require camera calibration. The first system computing an action from matched image features between current and stored images is proposed in [15]. Further, wheel odometry information was included to improve a place-recognition [4]. A limitation of the system is a requirement to start navigation close to an initial position. Similar system built on SURF [1] features is presented in [6].

2.3 SSM-Nav

SSM-Nav [3] is a new teach-and-repeat appearance-based system. It uses a visual place-recognition method [2] to localize a robot on a trajectory. It is based on robust features extracted from a convolutional neural network. An estimate of a current position is used as a sensor model in a particle filter, which returns a corresponding frame for computing a heading correction. A steering commands control a robot to minimize the horizontal displacement between current and matched images.

In a place-recognition [2], a corresponding frame is determined in two stages - filtering and spatial-matching. In a filtering step, the features from high levels of a network are used as descriptors. The dimension of descriptors is reduced by applying a principal component analysis (PCA) model. A selected number of the best matching images is returned based on a nearest-neighbor search and a histogram voting for particular places represented by images. Further, the selected images are processed in a second stage. The features are extracted from lower layers than in the previous step, and the dimension is also reduced by the PCA model. Each candidate is spatially matched with the input image. Particular features are matched to the closest features from the input image, and they are used as anchor points. The features in a region around the anchor points in both images are matched, and they are considered to be a correct match if they have the same mutual shift between images as the anchor point. The images are ranked according to a score of its best anchor point.

During the teaching phase, images are processed by the neural network, and extracted features for both visual place-recognition stages are stored to the database. Images are captured based on a driven distance, which is updated online according to a current angular velocity to increase a frame rate in curves. Further, the system stores odometry information for each image.

Navigation is split into determining a current position and computing an action command. The final position is gained from a particle filter. Particles represent positions along the path, and at the beginning, they are distributed uniformly. Particles' positions are updated by the driven distance between two subsequent frames. The current frame is processed by the visual place-recognition, which returns N best matching places. Weights of the particles matching to the best N frames positions are increased. Afterward, the weights of all particles are normalized. 10% of particles with the lowest weights are replaced by new uniformly distributed particles. The final position is the average position of the selected number of particles with the highest weight. The image matched to the computed position is used for the computation of a horizontal offset. The angular velocity is controlled by action commands to decrease the offset. Besides, a linear velocity is adjusted based on an angular velocity applied in a previous step. It is increased on straight paths and decreased in curves.

3 Teach-and-Repeat Method

The presented method is based on ORB-SLAM2 system [9] (Sect. 2.1). It is used to create a map of an environment and to localize a robot in a map. The system is used with a monocular camera only. The mapping is performed only in a teaching phase. It is not used for navigation to prevent recomputing of a map, which can cause shifts of stored positions.

In a teaching phase, a robot is manually driven along the desired trajectory, and SLAM is performed. Robot positions on a trajectory are not stored directly based on localization, but the trajectory is represented by keyframes' positions. It is advantageous because keyframes' positions are updated whenever the map is refined. Another possible approach is to create a map first and afterward, use localization only and record the trajectory positions. This approach was not used because it requires at least two runs during a teaching phase. Using keyframes' positions only was verified experimentally to be sufficient for a trajectory representation. At the end of the teaching, the map and positions of keyframes are saved.

A steering command is computed during a navigation phase whenever the ORB-SLAM2 sends a robot position. The position is represented in 3D Cartesian coordinates by six parameters: spatial coordinates x, y, z, and three Euler angles. The linear velocity is constant, and the position is corrected by controlling a robot's angular velocity. At first, the closest point p_i on a trajectory to a current position is found. Further, only planar movement in xy plane is considered and points are represented only in 2D homogeneous coordinates - $p_i = [x_i, y_i, 1]$. A line passing by p_i and subsequent point on a trajectory p_{i+1} is computed (Eq. (1)). P regulator only is used to control the angular velocity, and it is proportional to a distance of a current position p_C from a line l. The formula for a distance d computation is presented in Eq. (2). The representation of a line incorporates information about the orientation of reference points, so negative distances on one side from a line are computed, and it can be directly used as an input to P regulator. The distances are normalized by a distance between the first two points in a map to reduce a possible difference in dynamics caused by the scale ambiguity of monocular SLAM.

$$\vec{l} = \vec{p_i} \times \vec{p_{i+1}} \tag{1}$$

$$d = \frac{\vec{l} \cdot \vec{p_C}}{\sqrt{l_x^2 + l_y^2}} \tag{2}$$

The comparison of current and closest point's orientations is used to prevent a navigation failure of a robot. If an angular difference is larger than 30°, only constant action is applied. The large difference in orientation can lead to a failure because the robot can see an unknown part of a scene, and it can not be localized. In addition, it prevents a robot from getting perpendicular to the trajectory, which is difficulty reduced by a regulator based on a distance.

4 Experiments

4.1 Hardware and System Implementation

The presented system is implemented on the same platform as SSM-Nav system presented in [3]. It is Clearpath Husky A200 UGV, which is a four-wheel robotic platform with differential driving. The robot is equipped with an external computer and Intel® RealSense™ Depth Camera D435 with maximal resolution 1920 × 1080 px, but the system uses a resolution 1280 × 720 px and frame rate 30 fps. The camera is calibrated from the manufacturer, and it has zero distortion coefficients. The computer contains Nvidia GeForce GTX 1070 graphic card, processor Intel® Core™ i5-7300HQ, and 32 GB RAM. The platform is depicted in Fig. 1. Vicon motion capture system was used to evaluate the precision of tested teach-and-repeat systems.

The presented system is implemented in ROS Kinetic. ORB-SLAM2 runs in a separate node, performs mapping during a teaching phase, and publishes messages with the localized position of a robot. A slightly modified version of ORB-SLAM2 for ROS was used because it directly supports saving a map [10]. The teach-and-repeat system was implemented in Python as a node. It computes action commands based on received robot's positions from SLAM and sends them to a robot. Besides, it includes a GUI for the whole teach-and-repeat system control.

Fig. 1. Robotic platform Husky used for experiments equipped with Intel® RealSense™ Camera

4.2 Tests of Systems

The presented ORB-SLAM2 based system (ORB2-TaR) was tested together with SSM-Nav teach-and-repeat system [3] to compare both approaches. The systems were tested in indoor conditions under the Vicon motion capture system. The tests were designed to compare the precision of navigation and the robustness to different lighting conditions. In addition, the systems were tested to starting in a middle of the taught trajectory, operating in a changing environment and passing through a corridor with doors.

The teaching was performed for both systems at the same time to ensure the same conditions. Two different paths were taught for testing precision and robustness to lighting changes. The shorter path was 21.8 m long without sharp turns. The second trajectory was longer, it was 37.9 m long, and it contained sharper curves. Both trajectories were closed so the robot can repeat navigation several times without interruption. The results from all experiments on the trajectories are presented in Table 1. Particular experiments are described below in detail.

The first experiment on a short trajectory tested a precision of the navigation during three consecutive repeats. The robot started with a zero shift from the initial position. The final trajectories are shown in Fig. 2. The ORB2-TaR was

Table 1. Average and maximal precision (error) of systems in cm in proposed experiments. The experiments were taken under three different lighting conditions - light, dim, dark. "Repeats" represents a number of repetitions, and "Shift" is an initial lateral displacement of a robot in cm.

Experiment				ORB2-TaR		SSM-Nav	
Trajectory	Lighting	Repeats	Shift	Avg. err	Max. err	Avg. err	Max. err
Short	Light	3	0	0.8	2.9	5.8	19.2
Short	Light	1	50	3.2	49.7	16.9	49.7
Short	Dim	1	0	1.0	2.7	4.8	15.4
Short	Dark	1	0	Fail	–	3.9	12.1
Long	Light	2	20	5.1	23.8	4.7	21.0
Long	Dim	1	0	4.9	23.6	3.0	10.0
Long	Dark	1	0	Fail	–	3.8	9.7

more precise than SSM-Nav with maximal error 2.9 cm compared to 19.2 cm for SSM-Nav. Subsequently, the same trajectory was tested with an initial lateral shift 50 cm. Both systems were able to recover from the initial error, but ORB2-TaR converged faster. The paths are depicted in Fig. 3. The SLAM approach benefits from higher image resolution and localization in space. The heading only correction is less sensitive to a lateral shift. A similar test was performed on the longer trajectory with two repeats and an initial lateral shift 20 cm - Fig. 4. Because of the more complicated trajectory, the SLAM could not perfectly optimize a map, which could lead together with the tuning of a regulator to higher average error (5.8 cm) compared to the shorter path (3.2 cm) even with the initial lateral shift. By contrast, higher differences between images caused by turning improved the precision of the SSM-Nav. All the experiments described in this paragraph were performed in an artificially well-lighted room. The measured errors are presented in Table 1.

The experiments on the shorter and longer trajectory were repeated under dimmed and completely turned off lights in the room. The room was not completely dark because lights in adjacent corridors illuminated it. The comparison of images taken in the tested lighting conditions is presented in Fig. 11. The automatic exposure setting compensated the difference between dimmed and normal lighting significantly. The tests were performed exactly from the initial position, and the robot repeated original trajectories captured under normal lighting. The results under dimmed lighting are similar to the original one presented in the previous paragraph, and both systems were able to repeat the shorter trajectory (Fig. 5) and the longer trajectory (Fig. 6) as well. Whereas SSM-Nav successfully repeated the trajectory with turned off lights on both trajectories without impacting a system precision, the ORB2-TaR completely failed. The failure was caused by a lack of stable features detected in the environment. The trajectories are depicted in Fig. 7 and Fig. 8. The measured errors are presented in Table 1.

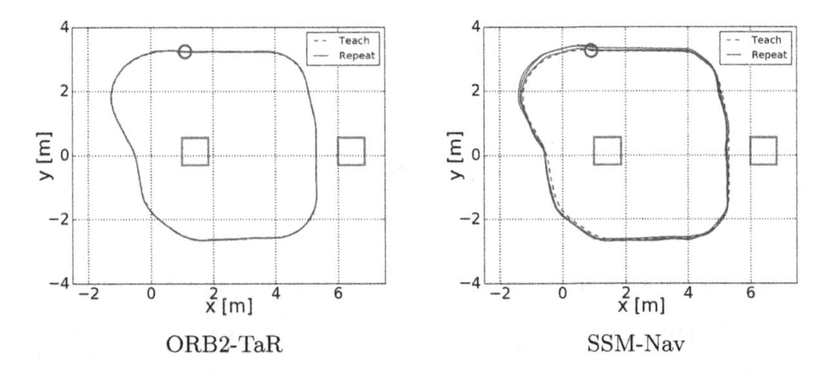

Fig. 2. Test of three times subsequently repeated navigation along the shorter trajectory from the initial position.

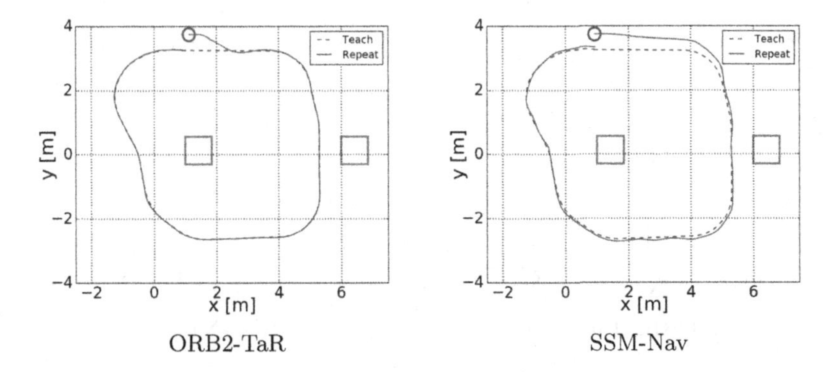

Fig. 3. Test of navigation along the shorter trajectory from a starting position with lateral shift 50 cm from the initial position.

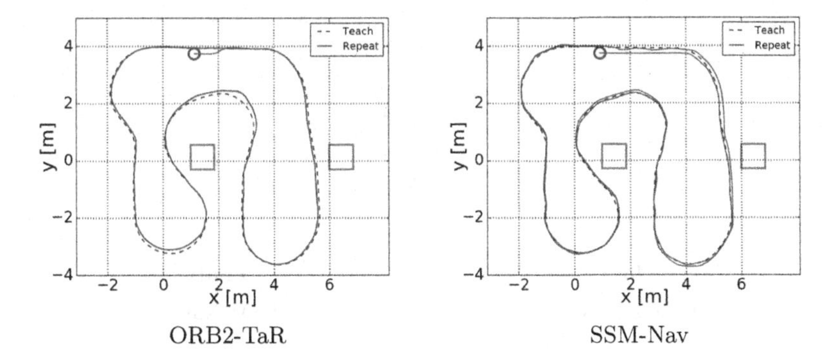

Fig. 4. Test of two times subsequently repeated navigation along the longer trajectory from a starting position with lateral shift 20 cm from the initial position.

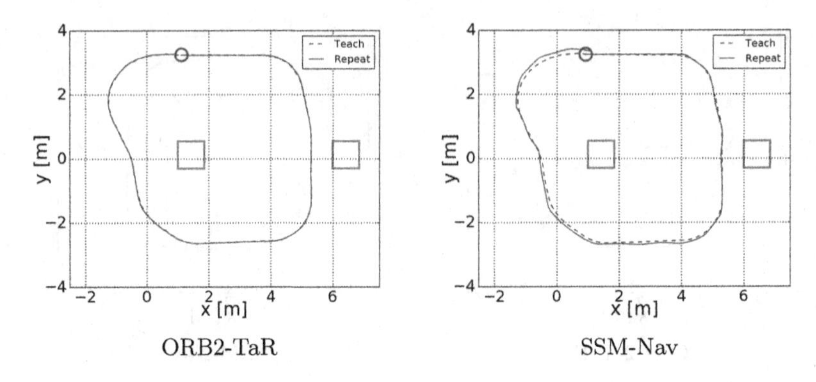

Fig. 5. Test of navigation along the shorter trajectory from the initial position under dimmed lights.

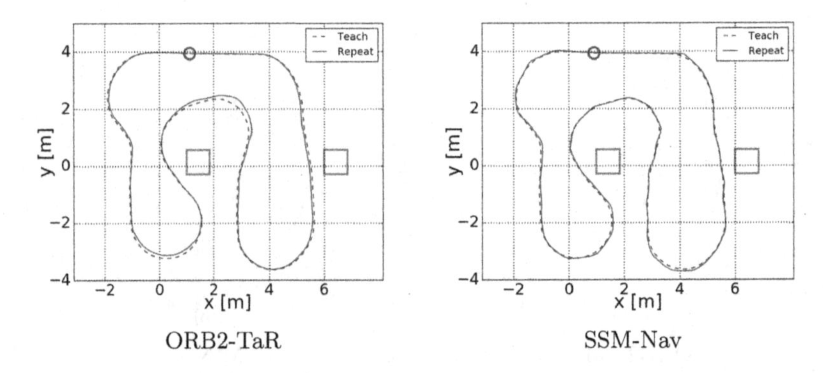

Fig. 6. Test of navigation along the longer trajectory from the initial position under dimmed lights.

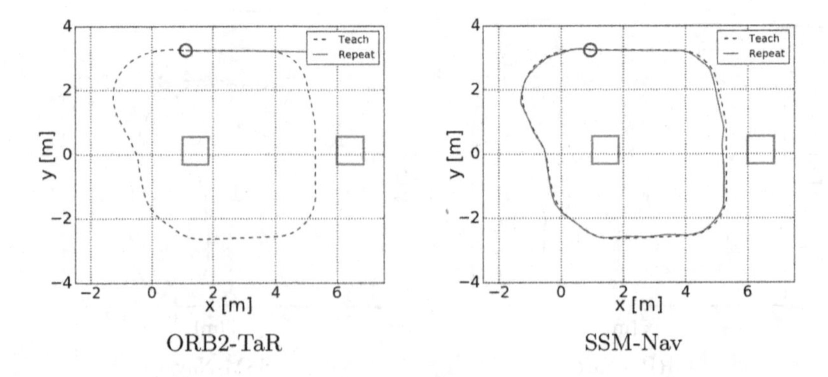

Fig. 7. Test of navigation along the shorter trajectory from the initial position under turned off lights with lighting from adjacent corridors. The ORB2-TaR failed due to a lack of detected features.

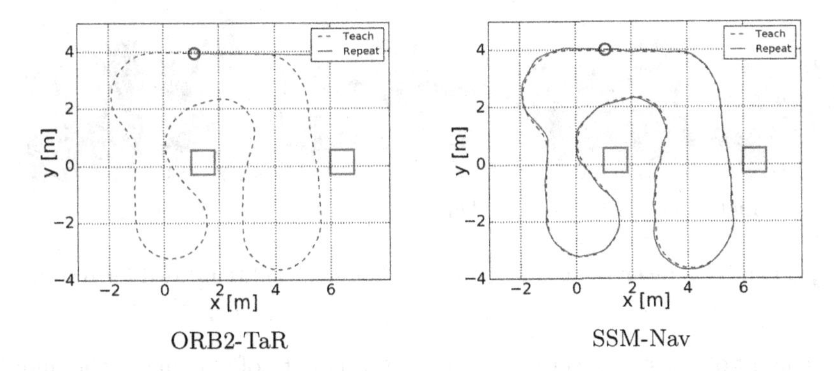

Fig. 8. Test of navigation along the longer trajectory from the initial position under turned off lights with lighting from adjacent corridors. The ORB2-TaR failed due to a lack of detected features as well.

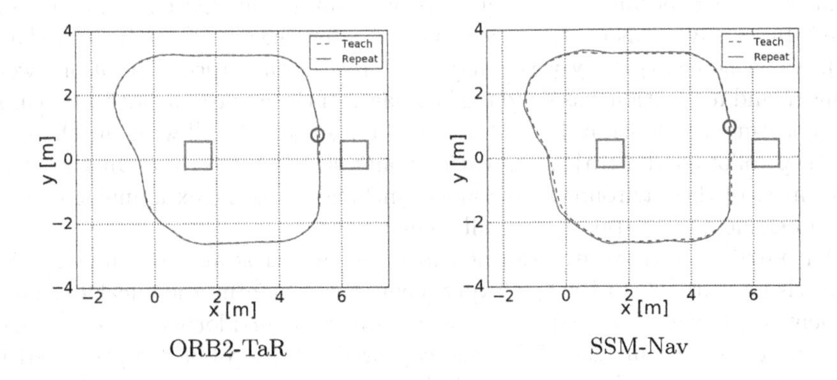

Fig. 9. Test of navigation along the shorter trajectory starting from a middle position.

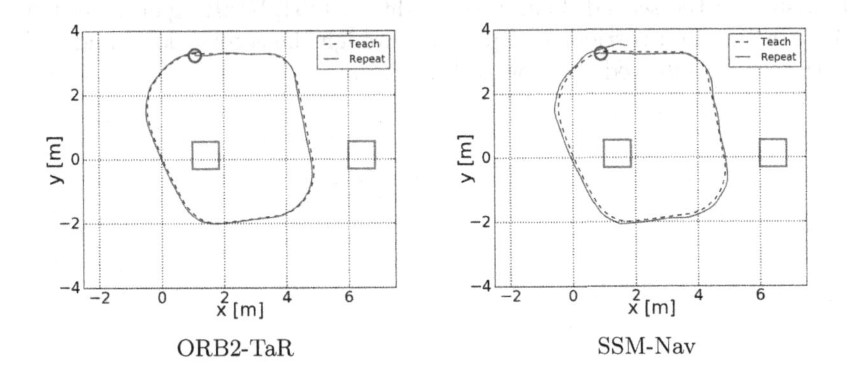

Fig. 10. Test of navigation in the environment with small changes.

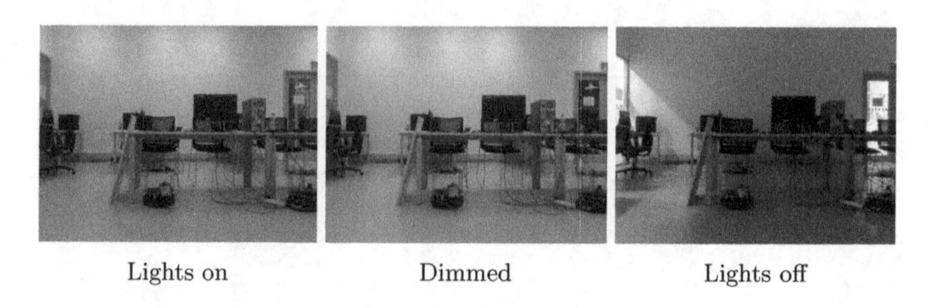

Fig. 11. Images taken under three different lighting conditions used for testing.

Further, both systems were tested to the capability of starting in the middle of the taught trajectory. Both systems were successful (Fig. 9), but it was possible to observe fluctuations of the position in the ORB-SLAM2 at the beginning. It was not caused directly by starting in the middle but by an initial orientation to a part of a scene with a lower number of features.

In the next experiment, the ability to work in a changing environment was tested. In an environment, there were placed a few extra objects (boxes, chair, etc.), and a new trajectory was taught. Afterward, the objects' positions were changed, and navigation was tested. The images of the environments for teaching and repeating are shown in Fig. 12. Both systems cooped well with the changes, and they followed the trajectory without any significant errors. On the other hand, most of the environment remained unchanged, so the experiment verified only an elemental resistance to small changes.

All previous tests were performed in a lab, which is an environment with many visual features. In the last experiment, navigation in a low features environment was tested. The trajectory was learned in a corridor with two passages through doors. The images of a scene captured by the robot are presented in Fig. 13. The ORB2-TaR failed already in a teaching phase, because of a sharp turn close to a matt glass wall. The SSM-Nav repeated the whole path with passing through doors successfully. The shorter path starting in the corridor and passing by the second doors was taught for ORB2-TaR again to test navigation starting in a corridor and passing through the second door, and in this case, the system followed the new path correctly.

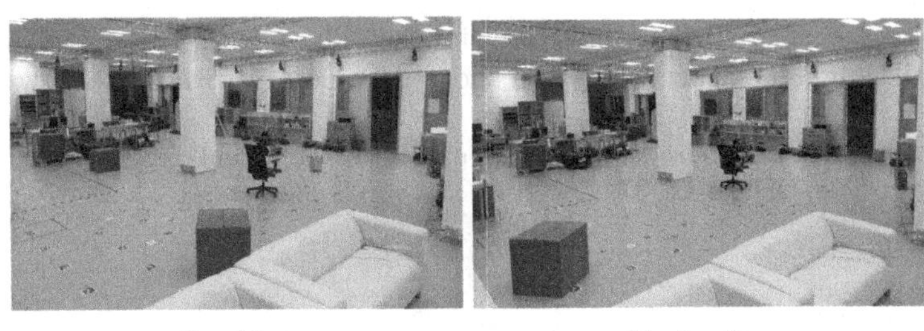

Fig. 12. Pictures of a scene for teaching and repeating with changed positions of some objects.

Fig. 13. Pictures of a scene from the corridor test captured by the robot camera.

5 Conclusions

The position-based teach-and-repeat system using ORB-SLAM2 [9] was proposed in this paper. In the experimental part, the system was compared with the novel appearance-based teach-and-repeat system SSM-Nav [3]. Both the systems were implemented on the same platform; therefore, their properties could be directly compared. Besides the presented system, the comparison of SSM-Nav with SLAM-based approach is the main contribution of the paper, and it extends experiments presented in [3].

The main advantage of the SLAM based system was higher precision and better resistance to a lateral shifts. But at more curved trajectory the precision decreased, which can be caused by worse optimization in a map. The map was not perfectly refined and it contained duplicated points of the same place. On the contrary, the SSM-Nav was much more precise on a curved trajectory, because there are more significant shifts between images and images of a places are more distinctive. The SSM-Nav is also more robust the changes of lighting and it is able to work in darker environment, where the ORB2-TaR does not have enough stable features.

The lab where the tests were performed was an almost ideal environment for a feature based method. Performance of the systems in a more challenging environment was tested on a setup with a corridor that contained matte glass walls and a lot of repetitive structures. The taught path included two doorways.

The SSM-Nav succeeded without any failure. The ORB2-TaR failed with the glass wall, but it was successful on a shorter trajectory with one doorway only.

Acknowledgements. The work is supported by the Grant Agency of the CTU (No. SGS18/206/OHK3/3T/37).

References

1. Bay, H., Tuytelaars, T., Van Gool, L.: SURF: speeded up robust features. In: Leonardis, A., Bischof, H., Pinz, A. (eds.) ECCV 2006. LNCS, vol. 3951, pp. 404–417. Springer, Heidelberg (2006). https://doi.org/10.1007/11744023_32
2. Camara, L.G., Přeučil, L.: Visual place recognition by spatial matching of high-level CNN features. Robot. Auton. Syst. **133**, 103625 (2020). https://doi.org/10.1016/j.robot.2020.103625
3. Camara, L.G., Pivoňka, T., Jílek, M., Gäbert, C., Košnar, K., Přeučil, L.: Accurate and robust teach and repeat navigation by visual place recognition: a CNN approach. In: ResearchGate, March 2020
4. Chen, Z., Birchfield, S.T.: Qualitative vision-based path following. IEEE Trans. Robot. **25**(3), 749–754 (2009). https://doi.org/10.1109/TRO.2009.2017140
5. Furgale, P., Barfoot, T.: Stereo mapping and localization for long-range path following on rough terrain. In: 2010 IEEE International Conference on Robotics and Automation, pp. 4410–4416, May 2010. https://doi.org/10.1109/ROBOT.2010.5509133
6. Krajník, T., Majer, F., Halodová, L., Vintr, T.: Navigation without localisation: reliable teach and repeat based on the convergence theorem. In: 2018 IEEE/RSJ International Conference on Intelligent Robots and Systems (IROS), pp. 1657–1664. IEEE (2018). https://doi.org/10.1109/IROS.2018.8593803
7. MacTavish, K., Paton, M., Barfoot, T.D.: Visual triage: a bag-of-words experience selector for long-term visual route following. In: 2017 IEEE International Conference on Robotics and Automation (ICRA), pp. 2065–2072, May 2017. https://doi.org/10.1109/ICRA.2017.7989238
8. Mur-Artal, R., Montiel, J.M.M., Tardós, J.D.: ORB-SLAM: a versatile and accurate monocular SLAM system. IEEE Trans. Robot. **31**(5), 1147–1163 (2015). https://doi.org/10.1109/TRO.2015.2463671
9. Mur-Artal, R., Tardós, J.D.: ORB-SLAM2: an open-source SLAM system for monocular, stereo and RGB-D cameras. IEEE Trans. Robot. **33**(5), 1255–1262 (2017). https://doi.org/10.1109/TRO.2017.2705103
10. Orb_slam_2_ros. https://github.com/appliedAI-Initiative/orb_slam_2_ros
11. Paton, M., MacTavish, K., Warren, M., Barfoot, T.D.: Bridging the appearance gap: multi-experience localization for long-term visual teach and repeat. In: 2016 IEEE/RSJ International Conference on Intelligent Robots and Systems (IROS), pp. 1918–1925, October 2016. https://doi.org/10.1109/IROS.2016.7759303
12. Rublee, E., Rabaud, V., Konolige, K., Bradski, G.: ORB: an efficient alternative to SIFT or SURF. In: Proceedings of the IEEE International Conference on Computer Vision, pp. 2564–2571, November 2011. https://doi.org/10.1109/ICCV.2011.6126544
13. Sivic, Z.: Video google: a text retrieval approach to object matching in videos. In: Proceedings Ninth IEEE International Conference on Computer Vision, vol. 2, pp. 1470–1477, October 2003. https://doi.org/10.1109/ICCV.2003.1238663

14. Vardy, A.: Using feature scale change for robot localization along a route. In: 2010 IEEE/RSJ International Conference on Intelligent Robots and Systems, pp. 4830–4835, October 2010. https://doi.org/10.1109/IROS.2010.5649557
15. Chen, Z., Birchfield, S.T.: Qualitative vision-based mobile robot navigation. In: Proceedings 2006 IEEE International Conference on Robotics and Automation, 2006. ICRA 2006, pp. 2686–2692, May 2006. https://doi.org/10.1109/ROBOT.2006.1642107

Author Index

Printed in the United States
By Bookmasters